# The
# Catapult

# The Catapult

## *A History*

Tracey Rihll

WESTHOLME
Yardley

Title page: Kheiroballistra in a "nest," Scene 66, cast 166, Trajan's column.

First Westholme paperback 2013


Westholme Publishing, LLC
904 Edgewood Road
Yardley, Pennsylvania 19067
Visit our Web site at www.westholmepublishing.com

ISBN 978-1-159416-103-2
Also available in hardback and as an eBook.

Printed in the United States of America

# Contents

## Abbreviations

Baatz 1994 = D. Baatz, *Bauten und Katapulte des römischen Heeres*

BAR = *British Archaeological Reports*

BAR IS = *British Archaeological Reports* International Series

CIL = *Corpus Inscriptionum Latinorum*

DS = Diodorus Siculus

GSW = W K Pritchett *Greek State at War*

IG = *Inscriptiones Graecae*

JHS = *Journal of Hellenic Studies*

JRS = *Journal of Roman Studies*

Ktesibios *Bel.* = *Belopoiika*

LSJ = Liddell, Scott and Jones, *Greek-English Lexicon*

Marsden 1969 = E. W. Marsden, *Greek and Roman Artillery: Historical Development*

Marsden 1971 = E. W. Marsden, *Greek and Roman Artillery: Technical Treatises*

Philon *Bel.* Bel = *Belopoiika*

Philon *Pol.* Pol = *Poliorketika*

R&O = Rhodes and Osborne *Greek Historical Inscriptions 404–323 B.C.*

RE = Pauly-Wissowa *Realencyclopädie der classischen Altertumswissenschaft*

REG = *Revue des études grecques*

SEG = *Supplementum Epigraphicum Graecum*

ZPE = *Zeitschrift für Papyrologie und Epigraphik*

## *List of Figures and Maps*

# Preface

*Antiquities are history defaced, or some remnant of history which has casually escaped the shipwreck of time.*
—*Francis Bacon* The Advancement of Learning *(1605) 2.2*

I EXPECT THAT MOST IF NOT ALL of the people who read this book already know a thing or two about catapults, some know a great deal about them, and some have even built one. Therefore, I have tried throughout to provide the evidence on which my opinions are based, so that their strengths and weaknesses are clear. I have also tried to lay out the arguments that other people have used to reconstruct different episodes in the history of the catapult where these are critical. Last but not least, as well as telling the story of what was invented and how it was developed, I have turned aside here and there to consider the catapult as merely an example in a larger story, that of the history and philosophy of technology—to consider why and how things are invented and developed.

### Catapult Types and Terminology

Although it seems to be general opinion that the names for different catapults are confused and confusing in the ancient sources, the situation is not as bad as some suppose. Technical terms were occasionally used by authors of technical handbooks but by no one else. Most writers most of the time were less particular, but I believe that there is a lot of order underlying ancient usage. Variant spellings are to be expected in the pre-modern world, and some flexibility in interpretation is often called for with rare words. Words also change their meaning over time, and we will be looking at a thousand years of history. But it seems to me that moderns cause some of the confusion by interpreting these unfamiliar terms with a good deal more flexibility than is warranted. I have tried to offer some clarification on this issue as it arises.

### Translations

Few now can read the sources in the original, and in my experience, those who are dependent on translations find it disconcerting, if not frustrating, to be faced with two significantly different translations. Literal translations often fail to communicate well the meaning of the text, while paraphrases may convey the meaning at the expense of accuracy, strictly understood. In both cases, how the translator understands the text influences the precise choice of words and phrases, as does the scholarly style current at the time of writing. I often find myself, therefore, adopting a compromise position, giving a convenient published English translation (Loeb, in most cases) and modifying it if it does not catch what I think is the correct meaning on topics that are central to this book. This is especially the case with respect to technical terms, where the translator's knowledge, interests, or assumptions can influence the choice significantly. One term I have avoided altogether is "arrow-firer" for the Greek *oxybeles*, because it is not just ugly but misleading in both of its halves. In the first place, catapults do not shoot arrows; the missiles shot by these machines varied in size from dart to twelve-foot stake (for which the word "arrow" is hopelessly inadequate), and in the second, no fire was involved in the process. I use instead the term "sharp-caster" for catapults that shot sharp projectiles.

## *Timeline*

| B.C. | |
|---|---|
| 490 | Battle of Marathon |
| 480 | Battle of Himera; battle of Thermopylai; battle of Salamis |
| 479 | Battle of Plataia |
| 431–404 | Peloponnesian War |
| 424 | Battle of Delion |
| 415–413 | Sicilian expedition |
| 409 | Siege of Selinos |
| 405–367 | Dionysios tyrant of Syrakuse |
| c. 400 | Invention of quinquireme |
| c. 399 | **Invention of catapult: mechanical bow** |
| 398–397 | Siege of Motya |
| 396 | Fall of Naupaktos |
| 390 | Rome sacked by Gauls |
| 388 | Plato of Athens in Syrakuse |
| 386 | Peace of Antalkidas |
| 373 | Jason builds up Pherai's arsenal |
| 371 | Battle of Leuktra |
| 370 | Liberation of Messenia |
| 369 | Invasion of Lakedaimon, foundation of Messene, construction of Messene's fortifications |
| 368/367 | Alliance between Dionysios and Athens |
| 366 | Dion of Syrakuse in Athens |
| 362 | **Catapult bolts on the Athenian acropolis** |
| c. 360 | Aineias Taktikos' *How to Survive Under Siege* |
| 360 | Timotheos nearly hit by catapult |
| 359 | Accession of Philip II of Macedon |
| 358 | Siege of Daton |
| 357–355 | Athens' Social War |
| 356–346 | Third Sacred War |
| 354 | Siege of Methone |
| 353 | First use of torsion stone-throwers? |
| 348 | Siege of Olynthos |
| 341 | Athens attacks Eretria; Demosthenes' 3rd Philippic |
| 340 | Siege of Perinthos |
| 338 | Battle of Khaironeia |
| 338–326 | IG $II^2$ 1467 = **Earliest concrete evidence for torsion catapult** |
| 336 | Death of Philip II; accession of Alexander the Great |
| 334 | Battle of Granikos; Siege of Halikarnassos |
| 333 | Battle of Issos |
| 332–331 | Siege of Tyre |

| | |
|---|---|
| 331 | Siege of Gaza |
| 330 | **Eretrian catapults in Athenian store** |
| 329 | Siege of Marakanda; siege of Kyropolis; Alexander crosses the River Iaxartes |
| 322 | Macedonian garrison installed in Athens |
| 323 | Death of Alexander |
| 319 | Siege of Kyzikos |
| 317 | Siege of Pydna |
| 317–313 | Offense begins to dominate in sieges |
| 315 | Siege of Tyre |
| 312 | Battle of Gaza |
| 310 | Siege of Tunis |
| 307 | Siege of Utica; siege of Mounikhia (Peiraieus) |
| 305–304 | Siege of Rhodes |
| 305 | Ptolemy I Soter becomes king/pharaoh |
| 294 | Demetrios captures Athens |
| 285 | Accession of Ptolemy II Philadelphos |
| 277 | Siege of Lilybaion |
| 270 | Ktesibios active |
| 264–241 | First Punic War |
| c.260 | **Sounion catapult** |
| 258 | Destruction of Myttistraton |
| 250 | Battle of Panormos; Romans using stone-throwers against walls; siege of Lilybaion |
| 246 | Accession of Ptolemy III Euergetes |
| 238 | Attalos I proclaimed King |
| 225–222 | Celtic War |
| 224 | Earthquake shakes Rhodes; Syrakuse donates catapults to Rhodes |
| 219 | Siege of Saguntum |
| 218–201 | Second Punic War |
| 214–211 | Siege of Syrakuse |
| 214–205 | First Macedonian War |
| 211 | Siege of Capua |
| 209 | Capture of New Karthage |
| 207 | Battle of Mantineia |
| 205 | Accession of Ptolemy V Epiphanes |
| 201 | Philip V attacks Attalos I |
| 200 | Philon active; first reference to monagkon (one-armed catapult) |
| 200–196 | Second Macedonian War; siege of Abydos |
| 192–189 | Rome v. Antiokhis III |
| 191 | Battle of Thermopylai |
| 189 | Siege of Same |
| 187 | Aitolian and Kephallenian ordnance displayed in Rome |
| c. 180 | Manlius v. the Galatians |

| | |
|---|---|
| 172–167 | Third Macedonian War |
| 168 | Rhodes begins to decline |
| 167 | **Ephyra catapults** |
| 158–138 | Attalos II |
| 156–155 | Siege of Pergamon |
| c. 150 | **Pergamon catapult** |
| 149–146 | Third Punic War |
| 146 | Destruction of Karthage |
| 138–133 | Attalos III |
| 133 | Pergamon absorbed by Rome |
| 122 | Conquest of Balearic Isles |
| c. 100 | **Ampurias catapult** |
| 91–88 | Rome's Social War |
| 89 | Sulla's Siege of Pompeii |
| 88 | Siege of Rhodes |
| 88–85 | First Mithridatic War |
| 87 | Sulla's Siege of Athens |
| 86 | Battle of Khaironeia |
| 83–82 | Second Mithridatic War |
| c. 80 | **Azaila catapults** |
| 76–72 | Sertorian revolt in Spain |
| c. 75 | **Caminreal catapult** |
| 74–66 | Third Mithridatic War |
| c. 70 | **Mahdia catapults** |
| 58–50 | Caesar's Gallic Wars |
| 55–54 | Caesar's Invasion of Britain |
| 52 | Siege of Alesia |
| 51 | Siege of Uxellodunum |
| c. 50 | **Tanais catapult?** |
| 49–45 | Caesar v. Pompey |
| 49 | Massilia absorbed by Rome |
| 48 | Battle of Dyrrachium; battle of Pharsalus |
| 47–46 | Siege of Lepcis |
| 44–31 | Roman civil war after assassination of Caesar |
| 43 | Rhodes absorbed by Rome |
| 31 | Battle of Actium; Alexandria absorbed by Rome |
| c.25 | Vitruvius *On Architecture* |
| 16–15 | Augustus conquers alpine tribes |
| 17–12 | Conquest of Pannonia, part of Germania |
| | |
| A.D. | |
| 5 | Tiberius reaches the River Elbe |
| 6–9 | Revolt in Pannonia and Dalmatia |
| 9 | Revolt in Germania; three legions massacred in Teutoburg Forest |

| | |
|---|---|
| 14–16 | Germanicus in Germany |
| 43 | Claudius' invasion of Britain |
| 45 | Cremona battleshield made |
| c. 50 | **Xanten catapult** |
| 55–64 | Armenian/Parthian War |
| 58/59 | Siege of Volandum |
| 62 | Heron active |
| 66–74 | Jewish revolt |
| 67 | Siege of Iotapata |
| 68 | Death of Nero |
| 69 | Year of four emperors; Placentia amphitheater destroyed by fire; **Cremona catapults**; earliest (indirect) evidence for carroballistae |
| 70 | Fall of Jerusalem |
| 73–74 | Siege of Masada |
| c.75 | **Auerberg catapults** |
| 81 | Death of Titus |
| 85–92 | Domitian's campaigns into Dacia |
| c. 90 | **Elginhaugh catapult** |
| 98 | Accession of Trajan |
| 100 | Apollodoros active |
| c. 100 | **Bath catapult** |
| 101–102 | First Dacian War |
| 105–106 | Second Dacian War |
| 116 | Revolt of Hatra |
| 117–119 | Death of Trajan; accession of Hadrian; Dacian Revolt |
| 122 | Hadrian's Wall begun |
| 135 | Arrian's campaign against the Alans |
| 140–143 | Antonine Wall constructed |
| 162–166 | Parthian War |
| 167–180 | German War |
| 180 | Death of Marcus Aurelius |
| 193 | Death of Commodus |
| 194 | Battle of Issus |
| 195 | Battle of Byzantion |
| 197 | **Lyon catapult**; accession of Septimius Severus |
| 198 | Severus' first attack on Hatra |
| 200 | Severus' second attack on Hatra |
| 217 | Murder of Caracalla |
| 227 | Artaxerxes' attack on Hatra |
| 238 | Murder of Maximinus |
| 241 | **Hatra catapults** |
| 244 | Murder of Gordian III; accession of Philip the Arab, the first Christian emperor |
| 249 | Philip killed in battle by Decius |

| | |
|---|---|
| c. 257 | Dura-Europos destroyed |
| 260 | Valerian captured by the Persians |
| 268 | Murder of Gallienus |
| 275 | Murder of Aurelian |
| 282 | Murder of Probus |
| 284 | Accession of Diocletian; creation of the Tetrarchy, division of empire into East and West |
| c. 285 | **Volubilis catapults** |
| 306 | Constantine proclaimed emperor (in York) |
| 312 | Constantine defeats Maxentius and becomes Emperor of the West |
| 324 | Constantine becomes sole ruler of both halves of the empire |
| 337 | Death of Constantine |
| 337–360 | Persian Wars |
| c. 350 | **Pityous catapult** |
| 359 | Siege of Amida |
| 363 | Siege of Maiozamalcha; death of Julian; peace with Persia |
| 364 | Accession of Valentinian I (West) and Valens (East) |
| 378 | Battle of Hadrianople; death of Valens |
| c. 380 | **Gornea catapults; Orşova catapult; Sala catapult** |
| 395 | Death of Theodosius I; Empire divided into East and West; Arcadius emperor in East, Honorius in West; Goths invade Greece |
| 406 | Barbarian invasion of Gaul |
| 408 | Gothic invasion of Sicily |
| 410 | Britain told to defend itself; Sack of Rome; death of Alaric |
| 429 | Vandals invade Africa |
| 453 | Death of Attila the Hun |
| 455 | Vandals sack Rome |
| 469–478 | Visigoths invade Spain |
| 472 | Ricimer captures Rome |
| 476 | Last Western emperor, Romulus Augustus, deposed by Ordovacer, who styles himself King of Italy; Zeno emperor in the East |
| 502–506 | Persian War |
| 526–532 | Persian War |
| 527 | Accession of Justinian |
| 529 | Justinian's anti-pagan legislation |
| 531 | Accession of Khosroes to Persian throne |
| 533–554 | Justinian (via Belisarios and others) attempts to reconquer Italy and North Africa |
| 537–538 | Siege of Rome |
| 540 | Siege of Antioch |
| 542 | Plague in Constantinople |
| 550 | Battle of Petra |
| 565 | Death of Justinian |
| 597 | Siege of Thessalonike |

Map 1. *The Ancient World*

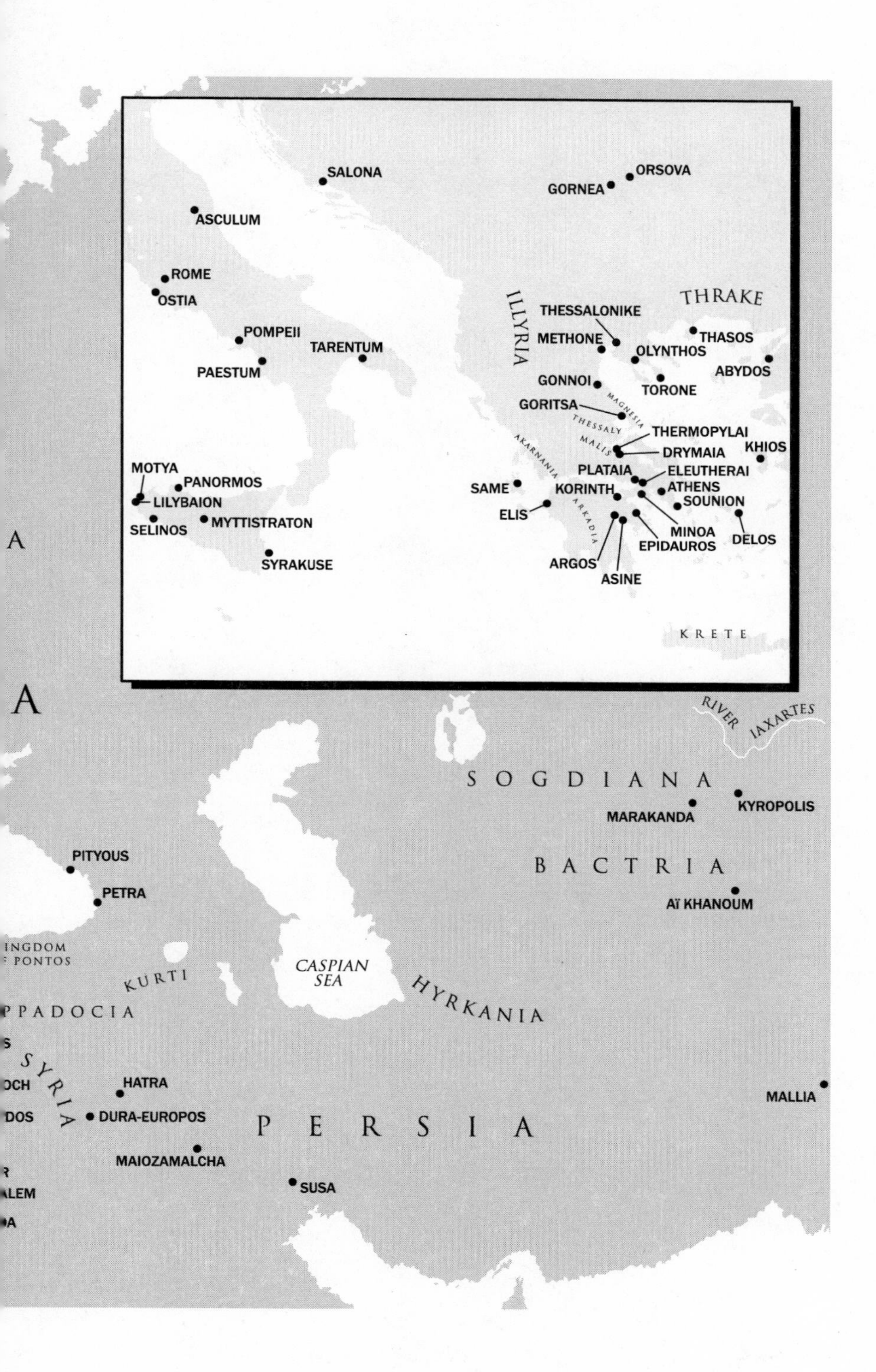
SALONA
ORSOVA
GORNEA
ASCULUM
ROME
OSTIA
THRAKE
ILLYRIA
THESSALONIKE
POMPEII
TARENTUM
METHONE
THASOS
OLYNTHOS
PAESTUM
ABYDOS
GONNOI
TORONE
GORITSA
MAGNESIA
THESSALY
THERMOPYLAI
MALIS
AKARNANIA
DRYMAIA
KHIOS
MOTYA
PANORMOS
PLATAIA
ELEUTHERAI
SAME
KORINTH
ATHENS
LILYBAION
SOUNION
ELIS
ARKADIA
SELINOS
MYTTISTRATON
MINOA
DELOS
EPIDAUROS
SYRAKUSE
ARGOS
ASINE
KRETE
A
A
RIVER IAXARTES
SOGDIANA
MARAKANDA
KYROPOLIS
BACTRIA
PITYOUS
AÏ KHANOUM
PETRA
CASPIAN SEA
KURTI
HYRKANIA
SYRIA
HATRA
MALLIA
DURA-EUROPOS
PERSIA
MAIOZAMALCHA
SUSA

# Introduction

The catapult is an amazing machine that captures the attention whenever it is seen. Rugged and primitive in materials but complex and polished in design, its performance is impressive. Well-made reconstructions hurl bolts or stones (or in more safety-conscious demonstrations, straws and onions) at great speed to thwack targets considerable distances away. The invention and development of this remarkable device is a story that spans about a thousand years, from about 400 B.C. to about A.D. 600, over an area of about a million square miles, and this book aims to tell that story.

The history of the catapult has an importance far beyond military and hence political history. The catapult is, without doubt, the best-documented technology from antiquity. We have several surviving treatises that explicitly aim to tell the reader how to build a variety of models and that are the source of all modern reconstructions that aspire to be genuine reincarnations of ancient devices and not modern hybrids or "improved" versions. We also have some briefer descriptions of them by ancient writers, and we have a large and growing body of archaeological evidence of machines and their missiles. Finally, catapults are mentioned in a large and diverse corpus of Greek and Latin literary texts in a variety of genres, prose and verse, history and comedy, and we thus have an extensive original popular as well as a technical literature on them.

Studying the story of the catapult, we can learn about the processes, means, and motives of technological development in the ancient world. As today, war or its prospect provided the strongest motive and prompted the biggest investment to develop technologies which had military use, but unlike today, the people who designed such technologies were not normally "employed" as such by the state and did not confine their activities to the military sphere. For example, one of the earliest engineers we know about, a man called Ktesibios, who designed at least two types of catapult, also designed the force pump, the water organ, and many other nonmilitary mechanical devices. Another, Arkhimedes, is as famous for his scientific discoveries and for shouting "Eureka!" as he is for his defensive siege engines that stopped the mighty Roman army in its tracks outside the city of Syrakuse, where he was born and had lived for the previous sixty-odd years.

One of my aims in this book is to bring the material up to date, since Marsden's magisterial *Greek and Roman Artillery*—which is essential reading for anyone interested in the topic and is the origin of most blueprints of ancient catapults to be found on paper or website today—was written more than thirty years ago, and we have a lot more archaeological evidence now. More important, I want to situate the technical developments more firmly in their historical contexts, so that alterations over decades or centuries are not compressed into single inventions and events. I also try to use comparative evidence to inform and widen the debates that exist over almost every single aspect of the topic. No other area of ancient history is blessed with legions of enthusiasts (pun intended) who examine the evidence in at least as much detail as academics (and sometimes a good deal more), and who have the benefit, moreover, of experience, building and using machines according to their preferred reconstructions on paper. No doubt, the debates will continue, but I hope this book will be found useful to the enthusiasts as well as informative to the uninitiated.

Let me then provide a quick and simplified overview of the story to serve as a map to the chapters to follow. In Chapter 1, "Sticks and Stones," I consider the catapult's antecedents. Since catapults could throw stones *and* sharps, these forerunners turn out to be the sling and the bow. In Chapter 2, I retell and analyze the story of the catapult's invention as a *gastraphetes*, belly-bow, a mechanized bow, a sort of crossbow, as told in antiquity. In Chapter 3, I examine the history of the early catapults, refusing to rush through some seventy-plus years of history in a hurry to get to the "classic" engines (as has previously been done). This careful review of the evidence from the period, together with the recollection that the catapult has *two* very different manual predecessors, sling as well as bow, leads to the argument that the classic two-armed torsion catapult has two, probably independent, origins. I develop the argument that the second origin—that is, the mechanization of the sling—occurs in about the middle of the fourth century, in the invention of the *monagkon*, or *onager*, the one-arm torsion catapult. The central argument of Chapter 4 is that the classic two-armed torsion catapult was a fusion of these two earlier catapults: the (two-armed) bow or tension catapult, and the one-armed torsion catapult. The classic machine marries the two arms from the bow catapult to the torsion power source from the one-arm to make a machine that can, with minor modification, shoot either sharps or stones. In Chapter 5, "Little Catapults," I focus on the catapult's primary role as an anti-personnel weapon. Chapter 6 then attempts to describe and explain the explosion of designs and sizes for such catapults that is attested in the literary sources of the Hellenistic period, comparing it with other similar episodes in the history of technology. On the basis of a careful examination of the historical sources, I argue that the *formulae* for the classic two-armed torsion engines were discovered about 320 B.C. In Chapter 7, I examine thoroughly the three principal technical treatises on the construction of war machines, which were written in the aftermath of that proliferation, when designs began to coagulate around standardized formulae that had been discovered in the previous age of experiment and now offered a sort of quality control for catapult builders. The focus switches to Rome in Chapter 8, as we follow their leaders' use and development of catapult weapons, and Rome's absorption of various different catapult-building traditions along with the territories whence they came. Chapter 9 pursues the story into the empire, with its continuing developments, not least the all-metal *kheiroballistra*, hand-ballista, and its debates, notably about whether or not on some models the arms became in-swingers. Chapter 10 concludes the story, not, as will be seen, because catapults died out in late antiquity, but because it would require a lot more research and another book to answer all the unresolved issues on catapults in the medieval, Islamic, and Chinese worlds.

A few themes reappear throughout as the evidence, organized in accordance with chronology (as any attempt to chart the rise and fall of a technology must be) dictates. One is variety. There has been an unfortunate tendency to try to squeeze all literary and archaeological evidence into a few configurations, notably those described in the surviving ancient technical treatises, even if a particular piece fits very badly indeed. While the creation of standards was probably the greatest single achievement of the ancient catapult builders, it is simply not reasonable or plausible to assume that everyone, everywhere, thereafter built to standard. People are not like that. Some, even if they know a standard exists, will deliberately depart from it, for one or more of a variety of reasons. Others will not follow through ignorance; and both qualifications can apply to a Roman just as much as to a barbarian.

Another regularly occurring theme is size. There has definitely been a tendency hitherto to emphasise size, probably unconsciously, perhaps partly as a result of using the modern word "artillery" or its equivalents in other modern languages to describe these machines. Often, there is absolutely nothing in the primary sources to indicate the size of a weapon, and by using the word "artillery" we think of a field gun, whereas if we had used, say, "gun," we could also have thought of a handgun. I am not for a moment suggesting that all catapults were hand weapons. But I argue repeatedly, as the evidence directs, that sometimes that is *exactly* what they were: personal weapons, carried by individual soldiers, shot from the shoulder, and shooting "bullets" or darts.

A few themes are picked out and developed at the end of chapters, each being assigned to what seemed to be the most appropriate chapter for that topic. Thus the nature of innovation and the issue of technology transfer, for example, form distinct sections within the narrative framework of the book. This seemed to me preferable to the more traditional practice of grouping the material into a narrative half and a thematic half, because important thematic questions arise naturally in the course of telling the story, and I prefer to deal with them as they arise, rather than later.

Catapults began as anti-personnel weapons. Over time, machines were developed that could shake walls; the largest known threw gigantic stones weighing some 75 kg. What all catapults have in common, and may be considered to define them, is that they project a missile by mechanical means. They may be powered by bows, or by skeins of rope (called torsion springs), they may discharge shot or sharps, they may be hand weapons or monsters that dwarf their human operators. "Shot" may be anything roughly spheroid that fits in the sling, for example, lead shot, a stone ball, a container filled with fire or other unpleasantries, or a human head. Sharps may be tiny darts or vicious bolts that could nail a man to a tree, or stakes with meter-long-bodkin points that could drive through several lines of defenses before burying itself in the ground. This variety grew over the course of the ten centuries that this book covers.

Those unfamiliar with ancient history might appreciate a little background on the world in which our story starts. When the catapult was invented around 399 B.C., the Parthenon was about forty years old. Athens had recently suffered defeat in the generation-long Peloponnesian War, and Sparta was adjusting (with difficulty) to her new status as the only superpower in the Greek world. "The Greek world" was wherever the ancient Greeks lived. The heartlands were modern Greece and the Aegean islands, Crete, and the coastal margins of the Aegean. In addition, in a more spotty fashion, the Greek world included areas of the Black Sea coast, southern Turkey, Cyprus, the Adriatic coast, South Italy, and Sicily, together with isolated sites further afield around the Mediterranean as far as Spain and Libya. Plato aptly likened the geography of Greek settlement to frogs around a pond, the pond being the Mediterranean Sea and the frogs the Greek settlements.

Other civilizations with which the Greeks clashed at this time, when not fighting each other, were the Persians in the East and the Karthaginians in the West. Nearly a hundred years earlier, in 480 B.C., when Leonidas with his 300 kamikaze Spartans and Themistokles with his cunning subterfuges famously fought the Persians at the battles of Thermopylai and Salamis in Greece, Gelon, the tyrant of Syrakuse, fought the Karthaginians at the battle of Himera in Sicily. And as the mainland Greeks defeated the Persians in the

East, so the Sicilian Greeks defeated the Karthaginians in the West. The Karthaginians did not invade Sicily again until the last decade of the fifth century, and during the course of that later war between the Greeks of Sicily and the Karthaginians, the catapult was invented.

ONE

# Sticks and Stones

*'As time passes human ingenuity naturally and perpetually keeps pace with it by innovating'*—Prokopios *Wars* 8. 11. 28

IN ORDER TO UNDERSTAND AND appreciate the invention of the catapult we need to reflect on war before the catapult: we need to understand the world in which it appeared, to seek the reasons for its invention, and the elements that were put together to create it. Thus, that is where our story begins.

Ancient catapults shot both arrows and stones. Therefore they had two ancestors: the bow and the sling. Curiously, neither bow nor sling normally leaps to mind when one conjures up an image of ancient Greece. Say "ancient Greek soldier," and one thinks of someone glaring out from the protective shadow of a Corinthian helmet, one continuous piece of beautifully crafted metal completely enclosing the head, except the eyes, with its bronze nose guard descending to the top of the lip where it meets cheek-guards descending from each side. The Greek warrior holds a large round shield called a *hoplon*, which labels him a hoplite, and carries a short thrusting spear. Below the shield, his lower legs are protected by bronze greaves, and he wears sandals on his feet. Atop his helmet, a crest of colored horsehair, running front to back, stands erect, and adds another six inches to his modest stature. It also exaggerates his movement. Massed together such warriors form a hoplite phalanx, a block of hard men, heavy infantry, bristling with murderous points and protected by strong armor.

But if we look a little harder we can also see slingers, stone throwers and archers lurking at the edges of the phalanx. The slinger is a slip of a man, who, if he wears any clothes at all, is lightly dressed in a wool or hemp tunic that is pinned at the shoulder and stops at mid-thigh; he may wear a felt cap. The weapons in his hand are a leather bag of stones and a sling. The archer is also without armor, and carries a bow and quiver of arrows. He is sometimes oddly dressed by comparison with the people alongside whom he fights, because a few places on the physical or metaphorical margins of the Greek world produced good archers, and their men were often employed as mercenaries. They brought their own habits of dress, as well as their weapons, with them;

hence their odd appearance. The Skythians are the most distinctive, because they wear trousers. But archers from Krete, Akarnania, and Arkadia are Greeks, so wear the tunic. Although the hoplites do not like to admit it (largely for social reasons), archers and other troops who use long-range weapons were useful in some circumstances, which is why they were employed.

Both sling and bow were old when the Greeks arrived on the ancient scene. Let us look first at the sling. The earliest evidence for the sling is from about 10,000 B.C., in the Neolithic or New Stone Age. Slingshot started as spherical in shape, but after about four thousand years men began to make it bi-conical and then roughly ovoid. Evidently they had discovered that the ovoid (egg, almond, olive, or acorn) shape helps the slinger to shoot accurately; it does this because it imparts spin, which assists the missile to fly true. The British rugby ball and American football exploit the same aerodynamic property. Slingshot was made from a number of different materials. The earliest slingshot was made of stone, which continued in use throughout antiquity. For example, marble shot found at Methone was probably discharged by Macedonian forces during its siege of 353 B.C.[1] Artificial slingshot, made of baked clay, was manufactured from about 5,000 B.C. Such sling bullets, of Neolithic date, have been found in Attike, the hinterland of Athens, and scattered across the Roman empire, for example at Caerleon in South Wales. Ceramic and stone shot had some special uses. For example, the Nervii used red-hot ceramic slingshot against Caesar's subordinate Quintus Cicero in Gaul in 57 B.C.

> On the seventh day of the attack, a very high wind having sprung up, they began to discharge with their slings hot balls of burned or hardened clay and heated javelins on the huts, which, after the Gallic custom, were thatched with straw. These quickly caught fire, and because of the violence of the wind, scattered fire to all parts of the camp. (Caesar *Gallic War* 5.4.3 Loeb)

Hot lead slingshot was considered one of the possible sources of the fire that burnt down what was, in A.D. 69, the largest amphitheater in Italy, at Placentia (Tacitus, *Histories* 2. 21). In classical times slingshot was sometimes made of lead as well as clay, and small water-worn pebbles identified as slingshot are known in quantity from Hadrian's Wall. A leather sling pouch was found in the extraordinarily preservative soil of Vindolanda.[2] The Greeks and Romans mined and smelted lead ores on a vast scale to make silver, which was contained within the lead ores. They wanted the silver principally to manufacture coins, but also for plate, jewelry, and other decorative purposes. Lead was generated in quantity as a by-product of this silver smelting—about two kilograms of lead for every four-gram silver drachma. Exploitation of the ores was on such a scale that the pollution from their furnaces reached the polar icecaps, where lead levels from Roman times are greater than at any other period in history until well into the Industrial Revolution.[3]

Lead was used for slingshot because it was easy to melt desilvered lead, and easy and quick to cast—even in something as simple as a hole made with the finger in sand—and it produced a dense "bullet" usable in as little as twenty seconds.[4] Lead shot could also be easily marked, either in the molding or after. For example, in 348 B.C. both sides were using slings in the siege of Olynthus, capital of the Khalkidian federation, on the peninsula of the same name in northern Greece. Molds were found inside the settlement bearing the inscription **ΧΑΛΚΙ**, an abbreviation for Khalkidian, and shot was found inscribed **ΦΙΛΙΠΠΟΥ**, "Philip's," or carried the names of some of his subordinate commanders, such as Perdikkas.[5] Some found hither and yon carry the name of the maker, "Timon's" or "Sokrates did it," or political slogans such as "Sertorius is pious," or sarcastic messages like "catch" (**λαβε**), or more insulting quips. Some are even addressed to the enemy, e.g., "hit Pompey" (*feri Pomp*), as apparently were some munitions discharged in World War II.[6] A few carry designs; five in the Greco-Roman collection at the British Museum bear scorpions; one also carries the Macedonian thunderbolt on the other side.[7] These will be discussed further in Chapter 5. An inscription or symbol, if present, is usually engraved in the mold, and thus stands proud on the shot; because lead is soft it is also possible to cut a message on the shot itself. It is reported in literary sources that slingshot were sometimes inscribed with treasonous messages, though this seems an intrinsically unreliable method of delivery for such missives; quite apart from the question of accuracy, it relies on someone in the enemy camp picking it up, reading it, and passing it on to the relevant authorities.[8]

A fairly common weight for slingshot, whatever it was made of, is around 30–40 grams. Shot lighter than this was perhaps designed to be cast in pairs or multiples to produce a shotgun effect.[9] The sling itself is essentially a piece of fabric, usually woven, or leather, with a cord attached at either end. One cord has a loop at the end, which is slipped over the second finger, and should be retained on the finger when the missile is cast. The other cord has no loop, is held between first finger and thumb, and is released to launch the missile. The slinger puts the missile in the material, whirls it around once or more to gain speed, and then lets go the appropriate cord, whereupon the missile flies off with centrifugal force. Usually a small shot is launched from a sling held in one hand; however, one can also launch bigger stones using both hands to wield a bigger sling. The movement then resembles a modern hammer throw, the whole body being rotated, usually a couple of times. (This is perhaps the origin of that sport—if it isn't, I don't know what is.) The slingers of the Balearic Islands reputedly trained to use slingshot weighing up to a *mina*, which was slightly less than a modern pound.[10] Irrespective of whether the sling is used with one hand or two, it takes a lot of practice to learn to release one cord at exactly the right moment for the missile to go off in the intended direction.

The staff sling is a sling on a pole about four feet long. One of the sling cords is tied to the pole. The other has a loop which slips on and off the end of the pole. The staff slinger holds the pole with both hands and in one swift movement whips it over his head. At a certain point, which varies with the precise form of the sling and the shot, the loop slips off the pole and the stone is discharged with greater velocity than from an ordinary sling—in Richardson's tests, the staff sling averaged 74 mph (33 meters per second) compared to the ordinary sling's 68 mph (30 meters per second).[11] The staff sling consequently has greater range than the sling. However, it is less adaptable than the sling; it always discharges its stone in a high arc, though the staff-slinger can adjust its precise shape. Roman legionaries were trained in its use, along with regular slings, and hand-thrown stones, darts, and javelins.

> Archers and slingers used to put up scopae, that is, bundles of brushwood or straw, removing themselves 600 feet from the target, to practice hitting it frequently with arrows, or stones aimed from a staff sling. This enabled them to do without nerves in battle what they had always done in exercises on the training field. They should also be accustomed to rotating the sling once only about the head when the stone is discharged from it. All soldiers also used to practice throwing stones of one-pound weight by hand alone. This was considered a readier method because it does not require a sling. They were also made to throw javelins and lead-weighted darts (plumbatae) in continual and perpetual exercises. . . . Technical skill is more useful in battle than strength. If training in arms ceases, there is no difference between a soldier and a civilian. (Vegetius *Epitome of Military Science*, 2.23 Milner)

Greek warriors disliked the sling for a number of reasons. One was because in a moment, unseen and unheard, slingshot could take a man's life. It is practically impossible to see slingshot coming, as Onasander observed, which is partly why it is so difficult, even now, to obtain reliable figures for range and velocity of slingshot. In any terrain, slingers could attack the enemy phalanx from more or less any position, and a distant one at that. Heavy infantry could not defend against slingshot as well as they could against arrows or javelins. A fully armed infantryman was more likely to be stunned by slingshot than killed by it—so Lucius Cotta was wounded, not killed, by a slingshot that hit him full in the face—but armor could be crushed in and teeth knocked out, and a stone to the head could cause a man to collapse, as Josephus says happened to him at the siege of Jerusalem in A.D. 69.[12]

> By hurling a shower of great stones, they wounded many and even killed not a few of those who were attacking, and they shattered the defensive armour of most of them. For these [Balearic Islands slingers], who are accustomed to sling stones weighing a mina [about 1 lb or 1/2 kg], contribute a great deal toward victory in battle. (Diodoros Siculus 19.109.2 Loeb)

Cavalry disliked the sling even more; the horse offered a larger target, the rider usually wore less armor and a more open helmet, and both horse and rider were easily startled by slingshot. Slingers, therefore, offered good cover from cavalry. That is why, according to Thucydides in his *History of the Peloponnesian War*, when, in 415 B.C., the Athenians launched their massive expedition against Syrakuse—a city renowned for its cavalry—the commanding officer Nikias asked for 700 slingers to be included among the expeditionary forces. And what was true of cavalry was also true of war elephants, though their use was much less common.[13]

> Scipio used the following technique to train and discipline his elephants. He drew up two groups for battle practice. One was of slingers, who were to act as the enemy and sling small stones at the elephants. Facing them he set first, the elephants in a line, with him and the whole army in battle array behind, so that when the enemy, by slinging stones, had frightened the elephants and made them turn toward their own men, they might be made to turn and face the enemy again by the volleys of stones from the army behind them. The training made slow progress, however, because these animals, even after many years' training, are dangerous to both parties when brought onto the battlefield. ([Caesar] *African War* 27 Loeb)

Another reason why the sling was unpopular was perhaps its unreliability, especially in inexperienced hands—and there are no shortcuts to experience, of course. Every article I have ever read on contemporary slinging is furnished with health warnings. When someone is learning how to sling, that person is probably the only person in the vicinity who is in a safe place—unless she is using the underhand method, in which case she is not safe either. Rammage estimates that some forty hours of practice is needed before the novice slinger gets beyond the "experimental" stage. He also notes that "rogue stones" (i.e., slingshots that don't go where the slinger intended them to) happen to everyone, even "experts." He advises that the safest (N.B. saf*est*) place for spectators is to the *sides* of the slinger (*not* behind).[14] The Slinging.org Web site's own safety advice includes the recommendation that a slinger of average height using a 36-inch sling "needs an open area of at least 27 feet diameter and height," at least when learning how to control it. Richardson observes wryly that "the job of sling instructor must have been one of the most dangerous in history."[15] I think that, if I were a hoplite, I would view the prospect of slingers lining up near me with some trepidation. It rather looks as if one had to brave not just the enemy slingers, but "friendly fire" from one's own.

Another use of slingers—and this was a role that the catapult adopted, supplementing rather than displacing the slingers—was against defenders on the battlements during an attack on the walls. For example, Hannibal (not the Karthaginian general who marched elephants over the Alps to attack Rome, but an earlier namesake) used slingers in this way at Selinous in Sicily in 409 B.C., Alexander the Great used them at the siege of Gaza in 331, and the Romans used them against Arkhimedes and his compatriots at the siege of Syrakuse in 214. Slingers were also employed alongside catapults and archers at the siege of Tyre in 332/331, the siege of Utica in 307 B.C., the siege of Same in 189 B.C., and the siege of Volandum in A.D. 58 or 59, where Corbulo demonstrated conclusively how it *should* be done.[16]

> [Corbulo] divided the army into four groups. One, massed in the tortoise formation, he led to undermine the rampart; another he ordered to bring the ladders up to the walls; a strong group were to shoot bolts and spears from the military engines; and the libritores and funditores were assigned a position from which to throw their shot at long range. The objective was to threaten danger equally in all places so that pressure at one point should not be relieved by reinforcements from another. . . . Before a third of the day had elapsed the walls had been cleared of defenders, the barricades broken, the fortifications stormed . . . all without the loss of one soldier, and with extremely few wounded. (Tacitus *Annals* 13.39 Loeb modified)

*Libritores* and *funditores* were evidently two different types of stone-thrower in the Roman army, perhaps "slingers of one-pound stones" with a two-handed sling (*libra* = pound) and regular slingers, or hand-throwers of one-pound stones and slingers.[17] Slings and bows were not replaced by catapults; they were supplemented by them. Catapults had slower rates of fire than either, but could instead deliver heavier missiles further and with greater impact.

Slingers were also employed against ships. At first sight, this might seem a bit bizarre: how could a slinger with his little bag of stones possibly harm a mighty naval warship? But the slinger was not aiming at the ship's timbers; he was aiming at its marines or its power source. Ancient warships went into battle under oars, not in sail. Propulsive power depended on all rowers working together as a team; in the case of the trireme, for example, 170 oarsmen moving in synchrony to raise, lower, and pull their oars. A rower hit by a slingshot would probably lose his rhythm at least, and that could disrupt the functioning of the entire ship, even diverting it from its course. (There were even more oarsmen on a quinquireme, though the problem would have been mitigated on the bigger polyremes, wherein oarsmen shared a sweep.) Thus we find slingers among the so-called "marines" on the deck of a warship, shooting without pause at any ship bearing down upon their own.

> All the time that another vessel was bearing down, the men on deck shot showers of javelins and arrows and stones upon the enemy, and when the two ships closed, the marines fought hand to hand, and endeavored to board. (Thucydides *History of the Peloponnesian War* 7.70.5)

Those oarsmen on the outrigger were afforded very little protection by the framework of the boat, and received higher wages than the other rowers on board, who were nestled in the comparative safety of the hull. Greek warships were manned overwhelmingly by the free, not slaves. Citizens and metics (resident aliens) who could not afford hoplite armor were expected (but not required) to do military service when necessary, either as light-armed troops or as rowers. In due course, other sorts of ships were decked, specifically in order to protect the rowers, and the platform thus created also made space for range troops and catapults to deploy.[18]

Dionysios Halikarnassos mentions another kind of slingshot. "Standing on the four-wheeled wagons were also many light-armed, archers and stone-throwers and slingers of iron *triboloi*." This is the "three-points," sometimes called caltrop, which resembled the device used now to puncture car tires, or a large sharp version of a jack from the game of jacks. Although the name means three-point, they are regularly made of four spikes joined together at their bases to make a three-sided (pyramid) shape (Figure 1.1). Dionysios' description implies their use individually by slingers: a particularly unpleasant sort of slingshot.

*Triboloi* are also mentioned by Polyainos, who says that Nikias on the Athenian expedition in Sicily (so 415–413 B.C.) strew *triboloi* over the ground to disable the Syrakusan cavalry, and then sent in light troops wearing shoes with stiff soles to dispatch the casualties. This is a much more common form of employment of these objects: as a sort of nonexplosive mine, rather than as slingshot. Philon, after describing circumstances in which *triboloi*, broadcast like grain, might be used for defensive purposes, says that once detected, they could be extracted from the ground with a two-pronged mattock, then swept up by others with a garden rake. This is a description of a minefield and its clearance technique, about 200 B.C. They might also be scattered in shallow water to hinder enemy movements. *Triboloi* were used, essentially unchanged in form and function, throughout antiquity, especially against cavalry and camelry. Vegetius recommended their use against scythed chariots.[19] Prokopios describes them fully for his noncombatant audience in the sixth century A.D.

FIGURE 1.1 *Tribolos. These are usually a couple of inches across or smaller. They were broadcast like grain to create non-explosive "mine-fields," and slung as missiles. (Author photograph)*

> These triboloi are like this. Four spikes exactly equal in length are fastened together at their bases so that their points form the outline of a triangle on every side. They throw these randomly on the ground, and because of their design, three spikes stick very firmly into the ground while the remaining one sticks up, and always present an obstacle to men and horses. As often as anyone turns over a tribolos, the spike which hitherto happened to point straight up in the air becomes stuck in the ground, but another one takes its place above, continuing to present an obstacle to those who want to attack. (Prokopios *History of the Wars* 7.24.16–18)

The anonymous author of a *Strategy*, who was writing at about the same time as Prokopios, advised laying a *tribolos* field twenty cubits wide beyond the camp ditch, as a regular defensive measure. He would make a mid-ranking officer responsible for gathering up the *triboloi*, partly to ensure their availability when required and partly to prevent "friendly fire" casualties among his own troops when they left the camp. Decades later, in Maurikios' time (c. A.D. 600), *triboloi* were not only regularly deployed around camps for defensive purposes, but were attached by small strings to an iron peg to facilitate quick recovery when the army marched on. The Anonymous' belt-and-braces approach also commends setting trip-strings with bells on beyond the *tribolos* field, to give advance warning of a stealth attack by the enemy at night.[20]

The Anonymous advised equipping the rear guard with *triboloi* to scatter during a retreat, so that enemy cavalry would be stopped in its tracks or risk greater injury than it might inflict. To counter such a move being employed by the enemy, he suggested fitting iron shoes to horses' hooves specifically to protect against *triboloi* "and suchlike."[21] These removable horseshoes are known as "hipposandals" (Figure 1.2) and one type was probably intended to defend against *triboloi*. (Other horseshoes were developed for other purposes, for example, to hold veterinary dressings in place, or with crampons for icy

FIGURE 1.2 *Hipposandal. These were designed to be fitted temporarily to horses' feet, strapped on via the loop and hook. There were several different types; this type would have prevented injury when crossing triboloi fields. (Sketch of Museum of London A1296 by T. D. Dungan)*

conditions.) These horseshoes were temporary fittings, slipped over the hoof, strapped or tied in place, and worn only as required.

Besides the practical reasons noted above for dislike of slingshot and slingers, there were social reasons for disliking them. Peoples with a reputation as good slingers often came from areas of Greece that a sophisticated Athenian like Thucydides might describe as a bit wild and woolly; mountainous and marginal are alternative descriptors, as are poor and pastoral.[22] There is probably a causal connection here. Of all military forces, the slinger costs least. He can make his own sling from readily available materials, and *in extremis* he may use any stone lying on the ground, though it was recognized that water-worn pebbles were particularly suitable for throwing. He can carry many missiles in a bag over the roughest terrain. He is well suited to warfare in mountainous areas, where there was no room for hoplites to follow their rituals of engagement.

For a variety of reasons, including economic, the sling was considered the weapon "most fitting for a slave," as Xenophon put it in the *Kyropaideia*. This was a society of masters and slaves. A free citizen who was wealthy enough to afford armor, or even more so, a horse, would ordinarily have been a slaveholder. He was used to dominating a slave or two. The heavily armed hoplites habitually fought each other face to face, looking straight at death or glory in the eyes of the man he would kill or be killed by. The idea that someone else's slave might injure or kill him on the battlefield, from afar, using the cheapest weapon available, was inglorious, to say the least. Range weapons such as stones, slingshot, and arrows were cowards' weapons, in these men's eyes. They acknowledged that the sling was a difficult weapon to master, requiring skill and training to use well, but it required no *courage*. There was, in short, no honor in the weapon; and so, by extension, there was no honor in the men who used it, nor in the death that was caused by it.[23] Nevertheless, the sling is a surprisingly powerful weapon, and its use in war continued at least until 1936, when it was used in anger in the Spanish Civil War.[24]

I said above that the slinger wears no armor: no shield, helmet, breastplate, or greaves. This is of course a generalization, and it is usually true for the ancient Greeks from the Peloponnesian War (431–404 B.C.) on. But before then, in sixth century archaeological material (in particular bronzes and images on pots), we find people who are wearing armor holding bows and arrows and stones; for example, one or more of helmet, cuirass, greaves, and even shields. Literary evidence suggests the same: the seventh-century Spartan war poet Tyrtaios depicts slingers with shields, acting also as spear-throwers, and apparently interspersed between hoplites. "But you, light-armed soldiers, crouching beneath your shields at various points in the line, hurl your great stones and cast your wooden javelins against the enemy, arrayed next to the heavy infantry."[25] We also have images of hoplites carrying javelins and holding stones, so the distinction between hoplite and light-armed was evidently fluid until about the time of the Peloponnesian War.

As mentioned, stones were not used just as sling weapons. Larger stones thrown by hand also played a role in ancient warfare. Figure 1.3 shows hand-stone throwers on the Nereid monument from Xanthos in Lykia, now in the British Museum.

Homeric gods and heroes such as Agamemnon, Hektor and Aias threw stones (suitably big ones, of course). About half a millennium after those epics were written down, i.e., around 200 B.C., Philon advised that houses near the city walls be equipped with stones for the hand, as well as spears, arrows, and large stones, as a secondary defense in the event of a breach in the walls. And when Khosroes of Persia attacked Antioch in A.D. 540, most of the young men, being unarmed, simply threw stones at the besiegers.[26]

According to Plutarch, Mardonios, who was made commander-in-chief of the Persian forces in Greece when King Xerxes returned to Asia, was killed at the battle of Plataia (479 B.C.) by a stone thrown by a Spartan called Arimnestos, which struck him on the head. This was a real triumph for stone-throwing, as well as for Sparta. Killing Mardonius was the key to the victory, for the Persian forces lost heart at that point in the battle, and this victory ended their invasion, resulting in the Persians being driven out of Greece. At the other end of antiquity, John, son of Thomas, commander of the Roman forces in the area now known as Georgia, was hit on the head and killed by a thrown stone at the battle for Petra in Lazika in A.D. 550. Another famous victim of a stone was Alexander the Great, who had a narrow escape (one of

FIGURE 1.3 *Hand-stone throwers on the Nereid monument from Xanthos in Lykia. (British Museum, author photograph)*

many) when he was hit hard on the back of the neck and head during the siege of Kyropolis in Hyrkania in 329 B.C.[27]

> The barbarians, although they saw that their city was already in the enemy's hands, nevertheless turned on Alexander and his force, and launched a vigorous onslaught. Alexander himself was struck violently with a stone on his head and neck and Krateros was wounded by an arrow, as were many other officers. (Arrian *Anabasis* 4.3.3)

Alexander's helmet presumably absorbed most of the force. The consequences are described in most detail by Plutarch and Curtius, the former when listing a variety of wounds he sustained on the expedition.

> On the banks of the Granikos his helmet was cleft through to his scalp by a sword; at Gaza his shoulder was wounded by a missile; at Marakanda his shin was so torn by an arrow that by the force of the blow the larger bone was broken and extruded. Somewhere in Hyrkania his neck was hit by a stone, whereby his sight was dimmed, and for many days he was haunted by the fear of blindness. Among the Assacenians his ankle was wounded by an Indian arrow; that was the time when he smilingly said to his flatterers, "this that you see is blood, not 'ikhor, that which flows from the wounds of the blessed immortals.'"[28] At Issos he was wounded in the thigh with a sword, as Khares states, by Dareios [III] the King, who had come into hand-to-hand conflict with him. . . . Among the Mallians [in India] he was wounded in the breast by an arrow three feet long, which penetrated his breastplate, and someone rode up under him, and struck him in the neck, as Aristoboulos relates. (Plutarch *On the Fortune of Alexander* 9 [*Moralia* 341b–c] Loeb)

Quintus Curtius Rufus tell a slightly different and more detailed version of the injury and resulting symptoms.

> No other city withstood siege more stoutly [than Kyropolis], for the bravest of his soldiers fell and the king himself was exposed to extreme danger. For his neck was struck with a stone with such force that darkness veiled his eyes and he fell and even lost consciousness; the army in fact lamented as if he had already been taken from them. . . . [Days later] Alexander, about to wage an unforeseen war with this race [of Skythians], when the enemy rode up in sight of him, although still ailing from his wound, and especially feeble of voice, which both moderation in food and the pain in his neck had weakened, ordered his friends to be called to a conference. It was not the enemy that alarmed him, but the unfavorable conditions of the times; the Bactriani had revolted, the Skythians also were provoking him to battle, he himself could not stand on his feet, could not ride a horse, could not instruct or encourage his men. . . . Even his own men hardly believed that he was not feigning illness. . . . So much had he spoken in a voice faltering, broken all the time and heard with difficulty by those who were beside him. (Curtius *History of Alexander* 7.6.22–7.20 Loeb)

He seems to have suffered short-term blindness and temporary speech dysfunction from the blow of this stone.[29]

Stone-throwing by hand required training, too, and advance preparation of ammunition to function well. Livy explains how the Galatians, who were used to fighting hand to hand with swords, used as missile weapons whatever stone came to hand in their desperate defense on Mount Olympus. "But like men untrained in their employment," they chose stones of unsuitable size, and threw them without skill or strength, so that even if they did hit their target they were ineffective.[30]

LIKE SLINGERS, ARCHERS ARE OFTEN overlooked in histories of warfare, but there are telling signs from all periods that they were common in battle. It has been estimated that more people have been killed by arrows than by any other weapon in the history of warfare, and it is still an effective weapon in certain circumstances. Some units in the Chinese army retained bows until the beginning of the twentieth century. Arrows killed about 5,000 soldiers in World War II, and they were used in the Vietnam War, too. In our own time, injury and death by arrow or slingshot is more likely to be accidental than deliberate (though murder by bow and arrow is not unknown), but the results are just as devastating. Moose, buffalo, bear, lion, rhinoceros, and even elephant have all been killed by bow and arrow in modern hunts.[31] The arrow kills, if it does so, by deep penetration of its sharp head; the victim bleeds to death. Or it may be a poisoned arrow, in which case it kills chemically. Use of poisoned arrows in ancient Greek culture gave us the word "toxic," for it comes from τοξικός, toxikos, which means "for the bow." Philon names "Arabian poison," a mussel poison, mistletoe, the venom of lizards, vipers and asps, naphtha, and fish glue as toxins to contaminate waters and anoint missiles.[32]

The origins of the bow are lost in time. It appears to us first in paintings from the Paleolithic, the Old Stone Age. Early bows were self-bows, made of one piece of wood, and the self-bow has a long and distinguished history that stretches down millennia to the English longbow.[33] Composite bows were invented as early as the fourth millennium B.C. in the Caspian Sea area. The Greeks appear to have adopted the composite Skythian bow from the early seventh century B.C., when Greek colonists first settled on the north coast of the Black Sea, though the role of the Kretans in preserving and disseminating native Greek reflexed self-bows may have been greater than is generally recognized.[34] The Skythian bow then diffused through Greek communities around the Mediterranean, and via them it spread as far west as the Seine River in France. The Etruscans then used the Skythian bow, too, as did the Romans and others. Other distinctly different types of bow developed and were employed at different times; for example, long-eared, short-limbed "Sassanid" bows appeared by the third century B.C., and the Hunnic (or "Qum-Darya") bow, found at a Chinese border post dating from the first century B.C. to the third century A.D., became popular in Roman forces in the fifth-sixth centuries A.D. Long ears, made of bone or antler and surviving in the archaeological record as so-called "bow stiffeners," give a mechanical advantage that allows a greater draw length while at the same time preventing stacking, so that the draw feels easier as it advances, instead of becoming harder and harder as it does with an English longbow.

It takes at least one year to make a good bow, though only about eight days' work on it is required during that period (so bows were usually made in staggered batches of fifty or more, depending on the number of people in the workshop). The especially season-sensitive and time-consuming parts of the operation involve the collection and preparation of components at specific times of year, and gluing, so that the glue sets slowly and thoroughly enough to ensure good adhesion of parts that are going to come under extreme stress.[35] The glue was commonly made from the liquor that results from boiling sinews; it needs to be as elastic as the rest of the bow.

The bow was a very effective weapon. It was good for the same sort of military purposes as the sling, and was sometimes subject to the same sort of prejudice in the ancient literary sources as was the sling. Training with the bow was part of the regular schedule for Athenian *epheboi*, young men undertaking two years' military training, from that institution's foundation in the fourth century B.C. Archers formed specific contingents in Roman armies, but other Roman troops sometimes used bows, too. Arrowheads are very common in the archaeology of border provinces such as Britain and Germany, where they often seem to indicate archers among auxiliary forces, mounted and foot. Archers were deployed regularly to screen an army in enemy territory, to provide missile support to legionary formations before and during battle, to protect the flanks on the battlefield, and to provide covering fire for river crossings and escalades of walls. Catapults cannot shoot downward, so could not be used from walls and towers for enfilading fire; consequently, archers continued to perform this defensive task on walls and fortresses, manning the parapets and shooting at those at the base of the walls throughout the centuries, even in Rome itself.[36]

Generally, armor would prevent an arrow penetrating very deep, and most arrow wounds that feature in the literary sources concern strikes to unprotected flesh. Achilles famously lost his life to an arrow in his heel. Philip II of Macedon, father of Alexander the Great, lost his right eye to an arrow at the siege of Methone in 354. Alexander the Great was several times hit by arrows, including one that broke a piece of bone from the tibia or fibula of his lower leg, but several times survived them.[37] Just what arrow wounds meant is well illustrated by Aulus Cornelius Celsus, who wrote an encyclopedia around the time of Christ of which only the medical parts survive, in which he describes various procedures for the removal of sharps and other missiles. I use the term "sharp" throughout this book for missiles that fit that description. Celsus observes that missiles are often very troublesome to extract, sometimes because of their shape, sometimes because of the position they are in. He recommends removing superficial and simple penetrations via the entry wound (extraction), after its enlargement, but if the missile has penetrated more than half way through the body, or has crossed blood vessels or sinews, he recommends cutting through the flesh from the other side and pulling it through (expulsion). He claims that this is an easier and safer procedure, and that the wound heals more easily because it can be dressed with a medicament at both ends, at least in the case of the limbs.

> In either case, the greatest care should be taken that no vein, nor one of the larger sinews, nor an artery, is cut. When any one of these is observed, it is to be caught by a blunt hook and held away from the scalpel. When the incision has been made large enough, the missile is to be drawn out, proceeding in the same way, and taking the same care, lest that which is being extracted should injure one of those structures which I have said are to be protected. (Celsus *De medicina* 7.5 Loeb)

According to him, arrows usually penetrate so deep that they should be expelled (extracted from the other side), which also minimizes damage from the barbs with which they are usually furnished. In the absence of X-rays and other modern methods, the missile's size, shape, and position would be established in each case by probing. He recommends nipping off small barbs and covering large barbs with split reeds during extraction. Sharps larger than arrows—which presumably means snapped-off spear heads or catapult sharps—were given different treatment, as he does not recommend making a second large wound in a patient who already has one. These were to be extracted via the entry wound, using some such instrument as the spoon of Diokles.

> The instrument consists of two [separate] iron or metal plates. One plate is turned down at the end, [rounded over] like a [broad] lip. The other has its sides bent up so that it forms a [shallow] groove, and its end is turned up slightly and perforated by a hole in the middle. The second plate is brought alongside the missile, then passed underneath it, until the point is reached. This plate is then jiggled until the point [of the missile] is engaged in the hole. Once the point has entered the hole, the other plate's lipped end is, with the fingers, slipped over the upturned end of the first's, and then the spoon and the missile are withdrawn simultaneously.[38]

In cases where the missile has forced its way into a joint between the ends of two bones, he recommends the following technique.

> The limbs above and below are encircled by bandages or straps, by means of which they are pulled in opposite directions, so that the sinews are stretched. The space between the ends of the bones is widened by this extension, and the missile is withdrawn without difficulty. If the missile is also poisoned, do exactly the same, but more promptly if possible, and give the treatment for poison drinks or snakebite. In all cases, care for the wound as any other.[39]

It is clear from this that, by the first decades of the first century A.D., a variety of strategies were known for a variety of types of wound caused by a variety of types of missile, some of them shot by catapult. Medical care had certainly improved a lot since the fourth century B.C., when a Hippokratic doctor lost a patient to a bolt (which we will discuss in Chapter 3), and when several casualties had the god Asklepios to thank for the miraculous removal of sharps that had been embedded, and sometimes suppurating, for months or years.[40] Celsus refers several times to the extraction of heads that had lost their shafts, and from what he says about such heads it is clear that he was not referring to the remnants of an arrow after the shaft had been snapped off deliberately in the field or suchlike. (In those cases, there would still be a small section of shaft protruding from the surface of the body, whereas Celsus is referring to foreign bodies buried in the flesh.) Socketed arrows are often designed and built to break just behind the socket on impact with a firm surface, since this prevents those that miss their targets from being shot back, and Celsus could be talking about them, or about short catapult darts.

Short catapult darts require some explanation. It has been suggested that some ancient arrows took the following form. The arrow had a softwood or reed shaft, into which an arrowhead was not inserted directly, but via a foreshaft, as it is called, of hardwood, about five inches (13 cm) in length. This was fashioned into two quite different shapes from midpoint to each end. The front half, which would carry the iron or bronze arrowhead, was lathe-turned and circular in cross-section, and formed a truncated cone. A hole, at least 1 inch (3 cm) deep, was bored into the center of the cone from its top, and this was designed to take the tang of an arrowhead. Once they were fitted together, this section of the foreshaft would probably have been bound tightly to hold the arrowhead firmly. The back half of the foreshaft, by contrast, was only loosely attached to the main shaft. One of the most sophisticated designs for this end of the foreshaft involved cutting the wood away, using a sharp blade, to leave three thin vanes, about 2 mm thick near the central axis line, and 1 mm thick at the edges, roughly equally spaced around the central axis line. The nearest analogy is the plastic flights on a British dart. And, like the modern British dart, these flights appear to have just slotted into the main shaft of an arrow; no trace of glue has been found on them, even on the incredibly well-preserved specimens of Roman date found at Qasr Ibrim in Egypt (Figure 1.4).[41]

Thus, the arrowhead was fitted to a short piece of hardwood, stronger and heavier than the softwood, cane, or reed of the main shaft. This would have improved the flight characteristics of the arrow, apparently. Then the foreshaft was fitted to the main shaft by vanes or, alternatively, a tenon that gripped without gluing or binding (so James and Taylor suggest). This could be expected to hold on launch, when the force was coming from behind and driving the shaft into the foreshaft, but, once embedded, to separate, with little encouragement, if someone pulled on the shaft to draw it backward—especially if the arrow head was stuck in bone. The easy separation of the foreshaft from the shaft was probably intended to inhibit if not prevent the enemy collecting and shooting back at their makers the many arrows that would inevitably have missed their targets. That they also made the victim's life more precarious and the doctor's task more difficult was probably an unintended consequence, if this is indeed a correct interpretation of the shortshafts.

It is possible that what might have begun as foreshafts were used in due course on their own as short darts; the shaft of an arrow is necessary only to span the distance between the hand holding the bow and the other hand holding the bowstring. The development of the catapult, with its slider to support the missile, removed the need for a shaft and allowed the head only to be discharged, like shot. The ancients certainly realized this, at least partially, over time, with shafts shrinking to 18 inches or less with the profile of the Dura bolt (and possibly much earlier, with the *kestros*, on which more later), which is the name given to a fantastically aerodynamic type of bolt found at Dura-Europos in Syria, dating to the second century A.D.[42] If some of the items currently considered "foreshafts" are in fact complete missiles, then they fully understood the point. Notice that the shaft on the left has been carved away at the "bowstring" end to reduce the width of the vanes; this additional and tricky work to the vanes would be necessary if this is, in fact, the end of the shaft and it had to fit between the fingers of a catapult claw. If, on the other hand, the vanes were supposed to slot into cuts in a long shaft, this extra work would be not just unnecessary but counterproductive. Certainly darts of little finger length existed from Byzantine times. Flight stability, which may be lost with very short sharps, is unimportant in an assault weapon, and a tumbling projectile usually inflicts more damage than a straight-flying one, which may pass right through the target doing relatively little damage en route.

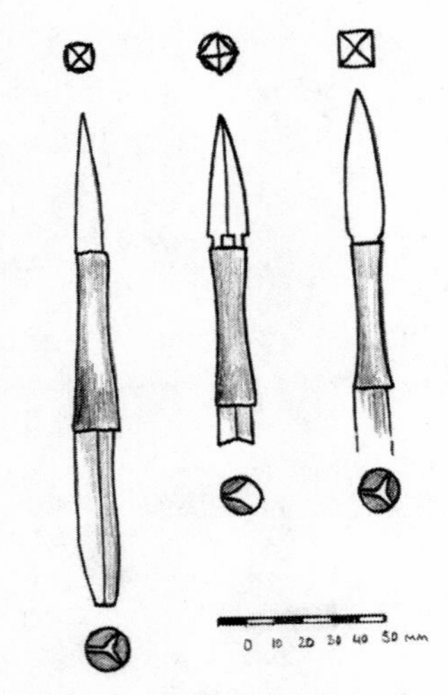


FIGURE 1.4 *Shortshafts. (Author sketch after Iriarte)*

TECHNOLOGICAL INNOVATION: WHY?

Both bow and sling were mechanized in the catapult. This was a momentous technological step. Both the bow and the sling had existed for millennia and had served their purposes well in hunting and in fighting. Why then did mechanization of the bow and sling happen? Before we tackle this issue in the next chapter, there is the general case to answer. Why was *any* war machine invented in antiquity? Why did the catapult happen *at all*? Was it *bound* to happen eventually? Is technical development "normal," like evolution in the natural world? Prokopios certainly seems to have thought so (see the epigram to this chapter), so did Polybios (see the epigram to chapter 2).

Crises, in the environment or society, have long been thought of as catalysts to innovation, as they seem to demand new responses. The notion that necessity is the mother of invention is present in classical authors. For example, Caesar noted that Trebonius' troops invented a completely new type of earthwork made out of brick because all the trees in the district had already been felled and used (*Civil Wars* 2.15). Thucydides points to other factors too: experience as a teacher of new ways, and more significantly, he seems to subscribe to the idea that, in technical areas at least, the new will always prevail over the old. His views are expressed in the form of a speech put into the mouth of a Korinthian ambassador trying to persuade the Spartans to go to war against Athens.

> This policy would hardly be successful, even if your neighbors were like you yourselves [Spartans], and in the present case, as we just pointed out, your ways compared with theirs are old-fashioned. As in the crafts (*technai*), so in politics, the new must always prevail over the old. In quiet times the traditions of government should be observed: but when circumstances are changing and men are compelled to meet them, much originality is required. The Athenians have experienced much, and therefore the administration of their state, unlike yours, has been greatly reformed. (Thucydides *Peloponnesian War* 1.71)

"Their ways," elaborated in the previous section, were described as follows:

> They are revolutionary, equally quick in the conception and in the execution of every new plan, while you are conservative—careful only to keep what you have, originating nothing, and not acting even when action is really urgent. They are bold . . . imprudent . . . hopeful . . . impetuous . . . always abroad . . . pursue victory to its end… fall back least. . . . This is the lifelong task, full of danger and toil, which they are always imposing upon themselves. None enjoy their good things less because they are always seeking for more. . . . They deem the quiet of inaction to be as disagreeable as the most tiresome business. In a word, if a man should say of them that they were born neither to have peace themselves nor to allow peace to other men, he would simply speak the truth. (Thucydides *Peloponnesian War* 1.70)

Now Thucydides' view of Athenian drive, ambition, and originality is very much tied up with the acquisition and maintenance of an empire (see also 6.18), but that should not be allowed to distract from the fact that he thought his people, and other imperialistic Greek states, were not motivated to innovate simply by crises in an external environment beyond their control. They themselves created the stimulus to change.

The assumption made by the technical artillery manuals, which we will examine closely in Chapter 7, is that people invent and develop tools to help them achieve some aim that they cannot otherwise accomplish, or to perform better some task that they can already do. "Better" might mean more easily, or more quickly, or more thoroughly, or more effectively, or whatever happens to matter to the person trying to innovate. The fact that some apes use things they find to hand, such as sticks or stones, as tools, suggests that this is a natural behavior, a result of brain size and function. Therefore, the "explanation" for people's use and development of tools *in principle* must lie in our genes. But that will not explain why any particular tool was invented, nor the particular form that it took, nor why it was invented at a particular time.

The reason for the first and last is a perceived need by a particular individual for a particular invention at a particular time in a particular place. The same need may be felt in different places or at different times, of course. That is why some things are independently invented in more than one place, as the wheel seems to have been, or are reinvented after being lost, as writing was, repeatedly. However, the reason why inventions take the particular *form* that they do is a much more difficult matter. One factor is the materials and techniques available; another is the inventor's knowledge of those materials and techniques, or their availability. This raises the issue that is at once the principal difficulty for the theoretician and delight for the historian: the idiosyncrasies of the individual. An invention is ultimately the result of a heady cocktail of perceived need, individual creativity and knowledge, and, if it is a material invention, the substances of its construction.

Historically, trial and error seem to have played a large role in the creative process. Education—by family members, co-workers, teachers, or rarely by written instruction—has then served to pass on the lessons learned by trial and error to others, so that they can reap the benefits without suffering as many losses that result from errors. However, education can also discourage further trials, although further trials in principle could produce even better results unless all *possible* trials had already been undertaken en route to the invention in question. And that, of course, could never have happened, because even if, by trial and error, the best solution was found with the materials of the time of the invention, new materials bring new opportunities and possibilities. There is a bias toward conservatism in human technology, which contributes to making change incremental rather than radical, usually. It is simply easier to try to improve an old design than it is to start from scratch with the proverbial blank sheet of paper or papyrus. Technological stagnation is what happens when education or dominant current interests block, rather than merely discourage, new trials.

Because people who use weapons of war entrust their lives to those weapons, they take the keenest interest in their technological capabilities, in themselves and in comparison with other, especially opposing, weapons. As

Quintus Curtius Rufus observed, in a speech he put into the mouth of Alexander, "Many inventions by which we have so far been victorious will recoil upon us [if we do not cross the river and defeat the opposing army now]. The fortune of war teaches its art even to the vanquished."[43] Over time, changes were required or desired by, for example, changes in circumstance of use, materials of construction, intended targets, or in the enemy's weapons (new weapons often appear in the hands of new enemies), and new designs were tailored to such requirements. These changes were generally triggered by contact with new people or new environments. For example, if an archer were to move from open country into jungle, then one would expect a reduction in the bow size to reflect the fact that a long-limbed bow might snag in the undergrowth. The same would be true of an archer who stepped onto a chariot, or who took to horse, for the same reason. Or, if the enemy changed the sort of armor that they used, then archers would probably need to change their arrowheads to suit. For example, chain mail is armor full of holes for a narrow-headed arrow or dart, which can fly straight through one of the holes. Plate armor can resist the average arrowhead, but the bodkin point can punch a hole straight through it (and through bone, as the Maiden Castle skull amply demonstrates).[44] There is an element of "scissors, paper, stone" to this sometimes, and it is clear from Roman reliefs such as Trajan's column that various types of defensive protection (chain mail, segmented armor, scale armor) co-existed, presumably because various sorts of offensive weapon co-existed, too. In other words, there was not a straightforward succession from one type to another. However, increased range was *always* advantageous; to be able to strike the enemy before he could strike back was the ideal situation, at least until his technology caught up.

There is a tendency in the literature to assume a significant level of inertia in ancient weapon development. "If it ain't broke, don't fix it" is a perfectly reasonable maxim to be held by a person whose life will depend on "it" working today just as it did yesterday. However, machines do not spring off the drawing board fully formed and fully functioning, and if modern experience is a good guide, that is more true of war machines than of any other type of machine. I imagine radically different designs on paper giving way to a few prototypes through the elimination of designs that, for one reason or another, were thought or shown to be unworkable. Prototypes may then have been constructed of a number of different designs, to compare performance. In ancient times, it is not obvious how the advantages and disadvantages of different designs could have been evaluated *except* by trial and error, and trial requires the device to be built, as a scale model if not full size.

Indeed, Philon specifically says that the calibration formulae (which are explored in detail in Appendix 1: "The Calibration Formulae") were discovered through a process of trial and error. There would be no point trying a number of identical machines. They *must* have varied, and why should the variation have been restricted—at least in the early days—to caliber and spring size? Surely the relationship between caliber and spring size was discovered after an earlier process eliminated other, less successful, structural variations or other features that could, in principle, and perhaps on some machines did, in fact, work. We might compare the warships known as the trireme, quadrireme, and the quinquireme (3-er, 4-er, and 5-er, respectively); all were developed, but the quinquireme and trireme were found to be better and became the mainstays of the Hellenistic and Roman fleets, while the quadrireme did not. I believe that there was significantly more variety than is generally allowed in the design of ancient catapults, both of bow and torsion types, especially in their early days, before torsion was adopted, and in their late days, when the formulae were apparently lost. Diversity and change seem to be characteristic of isolation (compare the homogeneity of modern life, which is becoming a global monoculture in the presence of saturated communication), and there may have been less technological variety during the High Roman Empire precisely *because* there was more communication. This seems to be a feature of the natural world, too.[45]

The torsion catapult is a completely different entity from the bow catapult. Indeed, they are *so* different that it is far from obvious how the former developed from the latter, though that is what is generally assumed to have happened. We will explore this further in Chapter 3. Having invented the "classic" two-armed catapult, the Greeks went on to try some radically new materials and new arrangements to produce other kinds of catapults. Ktesibios, who was active about 270 B.C., was particularly creative, inventing the compressed-air catapult and the metal-spring catapult. These were fundamentally different from any pre-existing catapult, and from each other, in their employment of utterly different materials for their sources of power: bow catapults were powered by a combination of wood, sinew, and horn; torsion catapults were powered by hair or sinew; Ktesibios' catapults were powered by air, and by metal.

Just because, more than two thousand years later, we only know about a few designs by some (extraordinary) mechanical engineers, it does not follow that only these people invented only these machines. Doubtless there were many more people and many more catapult designs in existence during antiquity: Greco-Roman civilization lasted about a thousand years and covered about a million square miles, and it would be absurd to think that we know about everything significant that happened in it. If we look at the history of technology in more recent times and places, the story is always much more complicated than it appears at a distance. Obviously, if evidence does not survive from antiquity then we cannot create it. But we can resist the temptation to squeeze all evidence that we have got into the corsets of a handful of well-known designs, some of which fit the evidence very badly indeed. This has happened in the past with the steel bow catapult, for example, on which more in Chapter 10.

Effort on designing and developing innovations was not confined to catapults, and sometimes catapult technology seems to cross over to and from other areas of military technology. For example, the Thessalians, early pioneers in catapult design, were credited with the invention of a particular kind of sling missile called a *kestros*, a type of bolt. It is described as having an iron head about two palms long (about six inches or 15 cm), on a wooden shaft about nine inches long (c. 23 cm) and a daktyl (finger) thick.[46] Instead of feather flights, it had three short wooden wings.

> The so-called *kestros* was a new invention at the time of the war with Perseus. The form of the missile was as follows. The pipe (?) was two palms long, the same length as the point. Into the pipe was fitted a wooden shaft a span in length and a finger's breadth in thickness. To the middle of the shaft were firmly attached three quite short wing-shaped vanes. The thongs of the sling from which the missile was discharged were of unequal length, and it was so inserted into the loop between them that it was easily freed. There it remained fixed while the thongs were whirled around and taut, but when, at the moment of discharge, one of the thongs was loosened, it left the loop and was shot like a lead bullet from the sling, and striking with great force, inflicted severe injury on those who were hit by it. (Polybios 27.11 Loeb modified)

This war is otherwise known as the Third Macedonian War, 171–168 B.C. Livy's version is more brief (42.65).

> [The Roman soldiers] suffered particularly from the *kestrosphendones* (dart-slings). This was a new kind of weapon, invented in that war. A sharp iron, two spans long, was set in a wooden shaft of half a cubit in length, a finger in breadth; around the shaft in the middle three short fir wood flights like those usual on bolts were attached. The sling had two unequal thongs; when the slinger swung and spun it by the strap with an extra effort, the missile, shaken loose, shot out like a slingshot. (Livy 42.65.9–10 Loeb modified)

As Livy tells it, this sounds even more dangerous than ordinary slinging. The description of the *kestros* as "cruel" on a gravestone at Gonnoi in Thessaly for a Greek called Dikaiogenes, who was apparently a casualty of this weapon, perhaps suggests that he was a victim of "friendly fire," perhaps during the war in which it was invented—the date at least is about right.[47] The Suda adds little more.

> *Kestros*: this was an invention made before the War with Perseus. The missile was two palms long, the pipe equal to the head. The shaft fitted into this, a span in length, a finger thick in diameter. In the middle of this were three wooden wings that had been made like wedges, really short. To shoot them, unequal cords of a sling, fitted to the middle, one of them easily released. The rest of them stretched in the spin. When the other cord is released, the projectile issues from the thong and is carried off exactly like lead shot from a sling, and strikes a forceful blow, bringing evil to those it happens to hit. (Suda s.v. κέστρος)

The Bertrand brothers in 1874 made a model of this sort of sling (Figure 1.5).

This *kestros* sounds remarkably like a Dura bolt. It has usually been assumed that these bolts were for delivery by catapult—they are totally inappropriate for launch by hand-bow—and so they may have been, but alternatively they may have been launched by sling. There is surely a connection between the *kestros* and the *plumbata*, the lead-weighted dart, thrown by hand and known best from Vegetius' work written in the fourth century A.D.[48]

But 171 B.C. (the Third Macedonian War) was not the first appearance of the *kestros*—the word, if not the missile. The term was also used of a type of sharp deployed by the Romans from ox-drawn wagons in a battle with Pyrrhos near Asculum a hundred years earlier, in 279 B.C. Dionysios Halikarnassos, describing the battle, says the *kestros* was "dagger-like." An ancient dagger was usually short, broad, and leaf-shaped. It appears from his account that these sharps were not shot into the air, but were fixed to some structure in the wagon. Indeed, these wagons were machines of a type that now seems quite bizarre, for use in the field, and they were being used, in effect, as armored cars. They bristled with sharps or fire, and carried some men who, together with the oxen that provided the motive power, were protected by surrounding screens of wattle. They also offered protection to more men taking cover alongside.

> Outside the line [of infantry] they stationed the light-armed troops and the wagons, three hundred in number, which they had prepared for the battle against the [19] elephants. These wagons had masts, on which were mounted [as on one-legged cranes] movable beams that could be swung round, in any direction desired, as quick as thought. On the end of the beams there were tridents, or *kestroi* that were dagger-like (μαχαιροειδεῖς), or solid iron sickles. Alternatively, they dropped down heavy beaks from on high. Many of the beams had attached to them and projecting in front of the wagons

> fire-bearing grapnels wrapped in tow that had been liberally daubed with pitch, which men standing on the wagons were to light as soon as they came near the beasts and then rain blows with them on their trunks and faces. Standing on the four-wheeled wagons were also many light-armed, archers and stone-throwers and slingers of iron *triboloi* (three-points); and on the ground beside the wagons were still more men. (Dionysios of Halikarnassos *Roman Antiquities* 20.1.6–7 Loeb modified)

According to Dionysios, these wagons were not adequate to the task. They worked initially, to bring the elephants to a halt, but then the men in the towers on the elephants' backs had the height advantage in the exchange of javelins, and once troops cut through the screens around the wagons and hamstrung the oxen, they were sitting ducks. The vehicles were then promptly abandoned by the light-armed stationed upon them; perhaps that is why we do not hear of such wagons again. Most innovations don't work, and deserve to be consigned to the dustbin.[49]

A diminutive form of the name *kestros* occurs even earlier, in an inscription from Athens, which is dated to 306/305 B.C. It appears amongst the treasures of Athena and the other gods, together with the first reference to a dual-function catapult ("a stone throwing and sharp-casting catapult," apparently designed by one Bromios).[50] Unfortunately, the damage to the stone increases as we go on, and although *kestrion* is mentioned in a military context, with generals before it, and Macedonians after it, and javelins, arrowheads, and catapult bolts—apparently lots of catapult bolts—what it means here is pure speculation. But diminutives do not normally precede substantives, and this suggests that the *kestros*, as a kind of sharp, goes back to the fourth century B.C. at least.

From present evidence it is impossible to decide whether the *kestros* was shot first from slings and was then used with catapults, or was developed for use with catapults and then slings were adapted to take it; it could even have begun life as a hand-thrown dart. In Athens, the military training officer called the *katapaltaphetes*, the catapult-discharger, was apparently replaced at the end of the first century B.C. by the *kestrophulax*, the *kestros*-guard.[51] Is this because the Athenians abandoned catapults, as has been suggested? I cannot imagine so. More likely, it seems to me, the change of title recognizes the military trainer's specialist knowledge of the latest addition to the broader range of catapult munitions.

Thus we see that the development of the catapult and its missiles did not take place in isolation from other munitions, and that a complex interplay between peoples and troop types was operating, rather than a linear progression. Change in one area could feed or prompt change in another, sometimes major, sometimes minor, sometimes transforming. The battlefield was a harsh testing ground that must have eliminated many more designs than those we know of, some deserving to fail, others victims of bad luck or bad management. We will see examples of these in later chapters.

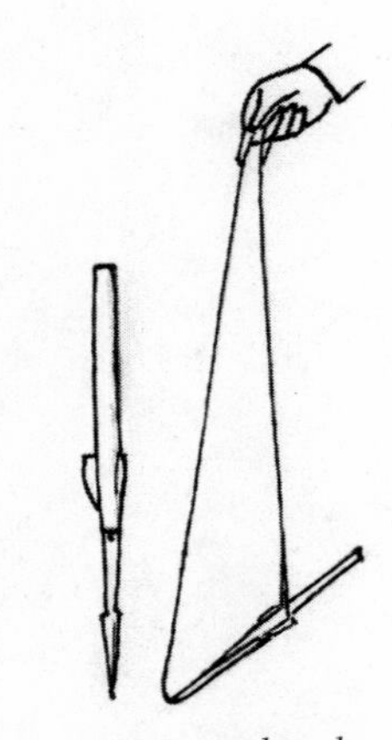

FIGURE 1.5 *Kestrosphendone. (Author sketch after Daremberg & Saglio 1877.)*

To conclude this discussion, my answer to the question "why was *any* war machine invented in antiquity?" is, in brief, because people naturally innovate, and they design and make tools to try to overcome what they perceive to be the constraints that prevent them from achieving their goal. In this particular case, the manual bow was developed to a point where it was recognized that the *human operator* was the limiting factor. The bow in itself had not reached its technical limits, and could be (and was) improved further by replacing those elements where humans had reached *their* limits with mechanical devices which were stronger. This meant, with a more powerful bow, the drawback—something was needed to replace the archer's arm and shoulder muscles, and the release mechanism—something was needed to replace the archer's fingers.

Let us then turn to the invention of the catapult.

TWO

# The First Catapult

*The love of innovation natural to man is, in itself, enough to produce any kind of revolution.*—Polybios 36.13

UNUSUALLY IN THE HISTORY OF TECHNOLOGY, the catapult seems to have a clear origin in time and space. To wit: 399 B.C. or thereabouts, on the island of Sicily, in the city of Syrakuse, which was then under the leadership of the tyrant Dionysios I. After setting the historical scene, I will describe this first catapult in detail.

Sicily was rich and fertile, but frequently disturbed by fighting between its various communities: Greek, Karthaginian, and native Sikel, Sikan, and Elymian. Syrakuse was a large and powerful Greek city that fifteen years earlier had destroyed an Athenian armada of two hundred ships; that fleet had been sent to conquer Syrakuse in particular and Sicily in general on the pretext of aiding another of the cities there (Segesta). Then in 405, Dionysios assumed dictatorial control of hitherto democratic Syrakuse during a Karthaginian invasion of the island. He bought time by gifting territory and possessions to the Karthaginians, and prepared thoroughly for a new war with them.

> He gathered skilled artisans (τεχνῖτες), commandeering them from the cities under his control and attracting them by high wages from Italy and Greece and Karthaginian territory. His intention was to make weapons in great numbers and every kind of missile, and triremes[1] and quinqueremes too, no ship of the latter size having been built before that time. After assembling many skilled workers, he divided them into groups in accordance with their skills, and appointed over them the most prominent citizens, offering great rewards to any who created a supply of arms. He distributed amongst them examples of each kind of armour, because he had gathered his mercenaries from many nations. For he was eager to have every one of his soldiers armed with the weapons of his [own] people, thinking that by such armour his army would, for this very reason, cause great consternation, and that in battle all of his soldiers would fight to best effect in armour to which they

> were accustomed. Since the Syrakusans enthusiastically supported Dionysios' policy, rivalry rose high in the manufacture of arms. Not only was every public space such as the porticos and back rooms of the temples (ἐν τοῖς προνάιος καὶ τοῖς ὀπισθοδόμοις τῶν ἱερῶν), the gymnasia and colonnades of the market place, crowded with workers, but the manufacture of huge quantities of arms went on in the most distinguished homes, as well as in public places.
>
> In fact, the catapult was invented at this time in Syrakuse, since the best artisans had been gathered from everywhere into one place. The high wages, as well as the numerous prizes offered to the workers who were judged the best, stimulated their zeal. In addition, over and above these factors, Dionysios circulated daily amongst the workers, talked with them in kindly fashion, and rewarded the most committed with gifts and invitations to his table. Consequently, the workers brought unparalleled devotion to the creation of many missiles and engines of war that were unfamiliar and capable of rendering great service. He also began the construction of triremes and quinqueremes, being the first to think of the construction of the latter. (Diodoros Siculus 14.41.3–42.2)

The result of all this effort was 140,000 shields, 140,000 swords, 140,000 helmets, 14,000 breastplates, "catapults of every kind and a large number of other missiles."[2]

Our source Diodoros was relatively well informed on the history of the island because he came from Sicily (hence his "surname" Siculus). He wrote in the first century B.C., so he was relatively close in time to these events as well. One of the sources Diodoros used was Philistos, who was a friend and courtier of Dionysios, so *his* account was that of an eyewitness to the extraordinary events of Dionysios' tyranny. When Diodoros talks about what was seen by a man looking around Syrakuse, astonished at the scale, vim, and vigor of the activities, it is probably through Philistos' eyes that he, and now we, look. For these and other reasons (to be discussed below), most historians have, after due and careful consideration, accepted Diodoros' assertion that the catapult was invented then and there. However, there are rival claims and we are duty-bound to review them.[3]

The first rival claim is that the catapult was invented in the Near East, in or before the eighth century B.C.; the second claim is more vague, that catapults were being used by Greeks east and west before 399. We will deal with the first and more concrete challenge here; the second was put without reference to, and apparently in ignorance of, the large body of scholarship on the history of the catapult, and does not therefore merit refutation. The first claim developed like this. Before the 1970s, most ancient historians followed

Diodoros Siculus' testimony that the catapult was invented in Syrakuse shortly before it was used in the attack on Motya, which took place in 398 B.C. However, at the same time, scholars of the Near and Middle East were following the testimony of the biblical Chronicles, which suggested that Uzziah[4] made sharp-caster and stone-thrower catapults somewhere around the generation-long period 780–750 B.C.

> And Uzziah prepared for them throughout all the host shields, and spears, and helmets, and habergeons, and bows, and slings to cast stones.
> And he made in Jerusalem engines, invented by cunning men, to be on the towers and upon the bulwarks, to shoot arrows and great stones withal. And his name spread far abroad; for he was marvelously helped, till he was strong. (2 Chronicles 26.14–15 [Authorized King James Version])

Now one cannot be too careful with literary sources, because they—just like the rest of us—can make mistakes, and it is very easy to misinterpret them, especially if one is dependent on only a brief piece of text. For example, according to Josephus, who was an aristocratic Jewish priest who went over to the Romans after the siege of Iotapata and who wrote an account (in Greek) of the siege of Jerusalem in A.D. 69 (in which he participated, on the Roman side), the Jews inside the city were inexperienced with catapult technology.[5] This might be taken to speak against a long history (never mind origin) of the catapult in this area of the world. But in fact artillery had been used in anger in this region since Alexander the Great passed through in the 330s B.C. (at the latest). So what is the correct interpretation of this comment of Josephus? He goes on to say that the Jews got better at using their catapults during the siege, because of their daily practice with them. Thus it transpires that, contrary to one's first impressions when reading Josephus' account, their inexperience was the result of prolonged peace and lack of practice, rather than unfamiliarity with catapults per se.

Something else is going on with Chronicles; the Chronicler was not writing history as we (and even Josephus) understand it. He "cares less about what happened than about the meaning of events. He is an *interpreter* of history . . . the Chronicler has a tendency to 'modernize'—describing events in terms the people of his own day would understand."[6] Thus Chronicles, which was probably written about 250 B.C., was probably anachronistic when making its brief assertion that Uzziah made machines to hurl missiles and large stones. Marsden and Garlan argue, rightly in my view, that it is simply not enough, quantitatively or qualitatively, to outweigh Diodorus' full, detailed, consistent, and coherent account of catapult invention in or about 399 in Syrakuse. On top of that, we cannot ignore the *absence* of *any* reliable evidence for the use of artillery *where we should expect it*, that is, at sieges, if it did, in fact, exist before 399.[7] It is hard to overestimate the importance of this point. If we drew a graduated timeline of the history of the catapult the length of this page, and Uzziah's putative mid-eighth-century catapults were at the top of the page and the first dated inscription mentioning catapults (i.e., bombproof evidence of their existence at the time it was carved) was at the bottom of the page, then the top 90 percent of the page would be empty and all the earliest evidence for the catapult would occur in the bottom 10 percent.[8]

It is not just the Chronicles verse that is rejected as unhistorical by the anachronism-plus-absence argument; three Greek and Latin passages are rejected too. Polyainos *Strategems* 7.9, which mentions sharp-caster catapults in Egypt during the Persian King Cambyses' siege of Pelusion in 525 B.C., is probably unconscious anachronism like the Chronicler's. Pliny *Natural History* 7.56, which credits the invention of a type of catapult called the scorpion to one Pisaeus son of Tyrrhenus and another type of catapult called the ballista to the Syrian-Phoenicians, is vague and lacks any supporting evidence. And the unique word *katapeltazontai* (καταπελτάζονται) in Aristophanes' *Akharnians* 160 (performed in 425 B.C.) is taken to mean "they will overrun with peltasts"—peltasts being the common name for troops equipped with a kind of shield called a *pelta* (as hoplites are troops equipped with the *hoplon* shield)—rather than a third-person-plural future of an otherwise unattested verb meaning "to catapult."

In addition to the evidence of absence before 399 B.C., there is other positive evidence for a Greek origin: *katapaltes* or *katapeltes* (variant spellings) is a Greek word, composed of the preposition *kata*, with the basic sense of "down," and either the noun *pelta* meaning "shield" or *palta* from the verb *pallein*, meaning to throw. A Greek name speaks against a non-Greek origin. When the Greeks adopted foreign things, they generally adopted the name too. This is a common phenomenon; for example, in our own time, "football" is "football" in places where the words for "foot" and "ball" are quite different.

But the hypothesis that the catapult was invented in the Near or Middle East was revived two decades ago when new material came to light in the archaeological record. In 1984, F. G. Maier and V. Karageorgis suggested that more than 450 stones found at Old Paphos on Cyprus are evidence for large stone-throwing engines before the Persian Wars, and that such artillery was used by the Persians.[9] These stones are broadly round, but are flattened on one side, and they weigh between 2.7 and 21.8 kilograms; most are 20–30 centimeters (8–12 inches) in diameter. They were found in what was described as a siege mound, apparently built by the Persians outside the walls of this city at the time of the Ionian Revolt, 499–498 B.C. Others were found outside the gate and its environs, leading some to suppose that they had been thrown against the gate. The flattened side of the stones remained unexplained. Then, in another development, it was claimed there was more evidence for artillery before 399: in 1994 Ö. Özyigit published an article in which he suggested that a hastily and carelessly made 22 kilogram stone ball found in the archaic gateway at Phokaia in Ionia was shot by Persian forces during the siege of the city in 546 B.C.[10]

In 2000, Isabelle Pimouguet-Pedarros reexamined the whole issue, taking into account all the literary and the archaeological evidence to date, in an article that includes photographs of the relevant stones.[11] She concluded that it is highly unlikely that these stones were designed for stone-throwing catapults. Stones were used in wars from earliest times in ways *other* than as shot for catapults, and much more commonly than as shot. We discussed hand-throwing in the first chapter; big stones were also dropped, tipped, and rolled onto enemy forces using a variety of manual and mechanized devices throughout antiquity. The fourth-century author Aineias Taktikos, in his treatise *How to Survive Under Siege*, describes some such methods in detail. For example, he retails a countermeasure against machines called wall borers that involves the use of a stone "large enough to fill a wagon." Philon describes a method involving three-talent stones (about 75 kg) being tipped off the end of a board, and one-talent stones being dropped from cranes and rolled out of windows.[12] Tacitus tells how Otho's troops rolled down millstones on Vitellius' forces at Placentia, causing them to be "crushed, paralysed, bleeding or mangled." Non-artillery stone use is also exemplified in the siege of Plataia in the fifth century, as told by Thucydides in *The Peloponnesian War*.[13] Thucydides was writing about a twenty-seven-year-long war, the "Great War" of ancient Greek history, which included a number of important sieges, and he was clearly interested in siege machines and unusual weapons. He provides us with the earliest record of a flame-thrower, for example, used by the Boiotians against the Athenians at the battle of Delion in 424 B.C. He also tells us of rams and tortoises (covered sheds to protect siege workers), sapping (undermining the walls) and counter-sapping (mining to intercept enemy sappers), among other siege techniques. That he says nothing of catapults, therefore, is significant. He makes no mention of catapults of any sort anywhere. If catapults had existed in the world that he knew, he surely would have told us about them. His silence is not easily dismissed. It is therefore much more likely that the Paphos and Phokaia stones were used with such manual devices and strategies than that they were shot by catapult. For these and other reasons, Pimouguet-Pedarros concluded that the traditional dating and context for the invention of the catapult appears to be correct. In this chapter I will offer additional arguments for accepting Diodoros' story that spring from the history and philosophy of technology; a variety of things he says are entirely consistent with what one would *expect* of a new technology. Let us then take up the story of the origin of the catapult.

CONFLICT BETWEEN THE GREEKS and the Karthaginians in Sicily was longstanding. When the Persians invaded Greece in 480 B.C., leading to the battles of Thermopylae, Salamis, and Plataia, the Karthaginians invaded Sicily, and fought the Greeks at the battle of Himera. According to Diodoros (11.1.4–5) this was not a coincidence; the Persians and Karthaginians planned to act in concert, the Persians attacking from the east, the Karthaginians simultaneously from the west. As fighting between the Greeks and Persians continued intermittently through the fifth century, either as direct antagonists or through proxies, so on Sicily fighting continued between the Greeks and Karthaginians.

Dionysios was of humble origins. He emerged in a time of great instability and insecurity and went on to rule Syrakuse for thirty-eight years. He was the real person who inspired Plato's and Aristotle's models of "the tyrant." Several people who were contemporary with Dionysios wrote about the man and his activities, and some of those writings have survived for us to read. One was Plato himself, who visited Dionysios in Syrakuse in 388–387 B.C. and later. Plato had ambitions to make the Syrakusan tyrant, or his son, a "philosopher-king," but the powerful pupil thought otherwise and put Plato up for sale, whence he was rescued by Arkhutas of Tarentum, general, statesman, mathematician, mechanic, and Pythagorean. Another important witness we met briefly above is Philistos, who was a friend of Dionysios from the beginning of his political career. Philistos wrote a history of Sicily, about half of which concerned Dionysios. One is inclined to assume that his account must have been more than usually biased, because the author was a friend and supporter of Dionysios, but the reality is not so simple. We have no reason to suppose that Philistos was even attempting to write an objective account—ancient historiography did not share that modern aspiration. But Philistos was also banished by Dionysios. So one could imagine that he might have been biased against, rather than for, Dionysios. However, instead of guessing, we have a good guide to the quality of a lost source like this in those portions that survive, when they can be reliably identified, and the person's reputation in antiquity, when his work was available in full and could be compared with other contemporary written and oral accounts. Philistos' reputation was good. His history was much admired in antiquity, and was used by later historians, notably Timaios of Sicily, whose work is lost, and Diodoros Siculus, whose work survives in large part and enables us to write this history. Ancient critics were every bit as clever as moderns are, and while the aims and objectives of historiography, the writing of history, have changed a lot over the past hundred years, Lucian's treatise on *How to Write History* is still a useful guide to the business, although it is now about 1,800 years old. It also indicates on what basis Philistos' work was judged then.

Our less intimate sources of information on Dionysios are relatively good, too. He lived at a time when many intellectuals were writing, in a variety of genres, and many of those works have survived. We have, for example, a public speech written by Lysias, a son of Syrakuse who was, at the time, living as an exile in Athens. He called on the Greeks, in his Olympic oration of 384 B.C., to unite and liberate Sicily *from* Dionysios. We also have an open letter written by Isokrates, an Athenian, which lauds Dionysios as *savior* of Greek culture and calls upon him to unite the Greeks, under his leadership, for an attack on the Persian Empire.[14] In addition, we have inscriptions from Athens recording their official relations with Dionysios.[15]

Over time, stories about the man became increasingly elaborate and extreme, so that by Roman times people alleged that he made his own daughters cut his hair and beard, because he did not trust anyone else not to slit his throat while they were at it. The trouble with being the archetypal tyrant, as Dionysios was, is that different people wish to hang different clothes on the model, depending upon their own politics and historical principles (history as truth versus history as moral tutor, for example). There is no touchstone upon which we can rub such stories, to discover which is twenty-four-carat gold and which is fake, which has some basis in fact, and which is pure fantasy. But this methodological handicap is not a license to reject those stories that don't fit one's predilections, only to replace them with one's own suppositions. There are many tests that a historian can and should impose on the sources, and one of the most important is chronological. As a rule of thumb, sources that are contemporary are to be preferred over those written later, as they come from the world they discuss, rather than from a later age when things inevitably would have been more or less different.

Dionysios was an autocrat who earned and kept his position through attacks on the rich among his own people and success on the battlefield against the Karthaginians. In the spring of 405 B.C. he was first appointed, along with nine others, to the board of ten generals (it had, until a few years earlier, been a board of three) in Syrakuse, which was then democratic. Selinos, the Greek city nearest to the Karthaginian stronghold of Motya on the westernmost tip of the island, had just fallen to the Karthaginians, who now led an army of perhaps 120,000 men toward Akragas, which lay half way along the southern coast of the island (Map 2). That city fell in December 406. Gela was next in line, and the Karthaginians began their march on it in the spring of 405. Dionysios, at the head of a small citizen army and a smaller band of mercenaries, was sent to help Gela. By April, he was the sole general, with a personal bodyguard—always a prelude to a tyranny in Greek thinking—and Syrakuse was filling up with mercenary troops whose pay he had doubled. The citizen army meanwhile was still in Gela, where the

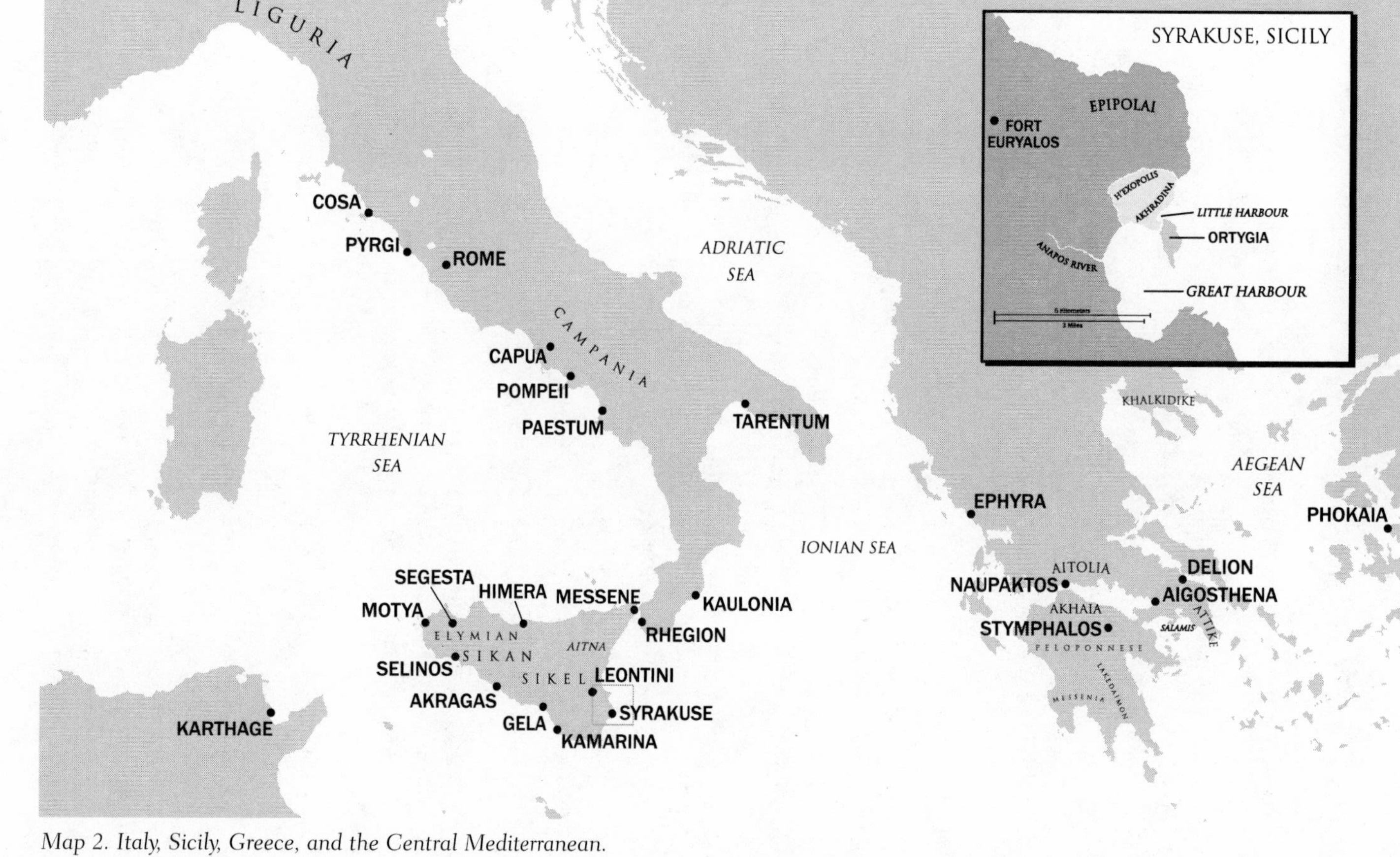


Map 2. Italy, Sicily, Greece, and the Central Mediterranean.

Karthaginians defeated it, together with contingents from Gela and south Italy, before Dionysios arrived back with his mercenary reinforcements. The Syrakusan cavalry attempted to oust him, but they failed. Some cavalrymen were killed, and others fled to the city of Aitna (Etna). Only Kamarina now stood between the Karthaginians and Syrakuse. Dionysios concluded a peace with Karthage, which in practice, if not in word, allowed him to continue ruling Syrakuse, while Karthage more or less ruled the rest of the island.

> To the Karthaginians shall belong, together with their original colonists, the Elymi and Sikani [native peoples of Sicily]. The inhabitants of Selinos, Akragas, and Himera, as well as those of Gela and Kamarina, may live in their cities, which will be unfortified, but they shall pay tribute to the Karthaginians. The inhabitants of Leontini and Messene and the Sikeli [native people of Sicily, after whom the island takes its name] shall all live under their own laws. The Syrakusans shall be subject to Dionysios. Whatever captives and ships are held shall be returned to those who lost them. (Diodoros Siculus 13.114.1)

He used the time thus bought by the peace to strengthen the defenses of the islet at the heart of Syrakuse, Ortygia, reinforcing its position as citadel. Walls and towers now enclosed this city within the city and the small northern dockyard that could hold sixty triremes. His supporters and troops were accommodated on the islet, and he himself took up residence on the acropolis. He gave the best of the property that had been abandoned by the cavalrymen—and we are here talking about at least several hundred large farms in the hinterland of Syrakuse and perhaps also town houses—to his friends and subordinates, and redistributed the rest of it in equal plots to citizens and foreigners alike. Diodoros claims that at this time he also enfranchised freed slaves as "new citizens" (*neopolitans*). The enfranchisement was presumably connected to the gift of land, for landholding was a prerogative of citizenship throughout antiquity and, indeed, most of history until very recent times.

Dionysios maintained his tyranny through the military. He obtained funds largely through military adventures in Sicily and Italy; for example, he won 1,500 talents in an attack on Pyrgi in central Italy. He spent those funds largely on military material—men and equipment. He employed those military forces to protect his position, which simultaneously protected theirs. His tyranny was quite different from those that had gone before, but in these respects was a striking precedent to the monarchy of Philip II of Macedon, which would begin about forty years later, in 359 B.C. But at this point in time, about 404 B.C., Syrakuse had been a democracy for two generations, and its citizens were not so easily persuaded to relinquish political and military power and become the subjects of one man. Just ten years earlier they had met, matched, and defeated the greatest force ever sent from Athens. So the next time Dionysios led them out to battle, some of them began an uprising, and when Dionysios fled back to Ortygia, they besieged him and his supporters on the island. They offered citizenship to any mercenaries that deserted Dionysios; some accepted the offer, and it was honored. They offered a large (unspecified) reward for his head. To counter this, and encouraged by Philistos (the historian-to-be), Dionysios sought more mercenaries, offering whatever reward the mercenaries sought. After 1,500 more mercenaries arrived, he led a successful attack on his disorganized besiegers (some of them had gone home, thinking the job done and the tyranny over), and the siege was lifted. The following harvest time, using what appears to have been a novel stratagem, while all hands of landholding households were in the fields gathering in the harvest, his troops entered their houses and confiscated their weapons. In most states landholders constituted the hoplite forces, the heavily armed infantrymen. Light-armed troops were typically artisans and slaves. Consequently, Dionysios' act would have been against the hoplite households, for those were the ones preoccupied with the harvest. After this he collected more mercenaries, says Diodoros. Through recruitment drives in the first few years he gathered a very large army, of which about 20–30 percent seem to have been mercenaries, and the rest were citizens, old and new, from Syrakuse and other Greek cities in Sicily. Dionysios was preparing for another war with Karthage. Preparations included the design and construction of novel weapons.

The catapult was invented in Syracuse by an artisan responding to the encouragement and incentives offered by Dionysios I. We do not know his name. This curious fact warrants and, I think, justifies some speculation on a topic for which there is no evidence but about which a reasonable circumstantial argument can be made. The Greeks and Romans liked to attribute inventions to named individuals—they would attribute to gods or heroes if necessary, for example mythological Prometheus taught men how to make fire—and the catapult was neither accidental nor prehistoric. It was invented in the full-blown historic period, in a major city, at a specific time, in the full glare of publicity, so to speak. If the inventor had been a native Syrakusan, I think our sources would have remembered his name; indeed, given that they do record the names of others who developed his invention, such as Zopyros of Tarentum and Poluidos of Thessaly, the anonymity of the inventor of the catapult is even more striking. So what is going on here?

Two obvious possible explanations suggest themselves. Often in the history of technology the first idea or version of a new invention does not work well, and the invention only really takes off when early snags, glitches, and teething problems are sorted out—and not necessarily by the same person who "invented" the original. As Polybios observed (29.17.2), many inventions that seem on paper to be plausible and likely to succeed do not live up to our expectations when tested. Something like that may have been the case here, so that the credit for this invention was not easily attributed to "an" inventor. If locals and foreigners, artisans, and amateurs really were jostling cheek-by-jowl in public and private spaces around Syrakuse, as Diodoros describes, then one can easily imagine people gathering round and joining in to develop an exciting new idea, so that the catapult that resulted was a group product, rather than an individual's. But then why was it not credited to X's group, since Dionysios is said to have put specific individuals in charge of teams of artisans?

So something else may be going on here, instead or in addition. This was a slave society, and the inventor may have been a slave. It is well known in the recent history of science and technology that dramatic advances have sometimes been made by people who moved between countries or disciplines. The resulting need to synthesize different approaches to similar problems, the frequent requirement to work on subjects tangential to the worker's original (home) speciality, and the liberation from "traditional" (home) constraints have all been identified as potential factors in this process. Hoch argued that, in the modern period, it was the very difficulty of integrating themselves that forced some migrants to make special efforts that led to, among other things, "the most successful new technologies in the atomic, aerospace, chemical, electrical and materials industries."[16] In many ways, skilled slaves in antiquity were like these free migrants of our own times; they were cultural hybrids, marginal people, whose potential to innovate has been as yet little explored. If the ancient inventor of the catapult was such a slave, then what caused his name to be lost was the general dishonoring of slaves. Slaves had no social identity; they could not even choose their own names.[17] This need not obliterate recognition, as there were enough creative slaves known in antiquity for someone to write a two-volume work called *Slaves Who Were Famous in the Cultural Domain* (lost, unfortunately), but it would depend entirely on the personality of the owner and the relationship between master and slave. Repressive owners probably did not nurture creative talents among their slaves—skilled slaves were usually motivated by carrot rather than stick—but it would be quite natural in such an environment for the maker, if a slave, to go uncelebrated.

We can also consider the alternatives. The inventor of the catapult, were he a free immigrant to Syrakuse who had moved there in response to Dionysios' call, might well have considered offering his new invention to other employers, as some of the mercenary forces that worked at one time for Dionysios offered their services, at another, to the Karthaginians. A free artisan could have taken his knowledge wherever he wished, and if he was famous for his skill, then he might have expected to earn very high wages.[18] This did not apparently happen with the catapult, for there are no *ancient* rival claims to this invention. There may be a number of reasons for that, but the possibility that the inventor of the catapult was a slave must be one of them.

IN ARCHERY, THE CHIEF DESIDERATA are the heaviest possible impact, with the greatest range, through a flat trajectory for ease of aiming. These ideals promote the use of a strong bow, the most powerful, in fact, that the archer can draw. The first catapult was nothing other than a bow stronger than any archer could draw in a normal way, with a mechanical device to take the place of the archer's arms, to draw back the bow, and another mechanical device to take the place of the archer's fingers, to hold and release the bowstring. A ratcheted stock and slider took the place of the archer's arms. The stock modeled the archer's bow arm, holding the bow firmly, and the slider moved backward and forward in the stock, modeling the archer's bowstring arm. In the braced position, the slider protruded out in front of the stock, and the trigger caught the bowstring. The trigger was attached to the slider so that the bow was drawn as the slider was pushed back. (This is the key difference between this and a crossbow.) The whole device was called a gastraphetes, which literally means "belly-bow," apparently because the slider was pushed back and the bow thus drawn by pushing the back end of the stock with the belly, while the front of the slider was jammed against a rock or the base of a wall (Figure 2.1). To facilitate this loading action, a beam of wood in the shape of a lazy omega (**Ω**) was added to the back of the stock. The user pressed with his stomach in the recess and his hands on the handles either side. A similar action can occasionally be observed today of someone using a manual lawnmower in an unkempt garden. Ratchets on the sides of the stock and a pawl on each side of the slider prevented the slider from slipping back. When the user had drawn the bow as far as desired or possible, he would lift the front end from the ground. The weight and cumbersomeness of the machine, as described in the one surviving source on it (on which more in Chapter 7), would have required it be rested on something to take aim.19 It could then be loaded. To keep the arrow on the slider the bow must have been laid horizontally or elevated; it could not be aimed downward.

The first catapults probably had bows of compound construction. The compound bow is made of three layers. It has a wooden core, glued to which is a thin layer of horn on the belly (the side facing the archer) and a thicker layer of sinew on the back (the side facing the target). Horn resists compression, and sinew resists stretching, so when the bow is bent, both materials exert pressure to restore it to its original shape. Ideally the horn glued to the belly should be a single lath cut from a single horn; animal horns reach a certain size depending on species, and do not grow larger. While a series of horn lathes can be stuck end to end on the belly, the more there are, the more difficult it is to ensure that they all curve in the same plane and do not twist. Indeed, it has been suggested that after the fall of the Roman empire, difficulty obtaining suitable horn may explain why western European peoples developed the self-bow, and why whale-bone was usually used as the belly-material in those composite bows that were made there (for crossbows).[20]

The materials actually impose a limit on the maximum size of a compound bow. Quite apart from trying to find, say, horn six feet long, as components get larger they become increasingly hard to move, and accelerate more slowly, which is not ideal in a catapult. As the length of the arm increases, assuming its proportions stay the same, its volume cubes, which means its weight cubes, which puts significant extra load on the arm itself, never mind on the stock to which it is attached, or the stand which supports the whole machine. If one built ever larger arms, at a certain point the arm would break under its own weight.

Other considerations also play a part in restricting bow size; for example, ordinarily a bow would have been unstrung when not in use, in order to preserve the elasticity of the sinew. A bow that was stronger than a man could draw presented a challenge to string. Did this put a limit on bow size? Or did they use a bastard string and windlass to string it? Or was the bow so constructed that it was not stressed in the brace position? That in turn would reduce the power that could be developed in the draw. With a hand bow a certain amount of tension is imparted in stringing the bow; more is stored drawing it.

Diodoros says that "catapults of all kinds were constructed, and a large number of other missiles" (14.43.3). This implies a lot of development and innovation hot on the heels of the invention of the *gastraphetes*. I take this as another indication of the reliability of his account, rather than, as some have, an indication that the catapult was invented somewhere else at some indeterminate time earlier. Early versions of new devices are normally problematic, and usually lots of tweaking (or something much stronger than is indicated by that word) is required to make the thing work properly or at all. In the worst case, the thing is sent back to the drawing board, for a fresh start. In the early life of a new design major changes may be made. As the design is worked on and tested, intellectually and physically, it stabilizes and congeals. Once one has a working device, the "if it ain't broke, don't fix it" mindset is more likely to dominate than the "back to the drawing board" one, especially if the device is actually required to work here and now. Consequently, "all kinds" makes more sense in the early history of the catapult than it would later on.

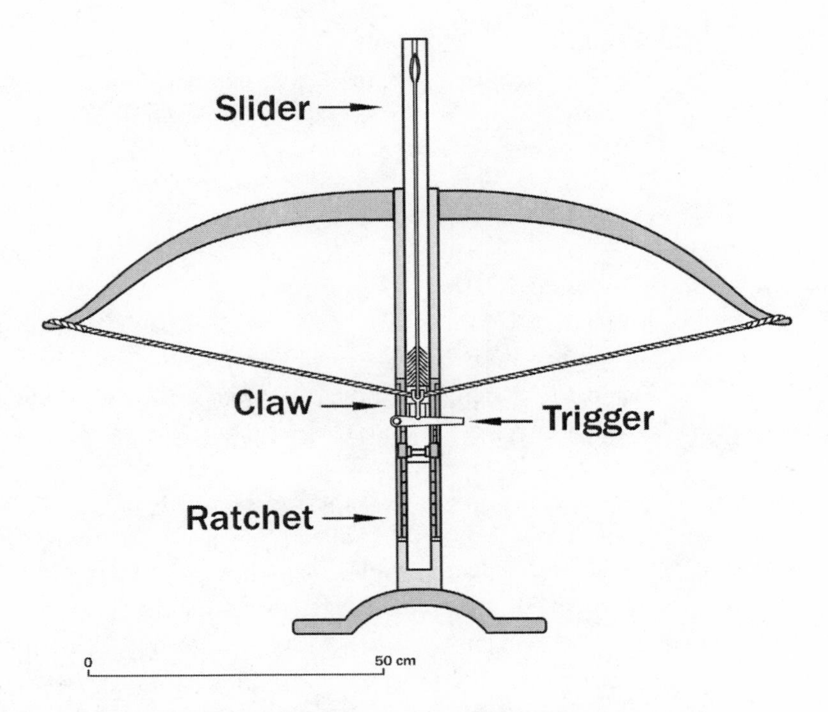


FIGURE 2.1 *The gastraphetes. (T. D. Dungan after Baatz 1991.)*

Regarding the missiles, I think it likely that, since the catapult was a mechanical bow, it was probably furnished, at least initially, with a variety of sharps, just as an archer would carry a variety of sharps to suit the different sorts of targets he might expect to shoot. The notion of caliber probably coagulated over time too, culminating in the experiments in Alexandria which Philon reports and which we will discuss in Chapter 7. We are often told, hereafter in Greek history, that a city expecting a siege prepared "all kinds of missiles," and there is no reason why a catapult could not have shot sharps of different design and length, especially before the formulae were discovered. The great variety of shapes and styles of sharps at Caminreal,[21] many of which were around 27 centimeters long, probably had something to do with presence of at least one sharp-caster catapult there. Which sharp was selected for use on any particular occasion would have depended upon what sort of target it was intended to destroy.

Dionysios' army was a diverse and polyglot force, composed mostly of Sicilian Greeks, but with significant numbers of others. There were native Sikels, and Iberians, Ligurians, and Campanians from what is now Spain, France, and Italy, respectively. From mainland Greece came Messenians, who in 396 sought a new home with him after the fall of Naupaktos in western

Lokris, where they had been living since the end of the siege of Ithome around 465 B.C.[22] There were also assorted Peloponnesians and Lakedaimonians, and no doubt others too, from diverse backgrounds. In keeping with tradition when employing mercenaries, Dionysios expected each group to fight in its own particular way. Some were renowned as cavalry troops; some, as light-armed infantry; some as slingers; some as archers. Diversity of troops therefore produced his multi-skilled force—an unusual feature in armies of his day.

Dionysios also recruited artisans from hither and yon. Diodoros' remark that Dionysios wanted to ensure that each group could maintain its own type of armor and weaponry to preserve the diversity of troop types should not be misunderstood. Ancient warriors were usually *self-arming*, and having bought their own equipment with their own money for their own use, they are not easily persuaded to give it up in favor of other, possibly inferior, and almost certainly less well fitting equipment. Consequently, when we do hear about troops of one type being armed and fighting as another type, they are usually troops who fight unarmed, such as slingers, being supplied with heavy infantry equipment and being asked to fight with the hoplites. Self-arming soldiers explain not only the diversity of Dionysios' forces, but also the diversity of weapons-making artisans. Some weapons-makers from the regions whence Dionysios' mercenaries were recruited might anyway have followed the mercenaries irrespective of Dionysios' inducements, because the mercenaries were the market for their wares.[23]

Now Dionysios disarmed the landed citizens initially by a ruse and later as a matter of course, and had to supply arms to those of his forces who were newly liberated and enfranchised slaves, so the issue was not just one of infrastructure building. He was supplying the weapons to at least some of his forces. Diodoros is surely correct to say that Dionysios wanted his troops to perform to their best, in whatever particular style and mode of fighting they knew. That is not incompatible with active encouragement of artisans of any type or background to innovate and develop weapons. There is no good reason to doubt Diodoros' claim that Dionysios offered material and immaterial encouragement and rewards for new weapons. As Diodoros tells the story, a given artisan seems to have been expected to be able to make a given type of war material, so long as he had a model to follow. But Diodoros thought that the most significant causal factor leading to innovation was the gathering together of skilled artisans from everywhere into one place. This could be interpreted as the creation of what we might now call a critical mass or a research hub, or more loosely the creation of an arena for stimulation and conversation between people from different craft-working traditions.[24]

Diodoros identifies a third stimulus to innovation which was very unusual for an ancient political leader, that Dionysios took a direct and personal interest in the work of the artisans—apparently on a daily basis, at least at some points in his career. Dionysios perhaps saw that new technology was a potential route to immortality for himself and his city, and he was right so to do: we are thinking about him now because of a military innovation born of this initiative, because the first catapult emerged from such an environment. As will be seen in later chapters, Diodoros appears to have put his finger on an important factor in innovation here; with respect to the development of weapons, real advances do seem to have been synchronous with genuine interest in technical development by the political leadership of the time.

All branches of warfare seem to have been grist to Dionysios' mill. One of the most dramatic of these was new, larger warships: shipwrights working in Syrakuse at this time invented the quinquereme, whose name implies rowers organized by fives. The Syrakusans had modified the design of their ships before the Athenian invasion of 415–413—they reinforced the prows and rammed the Athenian triremes head-on rather than side-on, stoving in their prows. Caven observed that "there must have been many unemployed shipwrights available in the dismantled Pireaus" after the end of the Peloponnesian War in 404 B.C. As the catapult emerged from a multicultural medley of artisans, so the quinquireme might have materialized amongst a multicultural posse of shipwrights, gathered where there was already a tradition of successfully altering existing designs. Diodoros' reference to Dionysios' costly new ship-sheds offers supporting evidence for the invention of the quinquireme, since the existing ship-sheds would have been too small to store the new larger ships. (Ancient warships were designed to ride high in the water, their timbers dry and light rather than wet and heavy, so they were stored out of the water whenever possible.) Historically, warships typically grow to the limits set by the physical environment in which they exist, and then stop. For example, the graving docks in British ports set a restriction on the breadth of British warships.[25] If Dionysios wanted to build warships bigger than triremes, new bigger ship-sheds would have been necessary too, so the fact that we are told that he built 160 of them—at great expense apparently—supports the assertion that he did develop the quinquereme, either *ab initio*, or to mass production.

Another war machine whose development he encouraged was the siege tower, which he took to Motya and used as a mobile platform for his catapults and other war machines. The description of fighting on and from the gangways laid down from the top of the towers again suggests a new technology whose strengths and weaknesses in combat had yet to be fully appreciated. These gangways in due course became proper skywalks, covered to protect the men inside from being seen or falling off, and with a front door, to provide protection until they were ready to burst out onto the enemy battlements, but at this early stage they had neither.

In 398 Dionysios assembled his new forces and weapons and marched on the Karthaginian stronghold of Motya, gathering Greek and Sikel forces as he advanced west across the island. The army thus marshalled finally amounted to 80,000 foot, more than 3,000 horse, and slightly fewer than 200 warships, the whole supported by at least 500 merchant ships "loaded with many war machines and all other necessary supplies." Motya, the Karthaginians' base at the western end of Sicily, was an island in a lagoon, connected to the mainland by a mile-long causeway. The Motyans destroyed part of the causeway as Dionysios approached, so one of his first tasks was to build a new mole.[26] "Architects," literally, "leaders of builders," often best translated "engineers" in ancient usage, accompanied Dionysios on his reconnaissance trips to decide where to build the moles, and it is likely that he employed technical people as well as fighters among his forces. Dionysios' new armaments, land and naval, were employed in this siege.

> Himilcon [the Karthaginian admiral] attacked the first ships, but was restrained by the multitude of missiles, for Dionysios had manned the ships with a great number of archers and slingers, and the Syrakusans slew many of the enemy from the land, by using the catapults that cast sharps (*oxybeleis*). Indeed, this weapon caused great dismay, because it was a new invention at this time. As a result, Himilcon was unable to achieve his plan and sailed away to Libya, believing that a sea-battle would serve no end, since the enemy's ships were double his in number. After Dionysios had completed his mole by employing a large force of labourers, he moved up against the walls war engines of every kind, and kept hitting the towers with his battering rams, while with the catapults he kept down the troops on the battlements. He also advanced against the walls his wheeled towers, six stories high, which he had built to equal the height of the houses. . . . They advanced the wooden [siege] towers to the first houses and provided them with gangways. Since the siege machines were equal in height to the houses, the rest of the struggle was fought hand to hand. For the Greeks would launch the gangways and force a passage by them onto the houses. . . . The Greeks found themselves in a very difficult position. For, fighting as they were from the suspended wooden gangways, they suffered badly both because of the narrow space and because of the desperate resistance of their opponents, who had abandoned hope of life. As a result, some perished in hand-to-hand combat as they gave and received wounds, and others, pressed back by the Motyans and tumbling from the wooden gangways, fell to their death on the ground. (Diodoros Siculus 14.50.4–52.4)

Motya held out for about six months. When it fell in 397, the destruction of people and property was extensive and ruthless. The Karthaginians in Sicily were reinforced from Africa and fighting continued all over the island in the following year. Just when Karthage appeared to have the upper hand, laying siege to Syrakuse and even penetrating the outer rings of defenses, "plague" broke out in the Karthaginian camp.[27] As Diodoros observed, there were natural reasons for this: besides the wrath of heaven that might have been expected (by the believers in supernatural interference in human life) in revenge for the Karthaginians' plundering of Greek temples, myriads of people were gathered in temporary accommodation, in close proximity, in the summer—and this was a particularly hot summer. Further, the spot the Karthaginians camped at was unhealthy: it was the same place that the Athenians had chosen a couple of decades before, and they had been struck with "plague" there too. Dionysios seized the moment, inflicted a significant military defeat, and accepted the terms which the Karthaginian general then offered: 300 talents in cash and evacuation of Karthaginian forces from almost the entire island. The native Sikels who had fought for Karthage were rounded up and sold as slaves, generating more funds, and some of the Iberian mercenaries that had formerly served Karthage were enlisted among Dionysios' forces. He then led two other successful sieges: Kaulonia and Rhegion. In these cases, as at Motya, he was dealing with coastal cities, and he had command of the sea as well as his catapults. But we must not lose sight of the important point. His catapults were *not* a transforming technology; Dionysios would have lost this war against Karthage had not their army been devastated by an epidemic disease.

Excavations at Motya in 1955 and 1961–1963 revealed seventy-seven bronze arrowheads. Arrows are easily lost and bronze heads are more likely to appear in the archaeology wherever they were used than are iron heads (because bronze degrades less quickly in the soil). An anonymous tenth-century A.D. treatise on *Tactics* anticipated losing most of its missiles and armament on any particular campaign, and so assigned two-thirds of the support troops that carried the material out, to carry something else on the return journey. Quite a large proportion of ancient military material should probably be considered as "consumables." These arrowheads at Motya were found scattered all around the site, every single one in close vicinity to the walls, either just outside or just inside. Some were bent or broken. Find spots, distribution, and condition all point to these arrowheads having been deposited in the archaeological record through use in battle, specifically, being shot at people on the walls and subsequently lost. The site was not reoccupied except by occasional squatters, thus it is highly likely that many if not all of these arrows were shot during Dionysios' siege of the city. Sixty-three of the seventy-seven bronze arrowheads found were so-called "Skythian" arrowheads, 1 inch (2–3 cm) long, which are usually three-edged and socketed to receive the shaft, but two-thirds of those found at Motya were a variant type, taking the form of an elongated three-sided pyramid, which is particularly suited to armor-piercing. This particular variant is characteristic of Greek sites of the fifth and fourth centuries B.C., especially Greek sites of Sicily and South Italy.[28] In one of the heads, a fragment of the wooden shaft remained; it was analyzed and found to be *taxus baccata*, yew, which is native to Sicily. The remaining fourteen arrowheads were of a different type entirely, a type that is exclusively Greek. They were two edged, leaf-shaped, tanged to insert into the shaft, and much larger—3.5 inches (9 cm) including the tang. This type of head is ideal for unprotected targets, causing large wounds and rapid blood loss. The same two types were found in similar proportions in the earlier excavations in the first decades of the twentieth century. There were also iron artifacts found which, though badly corroded, so strongly resembled the larger type of bronze arrowhead that A. M. Snodgrass thought they "may represent an attempt to copy them in forged iron."[29] He continued: "It is disappointing not to find any clear trace of the missiles fired from the newly-invented catapults of Dionysios, the ὀξυβελεῖς καταπέλται mentioned by Diodorus; but there are a few fragments, apparently from pointed iron bolts, which could be thus explained." Unfortunately, he does not make plain what sort of material he *would* consider to be a clear trace of Dionysios' catapults. Since the first catapult was essentially a very powerful bow, I would expect its missiles to be essentially either the same or perhaps slightly larger and heavier versions of missiles for hand bows. The large tanged heads in bronze or iron strike me as reasonable candidates for the first catapult warheads.

Dionysios and his engineers did not, apparently, think of putting catapults on board his ships, which is consistent with the idea, and yet another indication, that it was a new machine whose possibilities were not yet fully appreciated.

Dionysios' reign of nearly forty years was a lively one, punctuated by bloody and brutal battles with the Karthaginians, with other Greeks of Sicily and South Italy, and occasionally even with peoples further afield, though he never took his war against Karthage to Africa. Status and social and geographical mobility were fluid in a way that was rare before or since. Populations of cities he fought with or over were individually and collectively massacred, or sold into slavery, or liberated from slavery, or given Syrakusan citizenship, or moved to another city that he had depopulated and which he gifted to them and of which they then became citizens. His reign made Syrakuse one of the largest, wealthiest, and most powerful cities in the Mediterranean basin, with a population of more than half a million.[30] Its next great moment in the spotlight of history would come nearly two hundred years later, when another Syrakusan, motivated by another tyrant, would develop some extraordinary weapons, this time not to *take* a city by siege, but to *defend* one under siege: Arkhimedes, whose devices would stop the great Roman war machine in its tracks.

THREE

# The Development and Diffusion of Tension Catapults

*If history is stripped of her truth, all that is left is but an idle tale.*
—*Polybios 1.14.6*

THERE WAS A PERIOD OF ABOUT FIFTY to sixty years between the invention of the first catapult, a gastraphetes, a bow catapult that shot sharps, and the invention of the torsion catapult, which was powered not by a bow but by ropes made of hair or sinew. Because the torsion catapult overtook the bow catapult and became the dominant type of mechanized weapon, there is a wholly natural—but I think fallacious—tendency to overlook the history of the device that was superseded. This is a common phenomenon in the history of technology.[1] To put the argument in a nutshell: if one could choose any weapons from the arsenal of any day, one would not choose the same weapon to deal with all eventualities. A continuous row of one-talent stone-throwers, for example, would not constitute the ideal defense for a city any more than an armory equipped exclusively with cruise missiles would constitute a good arsenal for a state today. Even in cities where there were fortification towers big enough and solid enough to take such monster machines interspersed along the wall, there would still be a lot of curtain wall that could take men wielding gastraphetai as well as men wielding traditional unmechanized bows. As there was still a place and a role for the humble slinger in the most advanced armies of antiquity, so there was still a place for the simple machines alongside the more complex. Extensive trace walls on which nothing larger than a gastraphetes or hand-bow could have been used without hindering movement along the wallwalk at the Ionian cities of Kolophon, Priene, and Erythrai (Map 3), for example, were not obsolete in terms of siege warfare in the late fourth century B.C. when they were built. Rather, they were well suited to the terrain, and sometimes, narrow wallwalks were militarily sound even after the invention of torsion catapults and artillery towers.[2]

Economics should never be ignored, either. A gunpowder mill using medieval stamping technology to incorporate the ingredients was set up near

Bomlitz (in Lower Saxony) in 1814; this was not only long after edge-roller mills had been adopted by mainstream gunpowder producers, but after stamping powder mills had been *banned* in England (in 1772). Why was this obsolete mill set up in the area where European gunpowder mills had begun and many important breakthroughs were made? The millers in question knew all about edge-rollers and suchlike, so what is the explanation for their behavior? Their aim was to make gunpowder (albeit of lower quality) at much lower capital investment cost, and the old technology suited that better. A local carpenter could make a stamping mill with materials at hand inexpensively. This mill did not adopt edge-rollers until 1885, by which time they had become cheap. Nor should we overlook the employment of "outdated" technologies as contingencies, in the same way that many steam ships still carried sail, plus masts, rigging, and extra crew, well into the 1900s. As well as being relatively cheap, older technologies may also be relatively quick to bring into action. In 1814 at Waltham Abbey, which was then a government-owned gunpowder mill, eleven water-powered mills and nine horse-powered mills all worked side by side to produce gunpowder. The horses were not part of the establishment by this date, but had been brought in as a quick fix to supplement output during wartime.[3] Thus, there is no justification for passing over the tension catapult in a hurry to get to the torsion, and it is my intention in this chapter to try to restore the *gastraphetes* to its rightful place in the history of the catapult.

The name "belly bow," *gastraphetes*, originated with a crossbow-type weapon, as we saw in Chapter 2. It happened to be loaded by pushing with the belly, and Ktesibios says that is why it was so named.[4] Actually, he is one of very few people ever to use the word, and we do not know if his explanation of the term is any more accurate that the average ancient etymology. *Gastra* also refers to the *shape* of the belly, especially of a paunch, pot-, or, as we now call it, beer belly. The shape created by the arms of the bow, viewed in plan, resembles the shape of a beer belly, viewed in profile. This might explain why the name continued to be used for mechanical bows, even when they were very large and were drawn back by winch and not by pushing with the belly. The name now referred simply to the shape, and not to the method of loading. We know of a number of such *gastraphetai*, but only glimpses of them remain, and it is difficult to construct a continuous tale of the catapult's diffusion and development from the scattered and fragmentary evidence that survives from this period. Dionysios, remember, attracted to Syrakuse artisans from all over Sicily and south Italy, if not from the wider Greek world. Most of them would in due course have returned to their various homes or other places of employment, some at least taking knowledge of the catapult with them. Craftsmen have been the vectors for technology transfer in other, better documented cases, and they are among the most likely vectors in this case too.[5]

Our first scrap of concrete evidence for the bow-catapult's history, after its invention, is an apothegm preserved by Plutarch in both his *Sayings of Kings and Commanders* and his *Sayings of Spartans*, which are collected together with many other of his writings in a work called the *Moralia*. According to this, Arkhidamos III, King of Sparta, cried out in horror when he saw the catapult demonstrated for the first time.

> Arkhidamos, the son of Agesilaos, on seeing the bolt shot by a catapult, which had then first been brought from Sicily, cried out
> "O Herakles! Man's arete is destroyed."[6]

*Arete* means manly strength, valor, or prowess, especially in a fight; the word is derived from Ares, god of war (compare "martial" from the Roman war god Mars). Arkhidamos seems to have been referring to the fact that courage was no defense against the catapult. A Chinese observer echoed the sentiment but provided the view from the offense in 1044: "the crossbow is indispensable [for] overcoming men who do not lack bravery."[7]

Arkhidamos was born about 400 B.C., so this could have been said anytime after the invention of the catapult, but as told, the story seems to refer to its arrival in Sparta if not Greece, and Arkhidamos' comment sounds like that of a man not a boy. Therefore, I would date the story to about 380 at the earliest. Plutarch says explicitly that the catapult had been brought from Sicily, and he seems to have been tapping in to a reliable strand of information here. Sparta had a long-standing alliance with Syrakuse, and had helped Syrakuse resist and then defeat the Athenian expedition launched against them back in 415. They also helped Dionysios from early in his reign, and he helped them.[8] If Dionysios wanted to share his new technology, he would have shared it with his reliable allies—one of whom was Sparta.

Now Sparta was the dominant force in Greece for most of Dionysios' reign, but lost that position after being defeated by Thebes leading the Boiotians at the battle of Leuktra in 371. It took the Greek world, including the Thebans, a little while to realize just how significant that defeat had been and how weak Sparta had become, but the Thebans finally led an invasion of Sparta's territory a year later, liberating Messenia, the region which the Spartans had annexed centuries before, and then they invaded again in 369. The Spartans appealed to Dionysios for help, and he responded. Valor is all very well, but when the adult male population had shrunk to about twelve hundred, and tens of thousands of hostile men, nursing decades of anger and resentment, were on the warpath, I imagine that Dionysios might think that the surviving Spartans would embrace any assistance that technology could afford. Catapults would be jolly useful to the Spartans at a time like this. It would also be an appropriate point for the Athenians to obtain catapults (on

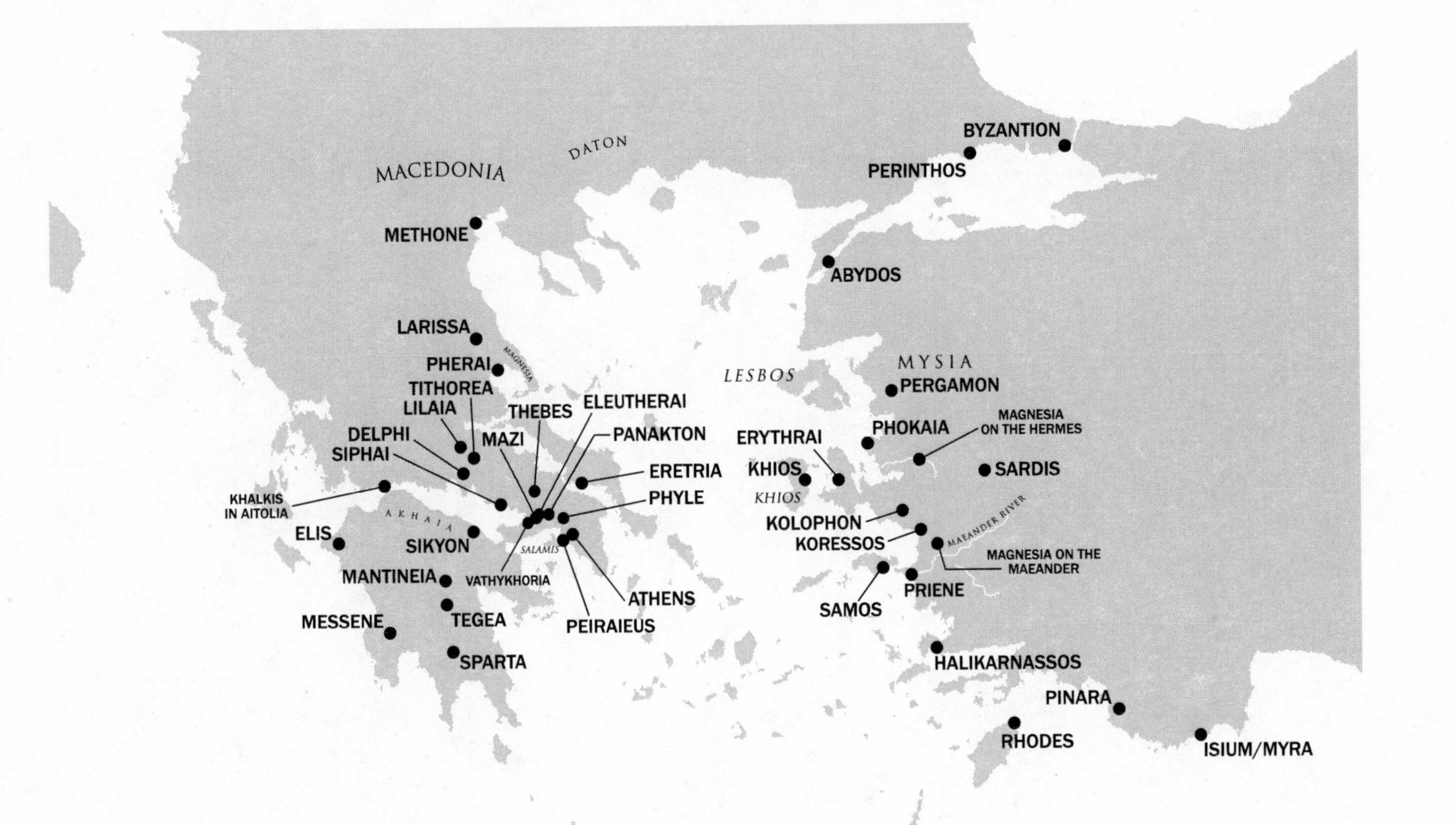


Map 3. *Greece, the Aegean, and Asia Minor*

which more below), since as soon as Thebes presented a greater threat to Athens than Sparta did, the Athenians abandoned their alliance with Thebes and allied with Sparta to oppose Thebes. Indeed, I can even imagine the Spartans, if they had it, quietly handing the stuff over to the Athenians—who were renowned for being ever à la mode—to use against the Boiotians. Xenophon, writing a history of his times in the fourth century B.C., reported on Dionysios' help. He says, "aid to the Lakedaimonians arrived, consisting of more than twenty triremes. They brought Kelts and Iberians, and about fifty horse;" he does not actually say that catapults came from Syrakuse. Later he observes that Dionysios' forces "fought the Sikyonians . . . and took by storm the fort of Dera. After that, this first aid from Dionysios sailed home to Syrakuse." The story of Dera may be an indication that they had catapults with them, for, as told, they seem to have taken the fort practically in their stride, whereas in previous Greek history strongholds are hardly ever taken by storm. Forts built for the pre-artillery world were abandoned or replaced in the fourth century, presumably because they offered inadequate defense against the new technology.[9] Sometimes this was because of their location or design. For example, any tower that has a gate within range and line of a natural feature that would be an ideal spot for artillery to shoot onto it, such as Rhamnous in Attike, can be assumed to have been built before the arrival of catapults in that area.[10] No one would knowingly build in such a weakness to what was meant to be a stronghold. Thus the catapult forced changes to military architecture and to choice of site for strongholds, and the torsion stone-thrower, when it came, would force further changes in due course. For now, we focus on the first generation catapults, the *gastraphetai*, the mechanized bow sharp-casters and stone-throwers.

Fortifications are typically very difficult to date; the structures in themselves, the stonework, the building style, these and similar things often give little or no indication of a precise date.[11] Most towers and walls are not excavated so lack stratigraphic data. Dates are usually estimated from any datable archaeological evidence that exists on the surface, such as associated pottery or inscriptions, and historical arguments about when whoever is thought to have built the fortification may have had the means and motive so to do. In such an uncertain area, it follows that opinions and dates are revised relatively often in light of new evidence or arguments. So this is a particularly fluid area, which does not make a good foundation for larger arguments built on them. Consequently, I report current opinion on what are called first generation artillery towers, but I will not rely on any of these dates in any other argument about the history of the catapult.

After the Thebans had invaded Sparta's territory and liberated Messenia, they founded a new capital for the region, called Messene, and fortified it.

The walls and towers of that site can therefore be dated very closely to 369 and the immediately succeeding years.[12] The Messenian fortifications include what must be some of the first "artillery towers." What characterizes towers as artillery towers are the embrasures (openings in the wall like small windows, but they sometimes have oblique sides, like loopholes or arrow slits, so that they are wider on the inside than the outside) in the topmost chambers. Archers can shoot through slits, which simultaneously offer them good security, and other rooms in these towers are usually so fenestrated. Little catapults could and did shoot through arrow-slits, but a larger aperture was necessary for larger models, and the original *gastraphetes* needed to be propped on a wall or sill to be loaded and discharged (later models had stands).

The earliest known artillery towers—even earlier than Messene—are perhaps those at Tithorea and Lilaia in Phokis. These are argued to be earliest because (1) they are not very high; the embrasures are only about 21 feet (6 1/2 meters) above the ground, despite being on easily approached ground. The windows of other towers in similar situations are 30 feet (9 meters) or more high—approaching 60 feet in some cases. (2) The Lilaia tower floor joists ran between the front and back walls, whereas in later towers they usually run between the sidewalls. Philon explains why the joists should run side to side and not front to back: the front wall was the one liable to be rammed, and if the joists of the internal floors ran front to back, and the rammed wall gave way, the floors would collapse too. If the floor joists ran side to side, by contrast, most if not all of that floor would hold even if the front wall fell down.[13]

According to Ober, other early towers are those at Siphai in Boiotia, apparently built by the Thebans before 362, and the Athenian towers at Eleutherai[14] (contemporary with Messene/Siphai?), Phyle (360s?), Aigosthena, Mazi,[15] and Vathychoria F and C (343 B.C.?); see Map 2 and Figure 3.1. Van de Maele is unconvinced of the argument for including Aigosthena this early, preferring Lawrence's link to Demetrius Poliorketes (whom we will meet in Chapter 6) and a date at the end of the fourth century, or even in the early third.[16] The recent excavation of Panakton, however, suggests that the 340s would be the latest for the Athenian towers at least. For although the walls and towers themselves at this fort have not survived well, an inscription has been found there that indicates categorically the presence of catapults. It is a list of the numbers and conditions of items in the arsenal when responsibility for them passed from one official to another. The beginning of the inscription is missing and the top few surviving lines have only odd words here and there. From line 7 it becomes meaningful and reads:

> Arrows (τοξεύματα) without heads and featherless 750; other arrows featherless without heads 1750; Skythian heads 1200+; three empty boxes

> unsound without lids; trigger (σχάσηια, from σχάζω, LSJ II.2) of a catapult; catapult bolts (οἱστοι) featherless 76; others without heads 16+; heads for those 13; a little wooden box, in which 11 catapult bolts, unsound; *saunia* (a type of javelin)124; *akontia* (another type of javelin) 28, without spearheads (λόγχια) 4, old 6; of these one is repaired another is broken. All of these things I handed over to Demokles son of Kleokritos of Akharnai in the arkhonship of Pythodotos.[17]

The last line is crucial: it dates this inscription to the year July 343-July 342 B.C., when Pythodotos was eponymous arkhon in Athens. The ceramic (pots) and numismatic (coins) evidence from the fort suggests that it was occupied throughout from about the beginning of the Peloponnesian War (431) until end of the third century (200 B.C.), when it was destroyed and then abandoned, with the most intense activity around 350–250.

Ober would date all his "first generation" artillery towers—designed to house first generation (bow) catapults—to between 375 and 325 B.C.[18] Fossey would date the Boiotian and Phokian fortifications to the period 370–347. Marsden notes similarities between the towers of Messene and those of Khalkis in Aitolia, and a tower still standing at Poseidonia/Paestum in Italy.[19] Ober thinks a tower on a relief from a tomb at Pinara in Lykia is similar to the Messenian, Siphai, Megarian, and Attic towers, and it is dated artistically to the period 370–350. Winter thinks that the fortifications of Rhodes that Demetrios Poliorketes assaulted and failed to take were of Ober's "first generation" type.[20]

Thus, various scholars have argued that the Phokians, Boiotians, Messenians, Athenians, Aitolians, Poseidonians in Italy, and Pinarans in Lykia added bow catapults to their arsenals sometime between about 370 and 325, on the basis of the existence of artillery towers. Whatever their actual dates, all these towers were probably designed to accommodate (and to face) bow catapults, not torsion catapults. Their walls are thin and high in comparison to later towers, which become increasingly robust and short as time moves on. The invention of the torsion catapult allowed attackers to throw stones big enough and to throw them hard enough to shake if not break single-thickness stone walls such as these towers have, requiring defenses to be strengthened or rebuilt. Torsion catapults also needed to be housed in more robust towers, as they were considerably heavier than bow catapults (and stone-thrower torsion catapults were considerably heavier than sharp-caster torsion catapults of analogous caliber).

Can we trace lines of contact between these places, to track the diffusion of catapult technology? In some cases, we can. The Thebans were responsible for the building of Messene's fortifications, and the towers at Siphai can

FIGURE 3.1 *Aigosthena tower. A first generation artillery tower: it has single-thickness stone walls, is relatively tall, and has small square embrasures on the top floor. Small calibre tension catapults would have been sited to shoot through these embrasures. (Bryn Mawr collection)*

be attributed to them too. But how did the Thebans get them? Thebes was perhaps courting Karthage (as Athens courted Dionysios), and Karthage, having been on the receiving end of catapults from the first, and laying close to their source so more likely to pick up willing or unwilling catapult-builders, probably developed them early. This is rather speculative, but there is (or rather, was; it is now lost) an inscription of the 360s or 350s naming a Karthaginian called Nobas as a *proxenos*, foreign guest-friend, of the Boiotians. This man's function was to defend the interests of Boiotians in Karthage,[21] the *proxenos* being the ancient analogue to the modern consul.[22] What were Boiotians doing in Karthage? Glotz and Roesch suggested that this might have had something to do with Thebes' development of a navy; perhaps it had something to do with artillery instead, especially if Dionysios were assisting Sparta and/or Athens. Although we are used to viewing Greek history from a point directly over southern Greece, wherefrom Sicily, Ionia, Macedonia, Krete, and so on are peripheral, it is possible that what was happening on the Greek mainland at this time was actually the continuation of the Siculo-Punic war, now being fought by proxies. The Persians had long since been successfully keeping the Greeks at bay by supporting Greek disunity (not fomenting; the Greeks needed no encouragement to keep fighting each other); perhaps the Karthaginians were doing it, too. The spectacular Tripylon gate to the Euryalos fort of Syrakuse, built at the end of the fourth century, was a response to the "menace karthaginoises," and was built to withstand artillery as well as house it, as Adam pointed out. If the history of weaponry at other times is a reliable guide here, Karthage had started her own program of catapult research and development immediately after Motya. As Alexander the Great is supposed to have observed, "the fortune of war teaches its art even to the vanquished," and my guess is that a second stream of catapult development began in North Africa immediately after the surviving defenders' return. Having witnessed the Syrakusan belly-bow(s) in action, it would be easy for them to copy the concept, even if they lacked any knowledge of the details of construction. Once the concept of the mechanized bow has been grasped, the construction can proceed independently. Detailed evidence is lacking because in 146 B.C. the Romans annihilated their city and dispersed their histories in an ancient act of genocide—though as Kiernan points out, "it was not a war of racial extermination. The Romans did not massacre the survivors, nor the adult males. Nor was Carthage a victim of *Kulturkrieg*."[23]

A tower at Khalkis in Aitolia may have been built by locals or by the Boiotians. It lies approximately midway between Naupaktos and Kalydon on the north coast of the Korinthian Gulf. Ober pointed out that Epaminondas captured Naupaktos and Kalydon from the Akhaians and allied with the Aitolians, so fortifying a site between them would make military and historical sense, though whether the Boiotians or the Aitolians undertook that fortification cannot be decided on present evidence.[24]

Our next piece of datable evidence is literary rather than archaeological and concerns the Samians; we consider it briefly in another context (sniping) in Chapter 5. Plutarch says: "When one of the reckless generals was showing the Athenians a wound he had received, Timotheos said, 'I was ashamed when a missile from a catapult landed near me, at the time when I was your commander in Samos.'"[25] The point Timotheos was making was that he, unlike the other general, prudently tried to stay out of range of the enemy's long-range weapons, and when he underestimated the safe distance, was ashamed of his error of judgment. The fact that he was nearly hit by a missile from a *catapult*, which had longer range than manual weapons, mitigates the self-criticism, which is clever if deliberate. So who shot a catapult at Timotheos, and when? Plutarch's evidence could refer to one of two sieges of Samos, with two enemies. The Peace of Antalkidas, otherwise known as the King's Peace of 386, handed the Greek cities of Ionia over to the King of Persia, but Samos and the other islands offshore were granted independence. She did not join the Second Athenian League, was still autonomous in 381, and probably stayed that way until shortly before 366. But by 366 Samos had apparently been annexed by the Persian Empire to defend it from a renegade Persian area commander, the satrap Ariobarzanes, and according to

Demosthenes a Greek called Kyprothemis was in charge of a Persian garrison there.[26] Timotheos led a ten-month siege to remove Kyprothemis and the Persian garrison. He then established a klerukhy, a colony of Athenian citizens, which was reinforced at least three times, resulting in the expulsion of almost all Samians from the island. (One of these Athenian settlers, incidentally, was Neokles, whose son Epikouros—Epicurus in Latinized form—born on Samos in 341, grew up to become the famous philosopher of atomism, friendship, and the good life. Lucretius' *De rerum natura* is Epikouros' philosophy done into Latin verse for sale to sophisticated Romans, and "epicure" remains to this day a label for a person devoted to pleasure.) Although Samos was not in the Second Athenian Alliance, the establishment of Athenian colonists there and the displacement of its people has been seen as a catalyst of the Social War between Athens and its allies of 357–355. In the course of this war Khios, Rhodes, and Byzantion besieged the Athenian settlers in Samos in 357/356.

If Plutarch's testimony relates to when *Athens* besieged Samos in 366–365, as Babbitt suggested in a note on the passage in the Loeb edition, then the Samians apparently had bolt-shooting catapults in 366. If, alternatively, it refers to Timotheos' defense at his trial and the *rebels'* siege of Samos in 357/356, as argued by Klee,[27] then Plutarch's testimony implies that some city in the Hellespont, probably Byzantium, had catapults, since that is where Timotheus was during the second siege of Samos. The latter scenario is easier to relate to the history of the catapult, since we know independently that Byzantium had catapults in 340, but Klee's interpretation of the text seems to me to be rather forced, and I prefer Babbitt's as a more natural reading. However, there is nothing else in the known history of Samos to connect her to known catapults or catapult-makers by 366. As with the historical contexts for the construction of the towers at Poseidonia and Pinaros, gaps like this in our knowledge serve as useful reminders of how much we do *not* know about the ancient world.

The next hint of the diffusion of catapult technology around Greece is two boxes of catapult bolts in storage in a building called the Khalkotheke on the acropolis of Athens from 363/362 or earlier.[28] How they got there is a bit of a mystery. We do not even know whether this inventory item represents first fruits, that is, the earliest and best of new produce, dedicated to the gods in thanks for past blessings and in anticipation of future goodwill, or whether a better analogy would be the store room at Buckingham Palace, which is reputedly full of unusual and unwanted gifts from domestic and foreign friends and diplomats. It has been argued recently that the building may have had a defensive function, serving as an arsenal, which, if right, undermines the argument of Keyser that by depositing catapult material on the acropolis the Athenians thereby took it out of service.[29]

A variety of reasons have been offered to explain the appearance of catapult bolts on the acropolis. One is that Dionysios may have given these bolts to the Athenians, together with (unattested) catapults from which to shoot them, as part of an alliance deal. This hypothesis stems from some surviving inscriptions. In the spring of 368 the Athenians awarded Dionysios and his sons citizenship, in gratitude for his help; what sort of help is unspecified on the stone. In another inscription from (probably) the late summer of the same year, they made a defensive alliance and promised not to attack each other. The other hypothesis is that the catapults may have been taken from one of Dionysios' warships that the Athenians seized, which implies a completely different relationship between Athens and Dionysios.[30] The idea that they were war booty is perhaps more plausible than that they were gifts, given that storage in a temple treasury is what one would expect to happen to booty dedications. On the other hand, enmities were sometimes resolved through alliances, and that was perhaps the case here, too.

Alternatively, or additionally, but perhaps less likely, Plato, or Dion (Dionysios' uncle and brother-in-law) may have been the vectors for this novel bit of hardware. They both moved (privately and socially) between the court of Dionysios and Athens, Plato going back and forth in the 380s, 367, and 361, and Dion moving to Athens in 366. Plato comments in the *Gorgias* (and the Gorgias after whom the dialogue was named was a philosopher from Syrakuse), that the machine-maker "at times preserves entire cities."

> The pilot is not accustomed to give himself airs, even though he saves us; no, my strange friend, nor the machine-maker (μηχανοποιός) either, who at times has no less power than the general or the pilot, or anyone else to save life, for at times he preserves entire cities. (Plato *Gorgias* 512b )

If this is not a reference to war machines then I do not know to what machines he could be referring.

Alternatively, catapults may have come to Athens as gifts from the Spartans, as suggested above. Ultimately, we can only speculate on the origin of these bolts and the reason why they were stored on the acropolis. But in 368/367 the Athenians were very grateful to Dionysios for something: in that year they gave him first prize for his tragedy *The Ransom of Hektor*. About twenty years earlier at the Olympic poetry contest, people had laughed at Dionysios' literary efforts, and no copy of his victorious tragedy has survived for us to assess it for ourselves. But that is, perhaps, beside the point. Dionysios had literary ambitions, as Nero did later, and like many other peoples, in many other times, the Athenians were not above letting politics pervert the rankings in an arts competition.

Our next witness is Aineias Taktikos, who about 360 B.C. wrote a book specifically about *How to Survive Under Siege*, and he mentions artillery *explicitly* just once in the surviving text. It is obvious from what he says, however, that he knows what catapult capabilities were. His advice is worth quoting, as it gives a real flavor of siege warfare of the time.

> Your opponents' assaults, with machines or troops, can be resisted in the following ways. In the first place, [sails] offer protection against missiles coming over the wall from towers or masts or the like. Cover them with something tear-proof, use capstans to stretch them taut, and once [they] are in position the projectiles will have to overshoot them. You should also start fires [which will emit] thick smoke from below and make as big a blaze as possible, and raise in defence wooden towers or other tall structures made of sand-filled baskets or stones or bricks. [Even] cross-woven wickerwork can stop the missiles. Be prepared in addition for rams and similar machines being directed against the battlements: protect them by hanging in front of them sacks filled with chaff, bags of wool, fresh ox-hides either inflated or filled with something, [and] other things of this kind. Whenever a machine is managing to breach either a gate or any part of the wall you should lasso its projecting end to stop its blows; and having ready a stone, large enough to fill a wagon, to drop on to and smash the drill. The stone should be held in place with clasping hooks and dropped from the forebeams. Make sure that the stone does not miss the drill as it falls by first lowering a plumb-line and dropping the stone after it as soon as it touches the drill. (Aineias Taktikos *How to Survive Under Siege* 32.1–6 Whitehead)

Drills were used against mudbrick walls, rams against stone, because, as Pausanias observed, mudbrick coped with the impact of a ram better than stone.[31] Stones are dislodged by rams, weakening the whole section of wall. Mudbricks just crumble, creating a hole little bigger than the ram head. Drills did the same job faster against mudbrick, but were useless against stone. Hence, well-equipped armies used rams against stone walls, and drills against mudbrick. We have a surviving description of how to build a wall-borer drill in Athenaios Kyzikos.[32]

Aineias continues with a highly risky strategy to be employed with mudbrick walls—the idea seems to be that a ram will smash a drill if the two come to blows, but it requires the defenders almost to puncture their walls themselves.[33] It is in this context that we find Aineias' one explicit reference to catapults, and he seems to mean catapults mounted in towers that are brought up to the wall, just as they were used in Motya; he is not talking about sending out saps to stationary catapults positioned at long range from the walls and shooting from ground level.

> Here is another excellent counter-measure against attempts to breach the wall. When you know the point against which the attack is directed, you get ready a counter-ram on the inside of the wall at that point and dig through part of the wall just as far as the outer layer of bricks, so as not to forewarn your enemy. Then, when their machine is close at hand you deliver a blow from inside with the counter-ram—which will be much the stronger of the two. As regards large machines used for bringing forward troops in quantity and for launching missiles, especially from catapults and slings, and shooting incendiary arrows on to the houses which have thatched roofs, you must counter them initially by having the men in the city excavate, in secret, beneath where the machines are to approach, so that their wheels fall through and sink into the holes. Then an internal barricade should be built, from sand-filled baskets and some of the available stones, which will overtop the machine and neutralise the enemy's missiles. At the same time hang out thick curtains or sails as protection against incoming missiles; any which get over the wall will be caught and easily collected, and none will reach the ground. And do the same anywhere else on the wall where missiles can come flying over and disable or wound those on duty or passing by there. (Aineias Taktikos, *How to Survive Under Siege* 32.7–10 Whitehead)

At this point in time, catapults still seem to be considered akin to bows and slings, and, most significantly, Aineias expects to counter catapults and slings by the same measures. The sort of catapult missile that he was expecting to meet was not a revolution in armaments, requiring a new type of response.

Our next witness is Xenophon, whose *Hellenika* tells of Greek history down to 362 B.C. The book ends with the battle of Mantineia and the death of Epaminondas, the great Theban general, where about 33,000 Boiotians, Tegeans, most of the Arkadians, Akhaians, Argives, and some other Peloponnesians fought about 22,000 Spartans, Athenians, Mantineians, Elians, and a few others. He does not mention catapults being used here, nor does Diodoros Siculus. Actually, Xenophon never mentions the catapult in any of his works. Some have suspected that he was prejudiced against it, but perhaps the explanation is that the catapult was neither common nor significant when he wrote. Both sides claimed the victory at Mantineia; the Greek states had fought each other to a standstill through decades of war and power struggles.[34]

For the next six years or so their maneuvers were political, and then a group that were usually marginalized, the Phokians, burst onto the pages of the history books by seizing the sanctuary of Apollo at Delphi.[35] Delphi is actually in Phokis, their own region, but for centuries the Phokians had been sidelined by stronger neighboring states who ran the sanctuary through a multinational organization called the Delphic Amphiktyony. Amphiktyones were literally "those who lived around," neighbors. A number of major sanctuaries had such organizations; an amphiktyony was not unique to Delphi. At times of international stress, such as in the middle of the fourth century B.C., this organization was liable to be perverted by the political interests of the state(s) that then dominated the Amphiktyony. And that was what the Phokians claimed had happened; they were being squeezed, via pronouncements of the Amphiktyony, by Thebes and Thessaly. In apparent desperation, they seized the sanctuary, smashed the stones on which the decisions and threats of the Amphiktyones against them were inscribed, and proposed to set up a new Amphiktyony to run the sanctuary, with membership from all over the Greek world instead of just Delphi's neighbors. They sought military and financial help from Sparta and Athens, neither of whom was happy with the orthodox Amphiktyony for reasons of their own (Sparta had been criticized and fined by it too) and both of whom were receptive to any request that involved challenging the power of Thebes.

The result was the Third Sacred War, between the Phokians and their allies on the one side, and the Amphiktyony on the other, which broke out in 356 and was waged on and off for ten years. The Third Sacred War is sometimes seen as the most important event in the rise of Philip II of Macedon, father of Alexander the Great, because it brought him into central Greece, it advertised his strength and his threat to the southern Greeks, and because by winning it he became in fact, if not in theory, hegemon of Greece. As Diodorus puts it, "Philip . . . added continually to his strength from that time on, and finally . . . was appointed commander of all Hellas and acquired for himself the largest kingdom in Europe" (16.64.3). It is also the first occasion on which stone-thrower catapults were apparently deployed—and not by Philip.

Having seized the sanctuary, the Phokians recruited mercenaries from all over Greece and soon had the upper hand over their opponents. Meanwhile one of those opponents, Thessaly, was rent with a civil war that had already been going on for two years. The developing Thessalian League wished to incorporate the city of Pherai, but the Pheraians, under their leader Lykophron, were having none of it. The League sought the help of Philip for an attack on Pherai to bring it into the fold. Philip was their neighbor to the north, and from 358 was tied by marriage to the Aleuads, one of the leading families in Larissa, one of the cities in the League. When Philip was first asked for help, he sent men but did not go himself; he was leading military actions north of Macedon, and reducing the Athenian outpost of Methone, where he lost his right eye, to an arrow according to Diodoros, and to a catapult bolt according to Strabo.[36] If Strabo is right, then the Athenian garrison of Methone in southern Macedonia had catapults before 353.

In 353 Philip came to help his allies the Thessalian League to reduce Pherai. He probably did not see his first intervention in Thessaly as part of the Sacred War, but as assistance for his allies in the Thessalian civil war. Pherai meanwhile sought help from the Phokians, who were already at war with the Thessalian League over Delphi, and who had a large and powerful army including a large numbers of mercenaries. Thus do civil wars become international wars.

> Philip was invited by the Thessalians and entered Thessaly with his troops, where at first he assisted the Thessalians and went to war against Lykophron the tyrant of Pherai. Subsequently, when Lykophron called in his Phokian allies, Phayllus, the brother of Onomarkhos, was sent with 7,000 men. Philip defeated the Phokians and expelled them from Thessaly. Onomarkhos, taking his entire army with him in the belief that he would secure control of the whole of Thessaly, came with all speed to help Lykophron and his supporters. When Philip and the Thessalians ranged themselves against the Phokians, Onomarkhos, who was superior in numbers, defeated him in two battles and destroyed many Macedonians. Philip was now in the midst of the greatest dangers and deserted by his despondent troops, but he encouraged the majority of them and with difficulty rendered them obedient. After this he returned to Macedonia. (Diodoros 16.35.1–3)

The exact location for either of these battles is never given; they remain battles without fields, somewhere in the vicinity of Pherai in Thessaly. According to Polyainos, one of these battles was on rolling ground, where Onomarkhos deployed stone-throwers (*petroboloi*) on the slopes of a blind valley, into which his infantry feigned retreat.[37] "He concealed men on the heights on both sides with stones and stone-throwers (πέτρους καὶ πετροβόλους)." Marsden misleads by referring to πετροβόλους *μηχανάς*, stone-throwing *machines*; the source is not that explicit and does not have the word *mekhanas*. However, like most scholars, I think Polyainos is talking about stone-thrower catapults; it is hard to believe that such destruction and dejection could have been caused just by slingshot and hand-thrown stones. The ensuing engagement seems rather to have resembled the Charge of the Light Brigade: "the men on the heights threw stones and crushed (συνέτρι–βον) the Macedonian phalanx." The verb *suntribein* implies a real pasting, smashing to bits, beating to a pulp, and to my mind clearly signals the use of machines, not just men. Onomarkhos' stone-throwers inflicted heavy casualties on the Macedonian forces, causing the despondency and desertion that Diodoros mentions. Philip was trounced, and the repercussions were tremendous: this was his first defeat in six years, and we will continue his story in

Chapter 4. Polyainos' evidence for catapults in Phokis complements the archaeological evidence of the towers that we examined earlier. Other scholars have argued on completely different grounds that Phokis has perhaps the earliest artillery towers anywhere. It should not then occasion surprise that one of the earliest testified uses of artillery in battle is by the Phokians.

We do not know where the Phokians got these catapults. We do not actually know even what sort of catapults they were, other than that they threw stones. This is, in fact, the earliest reference to *stone-throwing* catapults. Marsden guessed that they were probably small bow catapults capable of throwing a shot of about five pounds,[38] but it is equally possible that they were *monagkones*, one-arms, and if so, then *this* was not just the first stone-thrower, but also the first torsion catapult.

The one-armed catapult is a mechanized version of the staff sling (Figure 3.2), and it is, in fact, a completely different machine from either a *gastraphetes* or a two-armed catapult. It has a rigid oblong frame that sits on the ground, and a single arm (hence the name) inserted vertically into a skein of rope stretched tight across its width. The arm is pulled back and down, twisting the skein, and is held there by some trigger mechanism. Attached to the free end of the arm is a thick leather sling, one cord of which is fixed to the arm, and the other has a loop that goes over a pin in the end of the arm.[39] A stone is placed in the sling, and the trigger is released. The twisted skein of rope forces the arm rapidly up to the vertical. As the arm ascends and centrifugal motion pulls the sling-stone outward, the loop slips off the pin, releasing the sling, and the stone flies in a high arc over the machine toward the front. The arm may be restrained by a padded crossbar of some sort to stop it recoiling to the front of the machine, but this has nothing to do with releasing the stone, which should have left the sling before the arm meets such a crossbar. Figure 3.3 shows a large Roman variety of monagkon known as an onager.

Now as we have seen, the Phokian artillery towers might be the earliest known, indicating Phokaian interest in artillery decades before this battle. Alternatively, or additionally, the Phokians' ally in this episode, Pherai in Thessaly, might have been a locus of artillery development. About twenty years *before* this battle Pherai had an ambitious autocrat, Jason, who was, in some ways, very similar to Dionysios before him and Philip after; indeed, it has even been suggested that had Jason not been assassinated in 370, he might have pre-empted Philip and Alexander in dominating Greece and leading an invasion of Asia.[40] Jason employed large numbers of mercenaries, and it is possible that there was some crossover from Dionysios' forces, bringing knowledge of his arsenal with them from Sicily to Thessaly. Poluidos, Philip II's chief engineer and the teacher of those who built engines for Alexander, came from Thessaly.[41] If catapult technology *was* being developed in Pherai, then Onomarkhos might have had them to deploy at this battle in his capacity as commander of the *Pheraian* (as well as Phokian) forces, rather than having them in his own arsenal and used by his own troops. If catapult technology was not being developed independently in Phokis or Pherai, then one would expect the Phokians to have used sharp-casters (they were allied with both Sparta and Athens at one time or another, so their acquisition of bow sharp-casters is not difficult to explain). Indeed, some story is wanting to explain the adaptation of the sharp-caster *gastraphetes* to shoot stones. In the absence of such evidence, we are reduced to educated guesswork, and in my opinion the most plausible guess is that the Phokians or the Thessalians independently invented the torsion one-armed stone-thrower, as a mechanization of the staff sling, analogously to the invention of the gastraphetes as mechanization of the hand bow. I would guess that this took place with the inspiration, encouragement, and financial patronage of either Jason in Thessaly or Onomarkhos (or his predecessor Philomelos) in Phokis, by 353. The argument is developed in Chapter 4.

The next piece of evidence, hitherto overlooked, comes from a medical work that has been dated, for reasons that have everything to do with medical history and nothing to do with the history of the catapult, to between 358 and 348.[42]

> At the siege of Daton, Tukhon was struck in the chest by a catapult [bolt]. Shortly afterwards he was overcome by a raucous laughter. It appeared to me that the doctor who removed the wood left part of the shaft in the diaphragm, and he thought so too. The doctor gave him an enema towards evening and a drug by the bowel. He spent the first night in discomfort. At daybreak he seemed to the doctor and others to be better. Prediction: spasms would come on and he would die quickly. On the subsequent night discomfort, sleeplessness, he lay on his stomach for the most part. Convulsions began with daybreak the third day and he died at midday. (Hippokrates *Epidemics* 5.95 Loeb with modifications)

Daton was a colony of Thasos, founded at an unknown date on the Thracian mainland, some distance inland from the coast. In this area, in the decade in which this text was composed (358–348), either Athens or Philip could have been besieging the place, though Philip is more likely. The author does not remark on the novelty, or otherwise, of the missile, or the method by which it was delivered to his patient. That is, he takes it for granted that the reader will know what a catapult is. This is further evidence for the widespread occurrence of catapult technology by the mid-fourth century. Treatment evidently was a good deal more primitive than that offered to similar casualties in Celsus' time, about three hundred years later (discussed in Chapter 1).

FIGURE 3.2 (top) *Drawing of a staff slinger.* (Schramm) FIGURE 3.3 (bottom) *A reconstructed Roman onager, or monagkon, one-arm, in Greek.* (Courtesy of the Ermine Street Guard.)

About 350–340 someone set up in Peiraieus a gravestone for one "Heraleidas the Mysian, catapult-discharger."[43] This is the earliest evidence that handling a catapult was regarded as a specialty, and this man is sometimes said to stand at the beginning of the breed of artillerymen, which includes Napoleon Bonaparte among its members. That interpretation is based on the assumption that Athens had large catapults, requiring dedicated operators, and that this man was one such. However, many inscriptions from various places in the Aegean, especially Athens, dated between the fourth and second centuries B.C., list the winners of competitions in catapult-shooting, archery and javelin-throwing, and "catapult-dischargers" appear there alongside archers and javelin-throwers as individuals wielding (individual) range weapons, as we will see (Chapter 5). This man's name was probably Herakleidas, after the legendary hero and strong man Herakles, but whoever inscribed the stone omitted the "k." Misspelling apart, a Greek name with a Mysian ethnic suggests that our man was a metic, a resident alien, perhaps an ex-slave, living and working in Peiraieus, where so many metics lived. Mysia is that part of modern Turkey opposite Lesbos, slightly inland (the Greeks occupied the coast), where Pergamon would rise in due course.[44] By the 330s, if not before, the "catapult-discharger" was one of a group of trainers who instructed young men in Athens during their two-year spell of military training, and again he appears to be analogous to the archer or the javelin-thrower, rather than the different sort of soldier implied by the translation "artillery-man."[45] The Aristotelian *Constitution of the Athenians*, written about 325 B.C., refers to these officers.

> The Ekklesia also elects two trainers, with subordinate instructors, who teach the epheboi to fight in armour, to use the bow and javelin, and to discharge the catapult. ([Aristotle] *Athenian Constitution* 42.3)

When he wrote in the *Nikomakhaian Ethics* of an apparently accidental discharge of a catapult, Aristotle seems to be referring to an incident well-known to his anticipated audience that occurred around 335 B.C. "A man might be ignorant of what he is doing, as, for example, . . . a man might say he discharged it when he only meant to show its working, as the man did with the catapult" (1111a8–11). Clearly, catapults were familiar objects by the 330s, part of the standard equipment which Athenian soldiers might wield. In theory, catapult training ought to have required less time and effort than bow and javelin, since less skill is required to shoot effectively with a weapon that contains within itself the power to deliver, and which can be picked up, carried, and laid down when loaded.

In 330, we find reference to catapults "from Eretria" in a long and largely complete Athenian inscription, running to nearly two hundred lines of text in each of three columns. About ten years earlier, in 341, the Athenian general Phokion had attacked Eretria in Euboia and brought home war booty. About the same time, an Athenian comic suggested that Briareos "ate" catapults and pikes. In ancient Greek mythology Briareos was the name of a giant with one hundred hands, but a scholiast on this passage says it here means the Athenian politician and orator Demosthenes. The joke has been connected to the battle in Eretria and its war booty, but the reference and its significance remains obscure. According to the inscription, in "the large building near the gates" in Peiraieus, along with iron anchors and lead weights and other things that we might expect to find in a naval dockyard, were quite a lot of catapult bits.[46]

> . . . . πλαίσια (*plaisia*) of catapults from Eretria 11; σωλῆνες (*solenes*) of catapults 14; bases of catapults 7; bows in leather cases (τόξα ἐσκευτομε–να) 2; *solenes* of scorpions 6; ἐπιστύλια (*epistulia*) 5; windlass-drums 3 from the machine; catapult bolts headless and featherless 455; and with heads 51; staves for catapult bolts 47. . . . (*Inscriptiones Graecae* II$^2$ 1627 B lines 328–341)

I have left a number of terms in Greek transliteration because their translation is uncertain and needs discussion. Let's begin our discussion with the reference to Eretria. At this time, Athens and Philip were fighting a war by proxy. Marsden suggested that Philip had sent these catapults to Eretria when he supported the installation of Kleitarkhos as tyrant of that city. Macedonian troops arrived to reinforce his position twice, according to Demosthenes (who is a demonstrably hostile witness on Philip).

Marsden assumed that these catapults were torsion catapults, but if they were, then one must also suppose rapid diffusion of very new technology. It is often said that the first use of torsion catapults was at the siege of Perinthos in 340—which is a year after this Eretrian adventure, so consistency is not good here either—but in fact even that is uncertain. All Diodoros says is that Philip used *oxubeleis*, sharp-casters, at Perinthos, and that could apply to tension catapults, too. The earliest solid evidence for torsion catapults, because it mentions hair *springs*, is another Athenian inscription that is *probably* to be dated to the era of Lykourgos' dominance, that is, sometime between 338 and 326, which we will consider in a minute. The inscription from the dockyards currently under consideration might actually be earlier than that inscription, and thus it is important whether we think that the catapults listed in it were *gastraphetai* or torsion catapults. It warrants a close look.

Marsden gave three words in this short section, πλαίσιον, σωλήν, and ἐπιστύλιον, unorthodox translations. He took πλαίσιον, which normally means "an oblong figure or body" (LSJ) as a substitute for πλινθίον (whence English plinth). πλινθίον, with a basic meaning of "square," comes to be used as the technical name for the frame that holds the spring of a torsion engine. Thus, the first line of his translation is "Frames for the catapults from Eretria 11." His alteration of the meaning of *plaision* to *plinthion* is the only thing that qualifies this inscription as evidence for torsion engines. *Plaision* could equally well refer to, for example, the slider, which is oblong in shape. Next, Marsden made σωλήν (uniquely) the stock (case *and* slider), so that the *plaision* can be (uniquely) "frame." He further suggested that the term ἐπιστύλιον, which normally means "lintel," here, uniquely, means "hole-carrier." This strains one's credulity, as well as the Greek, too far. Giving unique meanings to words is, in principle, a dodgy procedure, and when someone does it three times on one short piece of text, albeit a technical and condensed text, it suggests that the text is being made to fit the vision, rather than the vision taking shape from the words that are there. Alternative, more orthodox readings can make just as much sense of this fragmentary text.

Marsden gave short shrift to the bows very clearly mentioned midway through the text. Contrary to Marsden, I think that this is a list of bow catapults components only, not torsion engine bits as well; it actually makes more sense wholly in relation to *gastraphetai*, and not to components of both bow and torsion engines muddled together. That the bows are said explicitly to be in cases shows that they were dismantled from the rest of the catapult and stored properly, presumably unstrung, as any bow should be. A composite bow must be unstrung when not in use to retain the elastic qualities of the materials of its construction, and that would have applied to catapult bows just as it does to manual bows.

The "*solenes* of the catapults" (second item) are probably the fixed part of the stock which Marsden calls the cases, not the whole stocks *pace* Marsden,[47] so that *plaisia* and *solenes* mean sliders and cases, which together constitute the stocks (see Figure 2.1). This word ordinarily means things that resemble pipes, tubes, gutters or channels, and both Heron and Biton employ it or its derivatives (σωλήνα and σωλήνια/σωληνίδια, respectively) for the case along which the slider travels. The case has a sort of open-topped triangular barrel, into which the triangular base of the slider fits and along which it slides. *Solenarion*, which is a diminutive of *solen*, is a term that survives as early as Galen and features in late antique sources (and beyond) as an "arrow guide," resembling a "barrel" unattached to the bow, which is used to enable an archer to shoot short darts from a hand bow. A similar sort of split barrel appears rarely on the crossbow, creating what is known as a slurbow (see Figure 3.4). [48]

The *solenarion* is as little known as the slurbow, so I will take a few minutes to explain it. It is described as a wooden tube hollowed out to the size of the dart but cut so that it resembles a reed split down the middle.[49] The split would need to be wide enough to allow free passage for the bowstring and long enough to extend (maximally) from full draw to the bow's brace point, or a little beyond. The tube would also need to be made of material sufficiently rigid that the split sides would not splay in use. It is, I think, depicted in the *Chronique de Pierre d'Eboli* of around 1197 A.D. (Figure 3.5).[50]

A small dart would be placed in the back end of the tube, the bowstring engaged with the dart's nock, and the dart inserted further in until the bowstring was engaged with the *solenarion*, passing through the splits top and bottom of the tube. The bowstring would then be drawn normally, bringing dart and *solenarion* with it. Some means is needed to secure the *solenarion*: either

FIGURE 3.4 *(top) The slurbow. A barrel is created but cut away to make space for the bowstring to pass. (After Payne-Gallwey, figure 84)* FIGURE 3.5 *(bottom) The solenarion (after Nicolle 1987)*

the left hand grasped its (unsplit) front, as in the Berne manuscript, probably with the index finger looped over to hold it, or a cord was attached to the rear of the device and held by the right hand, though this would require greater dexterity. On discharge, the dart is propelled through the *solenarion*. Multiple darts may be shot in this way; Arabic sources describe four or five being discharged in one shot, and two are depicted in the Berne illustration. The darts are called "mice" (μύας) in Byzantine sources, but this usage was much older: Paul of Aegina (seventh century A.D.) reported the same term in use for finger-sized small sharps among the Egyptians, and even Diodoros Siculus (in a section preserved by Constantine, so perhaps anachronistic) refers to "mice" (and something called "tragedies," τραγωδίαι) being used against the Rhodians in about 155 B.C. In due course this became confused with a very similar word, μυιας, meaning flies, which in Latin became *musca*, for which a diminutive developed, *muschetta*, which in the fullness of time became musket, which, via this extraordinary linguistic journey, became the term for the weapon rather than the little projectile it shot.[51] To return to our subject, *solen* probably means something like barrel, one rather larger than an arrow-guide for a hand bow. The case of a *gastraphetes*, with its central longitudinal groove, fits the image well. So the *solenes* were probably what we call the cases.

There is then mention in this inscription of *scorpions*: there were a further six "*solenes* of the scorpions" in the building. Now the best-known ancient catapult called a "scorpion" is probably the one described by Vitruvius and included in Marsden's collection, which is a small euthytone torsion sharp-caster with curved arms. But there were several other catapults that bore the same name, and like "ballista," the term "scorpion" appears to be used of quite different machines at different times and places, and it is also used of the missiles they shot; it was even used to refer to the man who operated it in one document dating to Roman imperial times.[52] The inscription that we are considering, written in about 330 B.C., shows that *scorpion* was also the name for one of the earliest types of catapult.

What links together the various catapults called scorpion is probably some analogy with the animal. Scorpions were ubiquitous in the ancient Mediterranean littoral and the Near East. Ancient natural historians, medics, and toxicologists preserve some details about this dangerous pest, including one sure to strike a chord with anyone fond of books: one of its favorite hiding places was in papyrus scrolls. Scorpions probably killed more people in the ancient world than did snakes and other venomous creatures.[53] Ktesibios, who was closer in time to this inscription than any of these other surviving sources, says, "Some people call euthytones scorpions from the resemblance of the shape" (74.6). This analogy is less obvious than one would like. He is perhaps referring to the pose—arms apart, making a headless, cruciform shape. Or the swollen-looking springs resembling the scorpion's enlarged claw "arms." Or could this even be support for the in-swinging arms theory, the arms' forward position resembling more closely the pose that the scorpion sometimes strikes with his claws? The key facts about scorpions are that they are small and have a ferocious, sometimes fatal, sting. Vegetius writing probably about A.D. 380 says that "they used to call scorpions what are now called *manuballistai* (hand-ballistas); they were so named because they inflict death with tiny projectiles" (4.22). It would appear that any catapult that satisfied the requirements of being small and shooting small missiles might conceivably be called a scorpion, just as *oxybeles*, "sharp-caster," covered a multitude of different designs that shot missiles equipped with sharp heads of one size and shape or another. This straightforward explanation has hitherto been overlooked thanks to Ammianus' comment that the onager, a one-armed catapult, was previously known as a scorpion, but he was confused (the depth of his confusion is revealed in Chapter 10 below). Onagers do not have *solenes* (cases), and we shall ignore this red herring.[54]

The windlasses may be explained as part of a large *gastraphetes'* loading mechanism or a device for stringing the gastraphetai. If they are stored, as they should be and as they appear to be, unstrung, and if they are too strong for a man to span, then stringing them would have been a significant challenge without mechanical aids. This windlass might be such an aid. The *epistulia*, which are lintels, architraves, or epistyles in their usual architectural context,[55] ought to refer to some piece that spanned something; what is known as the κανόνα καταγωγίδα or belly-bar[56] is the likeliest candidate in my view. If Heron's *kheiroballistra* is to be reconstructed with a *gastraphetes*-type belly-bar, then this component is called a "moon-like shape piece" there (σεληνοειδές τι σχῆμα, 124.8 Wescher).

It is frustrating to end the discussion of this important inscription on such an inconclusive note. What it shows, above all, is that the same evidence can be interpreted in a variety of ways. It is better method to state the possible options and the objections and limitations that have been or could be raised to those options than it is to try to force the evidence to fit one's hunches.

Bearing that caveat in mind, I would interpret the items on the inscription from the Peiraieus naval store as follows. First, we have an important negative conclusion: there is no clear evidence in this inscription of two-armed torsion catapults. The catapults from Eretria had sliders; *they* or other catapults also had cases, bases, and, most important, *bows* with their leather storage bags (I avoid the term "case" here to prevent confusion with the "case" of a catapult stock), and were therefore *gastraphetai*. There were, in addition, six "scorpions," which also had cases, and were perhaps a smaller type of *gastraphetes*. The *epistulia* were perhaps belly-bars, and the windlasses came from another machine, perhaps used to string the *gastraphetai* ready for action.

Drawing together all the threads of discussion, I would translate the text of this most important inscription, *IG* II[2] 1627, thus:

> sliders (πλαίσια) of catapults from Eretria 11;
> cases (σωλῆνες) of catapults 14;
> bases (βάσεις) of catapults 7;
> bows in leather bags (τόξα ἐσκυτωμένα) 2;
> cases (σωλῆνες) of scorpions 6;
> belly-bars (ἐπιστύλια) 5;
> windlass-drums (τροχίλοι) from the machine 3;
> missiles (βέλη) for catapults headless and featherless 455;
> and with heads 51;
> shaft missiles for catapults 47...

This is a tiny part of one of many large and largely complete inscriptions, running to hundreds of lines of text in several columns, listing naval equipment in Peiraieus, the port of Athens. It has bows in among the catapult bits, and there is no mention of springs; these catapults were surely *gastraphetai*. *IG* II$^2$ 1628 D 510–521 (another long and well-preserved naval inventory) dated 326/325, includes the same items, in the same quantities, as does *IG* II$^2$ 1629 E 985–998 (an even longer and well-preserved naval inventory) from 325/324, as does *IG* II$^2$ 1631 B 220–229 (another very long but not so well preserved naval inventory), from 323/322. This last inscription is different only insofar as the number of windlass drums seems to have grown from three to four, and bolts with heads from fifty-one to sixty. These differences may be explained as accounting errors in the original, inscribing errors in the original, or typographic errors in the transcription or publication of this inscription, or they indicate that Athens was re-arming, that is, stockpiling military material in preparation for a war planned or foreseen. Alexander the Great died in 323, the year of this last inscription, and the Lamian war, by which the Greeks hoped to throw off the Macedonian yoke, began shortly thereafter. Aristotle died a year later, and one of the minor treatises attributed to him but probably written by a member of his school (the Lykeion) seems to have referred to a bow (tension) catapult.

> What gives strength to the impact of the breath is that the lungs after wide distension contract and violently force out the air. This can be illustrated from the other parts of the body, none of which can strike a blow with any effect at a very close distance. It is impossible with either the leg or the hand to smite the object of your blow with any force or to hurl it far, unless you allow the limb a considerable distance in which to strike the blow. If you fail to do so, the blow is hard owing to the energy exerted, but it cannot force its object far. Under similar circumstances, stone-throwing catapults cannot shoot far, nor a sling, nor a bow, if it is stiff and will not bend, and the string cannot be drawn back far. But if the lung is large and soft and flexible, it can admit the air and expel it again in large quantities, regulating it at will thanks to its softness and the ease with which it can contract. ([Aristotle] *On Things Heard* 800b1–19, trans. Loveday and Forster)

This analogy seems to me to require that the author (whether Aristotle or not) is talking about bow catapults, for bows bend over a considerable distance, whereas springs twist over a little.

From *about* the same time (viz., about 330 B.C.) as the Peiraieus "scorpion" inscription comes an inventory of an entirely different kind: It lists two-cubit and three-cubit, complete and incomplete, hair-spring catapults in the bronze store (the Khalkotheke) on the acropolis.[57] (As it definitely does refer to torsion engines, we will discuss it at the beginning of the next chapter; all evidence we are discussing in this chapter does not certainly refer to torsion engines, though it has often been assumed to do so, so may concern tension catapults instead.) At *about* this time, Athens was home to two "armies": her own armed citizens, and a Macedonian garrison, which was installed in 322 after the Lamian war (Pausanias 1.25.5). We should not assume that the weapons stored in different military installations around Athens were all the same, especially at this time. It is likely that the militarily superior Macedonian forces, who initially occupied the ancient citadel of the acropolis but later moved into the Peiraieus as well, were better armed than the Athenian people's forces. But we have got a decade ahead of ourselves.

The Halikarnassans are said to have had *oxubeleis* back in 334, when they came under siege by Alexander: "Fierce fights occurred in front of the city, in which the Macedonians showed far superior prowess but the Persians had the advantage of numbers and of armaments (ταῖς παρασκευαῖς ἐπλεονέκ–τουν). For they had the support of men who fought from the walls using sharp-caster catapults (τοῖς ὀξυβελέσι καταπέλταις), with which they killed some of the enemy and wounded others." Diodoros uses a term (*oxybeleis*) that is often applied to torsion engines, but it does not follow automatically that this is evidence for torsion catapults in Halikarnassos at this date. For *gastraphetes* is a very rare term, used by Biton four times and Heron once, and a twelfth-century lexicographer seeking to preserve and explain it mentions it too, and it did not even appear in Liddell and Scott's standard lexicon of Greek until the ninth edition, which was published in 1940. For an ancient who did not have *gastraphetes* in his vocabulary (that is, almost all ancients, ever), *oxubeles*, literally, sharp-caster, would be a natural term for a device that threw pointed missiles, however they were propelled.[58]

We hear later that there were Athenians, specifically one Ephialtes and one Thrasyboulos, fighting on the Persian side at Halikarnassos, and thus it is possible that they could have brought catapult technology from Athens. Back at the siege, we then hear of a hundred-cubit- (about 200 feet, or 65 m) high wooden tower that was built by the defenders and "filled with sharp-caster catapults" (πλήρης καταπελτῶν ὀξυβελῶν). From this tower and the rapidly built inner brick wall, dense showers of missiles rained down on the Macedonians and killed many of them. Surprisingly, in Diodoros' account all catapults are on the Halikarnassan side; there is no mention of Alexander's forces using catapults, as distinct from other siege engines. Meanwhile another source on the siege, Arrian, has Alexander's forces using "stone-throwers" and "sharp-casters" but only hints at the Halikarnassans' equipment. Clearly, there was some confusion or ambiguity in the sources that Diodoros and Arrian used for this episode, because it is not just equipment that appears to swap sides. One Neoptolemos, who fell in the fighting in the first days of the siege, was said by Diodoros to be fighting on Alexander's side, and by Arrian to be fighting on the Halikarnassan side.[59] Diodoros also talks about siege-sheds to protect the workers filling ditches, where Arrian simply observed that the ditches were filled easily. These sheds were a relatively new invention. Now when there is a source conflict like this between the historians of Alexander, Arrian usually turns out to be more reliable than other authors, but he is not infallible. In any event, the problems are such that I think we cannot call the siege of Halikarnassos a key moment in the history of the catapult.

Perhaps we should not be looking to individuals or states with known connections to other catapult users by this time in order to explain their presence elsewhere; perhaps the diffusion of *gastraphetai* had grown beyond that which can be mapped like a family tree. One surviving treatise on catapult construction deals with *gastraphetai*: Biton, in his *Construction of War Engines and Catapults*. This is the most difficult of all the technical treatises: the descriptions are poor, critical figures are lacking, and the textual problems are so bad that some good scholars, including A. G. Drachmann, have written it off as incomprehensible. Marsden worked hard to make sense of it, but on some occasions his interpretation far outstrips mere translation of the Greek. There is always a danger in the history of technology that the scholar's knowledge will fill in original gaps, and anachronistically supply some detail that simply was not there, and the modeler's desire to make a physical reconstruction draws his or her attention to practical matters ignored or elided in the text. Marsden was both scholar and modeler. Supplying figures and ratios from other catapults, and his own knowledge and mechanical experience to compensate for omissions or clarify uncertainties in the text, Marsden made Kharon's machine a large version of the Dionysian *gastraphetes*, with a nine-foot bow launching stones weighing 5–6 minas (Figure 3.6). The caliber is probably about right, but many alternative reconstructions of the device are possible. In fact, Biton does not even explicitly call this catapult a *gastraphetes*. For him, the sling is attached to a beam, not the bow (46.11–47.1); and that beam can run up and down (not back and forth) in a *solenidion*, little case (46.3–4); these are just two elements that Marsden tweaks to make sense of the text. Until the problems are resolved, this particular machine cannot easily find a home in our story.

Kharon, who designed bow catapults, came from a place called Magnesia. Unfortunately, we do not know which Magnesia is meant, nor when this Kharon lived. Three places bore this name: a city called Magnesia on the Maeander (the form of which river gave us the word meander), which was about twelve miles southwest of Ephesos; a city called Magnesia by Sipulos (on the River Hermos) in Lydia about halfway between Phokaia and Sardis, and one region in mainland Greece, specifically, the east coastal part of Thessaly. If Kharon lived in the first half of the fourth century B.C., then the last Magnesia is perhaps the most likely of the three to have been his home. If, however, he lived in the second half of the fourth, or the third or second century B.C., then Magnesia on the Maeander is perhaps the most likely candidate. One could make arguments for any of these places and dates: Poluidos *of Thessaly* was Philip II's chief engineer, and the region Magnesia is in Thessaly, whereas neither of the towns called Magnesia have any known connection with war machine construction or use.[60] On the other hand, Biton says that Kharon designed his stone-thrower bow catapult in Rhodes, and Magnesia on the Maeander is a much closer and more likely origin for someone working in Rhodes. Further, Priene, whose tower could take a small arrow-shooter, is very close to Magnesia on the Maeander, and Kolophon, whose walls may have housed them too, is about half way between the two Ionian Magnesias.[61]

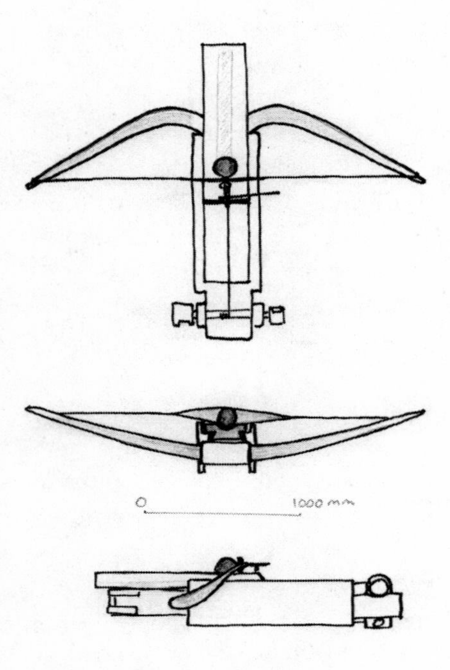


FIGURE 3.6 *Kharon's 5-6 lb stone-thrower gastraphetes or tension catapult. (Author sketch after Marsden)*

In the next century catapult technology really took off. By about a hundred years after Poluidos was active (that is, by 248 B.C.), little fortified towns in Sicily too insignificant to be named had "catapults and stone-throwers" available for passing Romans to commandeer. Even before then places otherwise unknown (and unlisted in many classics reference books and maps) were instructing their outgoing magistrates to ensure that when they handed over to their successors responsibility for the catapult on which ephebes trained in the gymnasium, there were hundreds of catapult bolts with it.[62] There is a tendency to assume that all catapults built by this time must be torsion machines, but we do not actually know that, and they may not have been.

## Technological Survival

The *gastraphetes* evolved into the crossbow. This evolution probably did not occur via the torsion catapult, or the large mounted bow catapults, for that matter. It probably evolved directly, albeit through a now unknown number of stages, from the *gastraphetes*. I guess that a hand-held individual weapon remained in existence from the time of its invention. We know that the Romans had crossbows in the first century A.D.: Arrian (who lived about A.D. 86–160) says that cavalry are trained to shoot missiles "not from a bow but from a machine."[63] The washer and ratchet found at the cavalry fort of Elginhaugh may just be parts of such a machine, though a torsion version rather than a tension version. From the second century A.D. we have illustrations of crossbows on reliefs from Solignac-sur-Loire and from Espaly. It is far more likely that, even though we have no record of it, this type of weapon—a personal compound bow with mechanical locking device and trigger—was being reproduced and developed over the centuries, than that it was lost and then reinvented in a slightly different form. Pliny described what is perhaps a genuinely independent large bow catapult being used by the Trogodytes of Ethiopia to hunt elephants, and it seems to be a significantly different device. "Others employing a safer but less reliable method fix great bows rather deep in the ground, unbent; these are held in position by young men of exceptional strength, while others striving with a united effort bend them, and as the elephants pass by they shoot them with hunting spears instead of arrows and afterwards follow the tracks of blood."[64]

There is no reason to assume that the torsion catapult heralded the demise of the *gastraphetes*, any more than to think that the *gastraphetes* heralded the end of the bow, or the *lithobolos* the end of the sling. Patently, the latter two did not; armies throughout antiquity employed bows and slings. In these cases, the initial technology retained its usefulness in some circumstances, while being superseded by a new technology or adaptation in other circumstances. The proportion of devices that historically became truly obsolete is probably very small, because rarely do successive technologies improve on old ones in most, never mind *all* respects. Therefore there is a continued role for the old technology. Technologies usually involve compromises, and new technologies may select for different outcomes from the same raft of options. The *gastraphetes*, for example, achieved increased range and velocity over the bow, but at the cost of reduced flexibility and maneuverability, and higher expenditure on manufacture. Any of these three reasons could have been critical at specific times and places in the past, and which would have favored the manufacture and use of the bow, rather than the *gastraphetes*. In other circumstances, the *gastraphetes* would have been the best choice, not just over the traditional bow but also over torsion catapults.

As van Wees observed, funds were critical in ancient warfare, and "the story is one of growing centralisation of financial resources and maximisation of revenue from the early fifth century onwards, a process which made new forms of warfare possible, even if it never produced quite enough to allow warfare on the scale to which the city-states aspired. Poverty may have stimulated Greek martial valour, as Herodotus had it, but ultimately it also sharply inhibited Greek military ambitions."[65] Even the simplest catapult cost more to make than the bow and a great deal more than the sling, the weapons which were its manual predecessors and which in some circumstances performed almost as well as it did. Big catapults needed specialist operators as well as specialist makers and cost a lot more. Thus, although—unlike today—there were no legal, political, or social reasons to stop anyone who wished commissioning one, for economic reasons big catapults were very unlikely to be bought by private individuals in the classical period. Social reasons explain why those individuals who could afford to buy a catapult would probably not wish to commission one for their own use: the wealthy went to war on horseback, and (having dismounted to fight on foot) fought courageously, staring death in the face—the face of their adversary—in the phalanx of heavy infantry. In their view, all long-range weapons were for cowards and slaves. Meanwhile, individuals who could afford to buy catapults and employ other people to use them did not do so for political reasons: they would have been assumed to be aspiring to tyranny. So (at least until small affordable ones became available) catapults were probably bought only by the political community (the "state," such as the Rhodians), or by individuals occupying an office that was, to all intents and purposes, the political state: autocrats, tyrants, and monarchs. Most ancient political communities were as reluctant to spend money on public services as most modern ones are; many of the "public works" of the ancient world, such as temples, fortifications, marketplaces, fountains, harbors, and so on were built by more or less politically illegitimate autocrats, rather than by democratic assemblies of upstanding citizens. Military installations and arsenals tended to be an exception to this pattern because if they were not kept up to date, said citizens might find themselves on the brink of enslavement if not extinction. But the traditional method of self-arming in ancient armies was not easily replaced (outside tyrannies, where the tyrant seized or stole the funds with which to achieve it), and the classical Greeks employed mercenaries on an *ad hoc* basis. The alternative was to create a standing supplied army. The first man to take that step was Philip II of Macedon, to whom we now turn.

FOUR

# The Sinews of War: Torsion Catapults

*As a rule, those who have the better equipment secure the victories.*
—*Cassius Dio 50.18.4*

IN DUE COURSE THE TECHNICAL LIMIT of the size of bows was reached: timber, horn, and sinew are natural products that have finite lengths and thicknesses. Bigger does not necessarily result in better performance, and while fifteen-foot bows were produced, bows half that size were perhaps more common. To move forward required a radical change: it required a catapult that drew its power from something other than a bow. Torsion catapults are powered by taut ropes of sinew or hair. They are not as dissimilar to bows as is usually supposed; the key difference is that an alternative resilient material replaces the bendy bit of the bow.

The "classic" two-armed torsion catapult cannot have appeared suddenly like a bolt out of the blue. As Barker observed in an apparently little-read paper, "Great inventions are never such leaps as they look: there has always been a tentative stage—preliminary tinkering, perhaps for a generation or so, perhaps more: hence we may be sure that before its complex application to the double-spring, torsion power must have been showing itself effective in simpler devices. In other words, the primary torsion-gun was the μονάγκ–ων."[1] I concur wholeheartedly with this view. I think that the *monagkon*, the one-arm, is a good candidate for the first torsion catapult because it is the simplest such machine, conceptually, and in terms of construction. It is simply a mechanized staff sling (see Figure 3.3). Today, children aged about seven can make one out of K'Nex and an elastic band, to shoot balls of paper over impressive distances. Further, people discover they can "twang" a length or skein of elastic material from a very young age (and the observant notice that the sound varies with the tension of the material), and begin to shoot acorns or whatever from such a skein held taut by the fingers. The Greeks twanged taut strings of sinew regularly—not just on hand-bows of a military and a carpenter's kind, but also on musical instruments, and even experimentally on a string across a sound-box, at least since Pythagoras began playing with sounds in (probably) the sixth century B.C. It may be objected that these things are easy when you know how, and that is, of course, true. Not many seven-year-olds would make an onager without someone else showing them one first. However, the two-armed catapult is in a different league altogether; one needs to do more than simply *see* a two-armed catapult in order to be able to make one. Even a simplified model is complicated and very time-consuming to make.

New inventions always have a history. They are a particular conjunction of bits and pieces, most, if not all of which, were known for some time before they were used in any particular construction. And herein lies the key to the probable origin of the "classic" two-armed torsion catapult: all the essential bits and pieces needed to make it can be found in either the *gastraphetes* or the *monagkon*. If the *monagkon* had been used by Onomarkhos in 356 (Chapter 3), then the key stage in the development of the "classic" catapult was to fit the power source of the one-arm mechanized staff sling, which lobbed stones in a high arc, to the two arms of the sharp-casting mechanized bow, which shot projectiles on a horizontal trajectory at high velocity, to produce a weapon that was both powerful and accurate, and could shoot sharps and stones. The torsion catapult has two very different antecedents and predecessors, remember: the bow *and* the sling. It makes sense for it to have emerged as a union of two machines, a mechanized bow, the *gastraphetes*, and a mechanized sling, the *monagkon*.[2]

The union of two distinctly different weapons, the tension bow catapult, the *gastraphetes* (see Figure 2.1) and the one-armed torsion catapult, the *monagkon* (see Figure 3.3), would have required rotating ninety degrees (from horizontal to vertical) the torsion spring power source of the one-arm catapult, and duplicating it. This might have been conceptualized as splitting the spring, for that would explain why each of the springs of the "classic" torsion catapult are called in the sources, at first sight bizarrely, "*half*-springs." In a similar fashion, modern French scholars refer to the frame of a palintone spring as the "demi-cadre," the half-frame, and a palintone is composed of two half-frames.[3] Thus a spring that was first laid horizontally was rotated to be held vertically. Then the spring was put in the place of the power source of the bow—the part of the limb that flexed when the bow was drawn. The monagkon's arm still needed to be perpendicular to the spring, so was now inserted horizontally, and a duplicate of it was made as a mirror image to stand on the other side of the stock to replace the other ear of the bow. The arms now took the place of the bow's rigid ears, transmitting and amplifying the power to the bowstring.

What sort of torsion catapult was this hybrid? Barker argued that the names suggested that the palintone came first, and the euthytone ("straight"- or "simple-spring") was a modification, a simplified construction for light catapults that did not require such a robust, but complex, structure as the palintone. (Figure 4.1)[4]

There are more arguments in favor of the palintone's priority, besides the linguistic ones Barker made.[5] First, its formula precisely determines the size of the machine from the missile it has to shoot, whereas the euthytone's is a simplified approximation in which missile length stands proxy for missile mass (length actually matters not a whit). Second, Ktesibios' historical account, starting with the *gastraphetes* and moving to early torsion frames, begins the torsion frame account with a palintone. Then, third, when he describes the machines of his own time, he begins with the palintone, describes it in detail, and when he moves on to the euthytone he describes only the differences in construction from the palintone—so Ktesibios considered the palintone the basic model, and the euthytone a modified version of it. Philon follows the same pattern; the palintone is described in detail first, and the euthytone more briefly, second. Fourth, there are historical precedents for the simpler version of a machine succeeding in time the more complex version.[6] It would indeed appear that, as Barker argued over ninety years ago, sometime after the palintone's invention somebody noticed that sharps made less demands on the weapon and could be launched as well or better from a smaller, simpler version. The euthytone was born in that moment, resulting ultimately in a catapult with shorter springs, mounted on one single frame with two holes in it (instead of two frames, each with one hole, that were then bolted together with more woodwork) and a shorter stock. This is *not* to say that stone-throwers preceded sharp-casters; as Ktesibios pointed out long ago, and others have since, while all stone-throwers were palintones and all euthytones were sharp-casters, not all palintones were stone-throwers and not all sharp-casters were euthytones. The pairs of terms are not interchangeable; palintones could throw stones *and* sharps.[7]

The two-arm catapult that resulted from the union of *monagkon* and *gastraphetes* embodied both the greater power of the torsion spring and the more manageable and more precisely aimed structure of the bow catapult. So when did the two-armed torsion device appear?

The conventional, albeit speculative answer is that engineers working for Philip II of Macedon invented two-armed torsion artillery in approximately 340 B.C. But the earliest surviving concrete evidence for torsion springs dates to sometime probably between 338 and 326. It is part of a large inscription which lists things in an Athenian store, together with the numbers of each item; that is to say, it is a sort of stocktaking register. The stone is damaged,

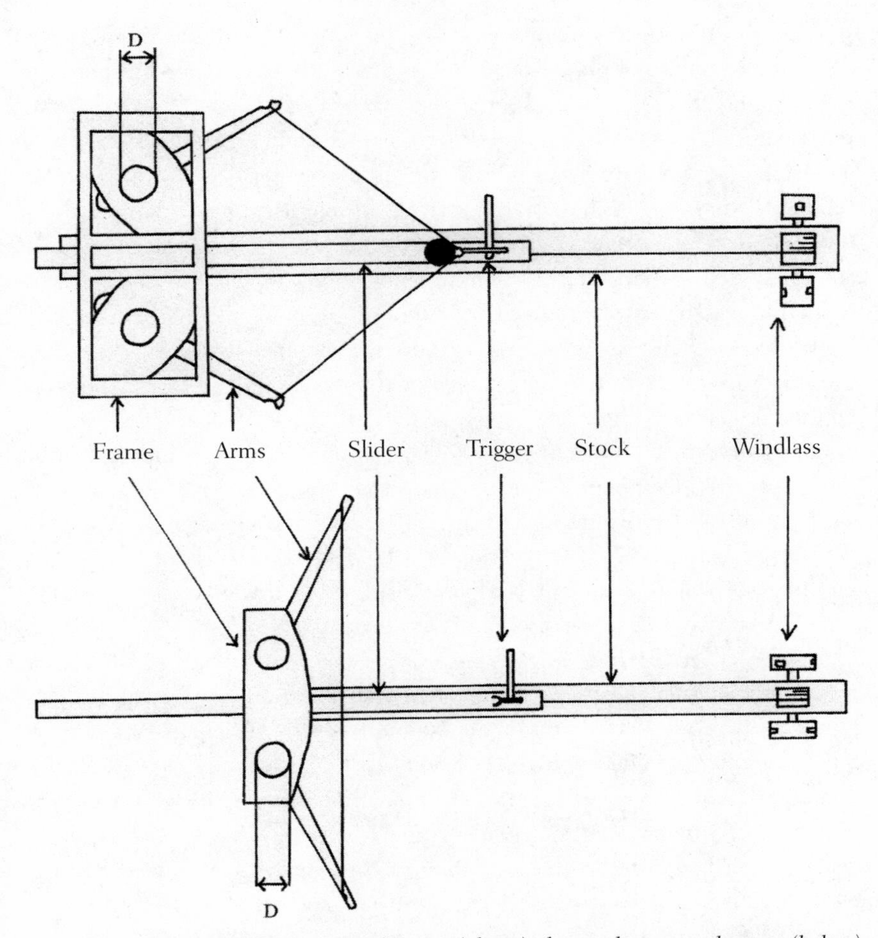


FIGURE 4.1 *Standard Catapults: Palintone (above) shown drawn; euthytone (below) shown braced. D is the spring diameter, which is the key parameter for the construction according to formula. (Author sketch after Marsden)*

and where, because of such damage, the letters are missing, probable missing letters, if they can be worked out, are enclosed in square brackets.[8]

> Ca[tapult]s two-cubit hair[springs] complete : 3 : cata[pults] two cubit hairspr[ings un]sound and incompl[ete : (a number) :] two other cata[pults th]ree cubit hairspr[ings unsoun]d and incomplete : [(7 spaces) catapults] ha[ir]sp[rings. . . . (IG II[2] 1467 face B column 2 lines 48–56)

This stone proves that the torsion catapult was in existence by (probably) 326 B.C., because that is the latest probable date for the stone's production. How much earlier the torsion engine was invented can only be guessed.

There were probably many years of trial and error, or research and development (the choice of phrase depends upon how systematic one imagines the effort to have been), before effective two-armed torsion catapults were being produced reliably.[9] Such a period of development and amendment is true of almost all new technologies—there are always imperfections or "teething problems" that may or may not get ironed out over time.[10] If such problems do get solved, the technology proceeds to the next stage or problem to be solved; if they do not, then it stalls then and there. Further, the people who make and use the new technology have to discover its scope and limits empirically, since by definition no one has experience of a new technology. This can be dangerous, even for the most skilled inventors and operators and the most confident patrons.[11] It is very noticeable that three of the eight catapult springs listed in this inscription, before the stone breaks off, are said to be unsound and incomplete. One assumes that some part has broken—the spring cord itself probably, or its frame. Breakage and damage are typical of early stage technologies, or of inexperienced operators. That is why, for example, new software usually needs a patch or three, and why we give our children inexpensive musical instruments, or cheap cars, to learn on. With a new technology, devices are metaphorically or actually fragile, and operators are metaphorically if not actually novices, so the likelihood of damage is compounded. The poor condition of the catapults in the Athenian store before 326 is completely consistent with the fact that this is actually the earliest evidence we have for torsion springs, and that they were at this time not only new to the Athenians, but absolutely new. It surely makes more sense to explain their condition in this way than to suppose, as some have, that the damage reflects their age or neglect, given that there is actually no firm evidence for the existence of torsion springs before this document.

So we have some idea when the two-arm torsion catapult was invented. But why, by whom, and where?

King Philip II of Macedon, the usual suspect in this case, was born into a family whose history revolved around the ruthless pursuit of power and was littered with battlefield deaths and domestic assassinations (only two of his predecessors on the throne died of old age). Philip emerged as a charismatic, clever, resourceful, and above all pragmatic leader of his people. He was responsible for many innovations in the military and political spheres. During his first year in power (359 B.C.), he reformed the army. In the battle with the Illyrians that gave him the throne, 35–50 percent of Macedonian men-at-arms had fallen, along with his older brother, King Perdikkas. To replace those casualties, Philip recruited heavily and he provided equipment for the troops—a real novelty at the time. This gave him the option of recruiting many more men into the armed forces than his peers could do, because they relied on self-supplying troops. It also gave him significant power and influence over the men enrolled: they owed their position to him.[12] This is not to say that Macedonian rank and file were not still levied and obliged to serve their king. He did not create a professional army overnight. But as pay for jury service allowed poor Athenians to participate in the running of their democracy, so Philip's equipping of his infantry allowed more men to fight in his army. It also heralded (in Europe; it already existed elsewhere) the creation of a military store where war material for land forces was produced—this is certain in the case of some metal items (such as lead shot and bronze bolt heads) that survive and are inscribed or embossed with Philip's name. Artisans must have been employed by Philip to make military equipment for him to issue to his troops.

Each man was equipped with a helmet, shield, greaves, and a sarissa. The sarissa was a pike about 14 cubits long, according to Polybios, which translates to 21 feet, and had a heavy butt, so that the center of gravity was set well back, and when held about that point, most of the pike projected forward of the infantryman, with only a couple of feet behind him.[13] It needed both hands to hold and wield, so the shield was light, and was supported by a thong over and around the shoulder and back. This equipment continued a process of changes to infantry gear that had been going on sporadically for decades, as hoplites with short spears and heavy shields were gradually replaced by peltasts with long spears and light shields (*pelta*). The *pelta* is the kind of shield that the *katapelta*, now spelled catapult, was, according to a widely retailed view, supposedly designed to oppose, though *palta* means "something thrown" and the machine is spelled *katapalta* as well as *katapelta* in early sources, thus the name may refer to "throwing down" rather than "against shield." (In addition, spelling with "e" or "a" is often a matter of dialect.) And instead of the breastplate, the infantryman now had a coat of mail.

But Philip did not stop with novel equipment. He put his men into the sort of training that we would recognize as such (a notorious deficiency in most Greek armies of the day) with route marches, carrying their own gear and rations—another novelty—and holding exercises under combat conditions—another novelty—throughout the winter—another novelty.[14] In close formation, the front man in each file would have four or five sarissa points sticking out in front of him, up to a distance of twelve feet. The hoplite, with a seven-foot spear, would meet this hedge of pike-points long before he came within striking range of, never mind face to face with, his opposite number in the new Macedonian phalanx. In battle, Philip often fought not with the cavalry, as was traditional for a Macedonian king, but on foot, leading his new infantry. For six years he was unstoppable. He fought and defeated armies large and small, Greek and barbarian, hoplite and light-armed, citizen and mercenary. Peoples hither and yon, notably those who were under threat or invasion by others, sought his alliance and protection. It was in the service of such allies—which promised him significant advantages too, of course—that he met his first defeat.

Philip's army was shocked and shaken when it found itself on the receiving end of Onomarkhos' stone-throwers in 353. It seems entirely natural to imagine him, on returning home, asking one or more of the architects who normally built siege towers and rams for him to try to build one ("establish an engineering corps" in the anachronistic language of some histories). The first time we know that he used catapults of any sort was five years later, in 348, at the siege of Olynthos. The surviving historians' accounts do not actually tell us that he used catapults at this siege, but sharps have been found at the site that are simply too big and heavy to have been shot by ordinary bow and arrow. Moreover, a couple of years later, an Athenian comedian, noting Philip's arms, armor, and distribution of equipment to all and sundry, joked about Philip's catapults as his crowning glory.

> D'you realise that in us you'll have to fight
> Men who at dinner make sharp swords their bite
> And lighted links their sup, who when they've dined,
> At head on shield and coat of mail reclined,
> At foot on slings and bows, and duly crowned
> With catapults for wreaths, have handed round
> Like nuts, for anyone to take who likes
> Fine Kretan darts or broken ends of pikes?
> (Mnesimakhos *Philip*, fr. 7, Edmonds modified)

Archaeologically, Olynthus is better known than almost any other Greek city.[15] This is largely because, when it was sacked by Philip following this siege, it had been extensively settled for less than a hundred years, and only a tiny fraction of it was resettled, so that the destruction level is clear. It records houses as lived in (not gradually abandoned), and there are no later layers complicating and obscuring the stratigraphy. Olynthus reveals a lot about Philip's military forces. Bronze bolt heads and lead shot (usually interpreted as sling "bullets") inscribed "Philippou" (Philip's), sarissa heads, and armor have been found. The bolts were probably shot from the *gastraphetes* or bow catapult, and the lead bullets may have been too (see Chapter 5). The gastraphetai were probably used to neutralize the defense forces on the walls, which remained catapults' principal function right down through history, while rams were used to try to break down the walls and ladders to try to get over them.

Seven years later, in 341 B.C., the great Athenian orator Demosthenes evaluated Philip's performance in a speech designed to motivate his compatriots to wage war on him. Demosthenes speaks to the traditional hoplite in the audience—all those who were, aspired to be, or sympathized with, the heavily armed citizen on whom Greek life and liberty had depended for at least 150 years (since the great Persian invasions with their battles of Marathon, Thermopylai, and Plataia), and who was celebrated in the literature and monuments of the city. In the good old days, Demosthenes imagines, disputes between communities had been settled by phalanxes of such warriors lining up with huge courage and discipline for a fair fight, by arrangement, over their possessions and lives, literally facing up to those who would take it from them. Nowadays, he laments, Philip is rolling up the country without fighting such battles and most cities fall because someone on the inside betrays them, either by persuading the people to accept an unequal alliance with Philip, or by opening the gates to let his army into the city during the night, or both.[16]

> Although nothing is the same now as it used to be, I think that nothing has advanced and innovated more than the military. . . . Now most disasters are due to traitors, and none is the result of a regular pitched battle. You hear of Philip marching where he wants, not because he leads a phalanx of heavy infantry, but because he is accompanied by light-armed, cavalry, archers, mercenaries, and troops of that sort. When, with these, he attacks a community that is internally divided [over what to do], and no one goes forth to fight for his country because of distrust [of his compatriots], then he brings up his machines and lays siege. I need hardly add that he does not distinguish between summer and winter and has no season set aside for rest. (Demosthenes 9 [*3rd Philippic*] 47–50)

It is striking how Demosthenes, specifically focused on Philip's innovations in the military, and listing many unconventional elements in Philip's army and tactics, says nothing about catapults. It suggests either that Demosthenes subscribed to an anti-catapult ideology at the time, as some scholars have proposed, or that Philip's success was not perceived at the time as being due to that new technology. I take the latter view. Demosthenes' diatribe is entirely consistent with Philip's reputation for winning over or politically disabling cities by bribing prominent people in them to promote pro-Philip policies, in preference to winning or weakening them via a bloody battle or long siege, which he would undertake only if the other tactics failed. As he is famously supposed to have said, is there any city so impregnable that not even a donkey laden with gold could approach it?[17]

Catapult technology was still not significant, I think, even in Philip's hands. According to Polyainos, Philip could not take a place called Karai, which was then and is still now essentially unknown, and had his engineers pretend at night to be building new machines when the noise created by them was in fact the sound of the machines being dismantled. When morning dawned on the bleary-eyed defenders, who had spent all night building counter-machines, the Macedonians had gone.[18]

Philip's next siege was in 340 at Perinthos. Here Philip's "artillery," which are specifically called sharp-casters, served the same purpose as Dionysios' had about sixty years earlier. That is to say, the catapults were still being used only against the defenders on the battlements. Further, once the defenders were reinforced "with men, missiles and catapults from Byzantion" they "became a match" for Philip's forces. This speaks strongly against the idea that Philip deployed catapults that were significantly different from or better than other people's were.

> [Philip] instituted a siege and advancing engines to the city [of Perinthos] assailed the walls in relays day after day. He built towers eighty cubits high, which far overtopped the towers of Perinthos, and from a superior height kept wearing down the besieged. He rocked the walls with battering rams and undermined them with saps, and cast down a long stretch of wall . . . both sides showed great determination. The king, for his part, rained destruction with numerous and varied sharp-casters upon the men fighting steadfastly along the battlements, while the Perinthians, although their daily losses were heavy, received reinforcements of men, missiles, and catapults from Byzantion. When they had again become a match for the enemy, they took courage and resolutely bore the brunt of battle for their homeland. . . . [Philip] had 30,000 men and a store of missiles and siege engines besides other machines in plenty, and kept up a steady pressure against the besieged people. (Diodoros Siculus 16.74 Loeb modified)

Perinthos apparently lacked its own catapults, but could borrow some from Byzantion if necessary. So Philip split his forces to attack the Byzantines, too, in order to divert them from continuing support for Perinthos. His siege of Byzantion was also unsuccessful because here, too, Philip failed to take the city by storm before one of two things happened that caused him to quit. Either reinforcements from Khios, Kos, and Rhodes arrived and forced him to break off the siege, or the opportunity to seize the Athenian grain fleet tempted him to raise it and take that instead, which he did. Although Diodoros gives the siege of Byzantion very short shrift (16.77), according to Athenaios Mechanikos, more and better machines were developed by Philip's forces during this siege—which was, as it happened, Philip's last.

> The making of siege devices progressed greatly in the reign of Dionysios, the tyrant of Syrakuse, and during the reign of Philip, the son of Amyntas, when

> he was besieging Byzantion. At that time Poluidos, the Thessalian, and his pupils, Diades and Kharias, took part in Alexander's campaign. (Athenaios Mekhanikos 10.1, Sage no. 221)

Now stone-throwers, although used by Onomarkhos in 353, are not specifically mentioned among Macedonian weaponry in Philip's time. They first appear explicitly at the siege of Halikarnassos in 334 under Alexander. Here, the stone-throwers were mounted in the siege towers.

> [Alexander] encamped about five stades from Halikarnassos. . . . He began to fill up the moat they had dug before the city, about 30 cubits broad and 15 deep, so as to facilitate the approach of the siege-towers, from which he intended to shower missiles on the defenders of the wall, and of other engines with which he intended to batter down the wall. The ditch was filled up without difficulty and he began to bring up the towers. . . . Again a sally was made by those in the city to burn [the siege engines]. . . . The besieged had the best of it, and they used to fire at the force protecting the engines not only from in front but also from the towers which had been left standing on either side of the breach and which made it possible to shoot from the flanks and almost at the back of those assailing the replacement wall. . . . Alexander again brought up his engines to the inner brick wall . . . and there was a sally from the city in full force . . . large stones were hurled from [Alexander's siege] towers by the machines, missiles were showered in volleys, and the besieged were easily repulsed and fled into the city. (Arrian *Anabasis* 1.20–22 Loeb modified)

Now we saw in Chapter 3 that our two principal sources on this siege contradict one another on who had what catapults (and on personnel matters), and there seems to have been some deep confusion somewhere along the line of retelling this Alexander tale. Assuming for the sake of argument that Arrian is correct in the passage just quoted, and Alexander did have stone-throwers in his siege towers, they must have been relatively small caliber machines (despite Arrian's comment that they threw μεγάλοι λίθοι, large stones), for a mobile wooden tower ascended by ladders would have had limited access, carrying capacity, and strength in the floors. About thirty years later, when Demetrios the Besieger built a monster nine-storey *helepolis*, city-taker, in 307 B.C. to attack Salamis in Kypros (Cyprus), Diodoros says he mounted "all sorts of stone-throwers" (πετροβόλους παντοίους), of which incredible three-talenters were the largest, on the lower levels. He mounted the biggest sharp-casters on the middle levels, "and on the highest levels he mounted his lightest sharp-casters and many stone-throwers" (*petroboloi*). Now Diodoros uses *lithobolos* and *petrobolos* alike (even within the same paragraph) to mean a stone-thrower catapult, and he uses *lithobolos* to mean a human stone-thrower as well as a stone-thrower catapult, but he appears never to use *petrobolos* for a human stone-thrower. It seems therefore that the *petroboloi* mounted at the top of his *helepolis* were small caliber stone-throwers, which would also explain why "more than two hundred men" were needed "to operate the machines on [Demetrios'] tower" (as distinct from the other men who moved the tower). Diodoros elsewhere calls (hand-thrown) one-mina stones μεγάλοι λίθοι,[19] so perhaps Arrian also meant stones of about this size, that is, one pound, or half a kilo. There are two (unpublished) shot of this sort of size in the collection at the British Museum, one weighing 439 g, the other 500, and unconsciously I mentally labeled them "big shot," which might be translated into Greek *megaloi lithoi*. A palintone built to formula for this caliber would have a stock a little over six feet long. A palintone built to formula for 15 drachma shot (65 g, big slingshot weight) would have a 50 mm spring diameter and a stock length of about a meter, or 39 inches. Some catapult washers that have been found are of even smaller caliber than this, at less than 40 mm; they would correspond to shot weighing about 30 g, discharged from a machine with a stock about 32 inches long.[20]

THE FIRST EVIDENCE WE HAVE for stone-throwers explicitly being used against the walls of a city (while sharp-casters were being used to clear the battlements of defenders) is at the siege of Tyre in 332.

> Alexander mounted the stone-thrower catapults (πετροβόλους καταπέλτας) in appropriate places and made the walls shake with the great rocks that they threw. With the sharp-casters on the wooden towers he kept up a constant fire of all kinds of missiles and terribly punished the defenders on the walls. (Diodoros Siculus 17.45.2.)

The Tyrians responded to this by filling sacks with seaweed and hanging them over the walls to cushion the blows. Wool came to be the favored material for such purpose, but at short notice and on a small island, the Tyrians showed great resourcefulness in this defensive measure, as they did in others. Again, this idea was probably a development of an existing practice: at the Tyrian colony of Karthage there was a tradition of covering the walls with sackcloth after a major disaster. If this tradition originated in Tyre, then the innovation was to make the existing sackcloth into bags and pad them with whatever resilient material was to hand—seaweed, in these trying circumstances. An alternative origin lies in Greece. Recall Aineias Taktikos' recommendation to hang "sacks filled with chaff, bags of wool, fresh ox-hides either inflated or filled with something, other things of this kind" in front of the battlements against which "rams and similar machines" were being deployed

(Chapter 3). About 130 years later Philon recommended, as an alternative to filled sacks, slices of date-palm trunks hung over the walls to help cushion them against the impact of catapult-thrown stones, and over the sides of mobile towers for the same purpose.[21] This expedient has an obvious geographic and botanic restriction (even more so than seaweed), and seems to be peculiarly suited to the Ptolemaic territories. Evidently, technologies were adapted to local conditions.

Both sides conducted the siege of Tyre vigorously and imaginatively. Alexander used the full range of weaponry available to him, and ordered an audacious engineering project to be undertaken that changed forever the coastline of the Mediterranean. The Tyrians used and devised an astonishing range of countermeasures, and this siege appears to have been one of the most remarkable in all of human history.[22] The first use here of stone-throwing catapults against walls, rather than people, did not have a dramatic impact, because ultimately the walls collapsed under the impact of good old-fashioned rams.[23]

Diodoros' next mention of stone-thrower catapults is on the offence at the siege of Kyzikos thirteen years later in 319 B.C., a couple of years after Alexander's death. Kyzikos, like Rhodes and Massilia (modern Marseilles, on the south coast of France), had a reputation for being very interested in military technology.

> Here [in Rhodes] too, as in Massilia and Kyzikos, everything relating to the architects, the manufacture of instruments of war, and the stores of arms and everything else, are objects of exceptional care, more so than anywhere else. (Strabo *Geography* 14.2.5, C 653 Loeb)

The defense were reinforced from Byzantion, as the defenders had been at the siege of Perinthos, and like that one, this siege failed, as did about 50 percent of all sieges in this period. Although some stone-throwers were now being used against walls, they cannot have been very effective for, as Keyser pointed out, they would have been preferred to rams if they were, for an obvious reason: they could be used in relative safety, at a distance, whereas the alternative, the ram, had to be used immediately below the walls, which was a very dangerous place to be.[24] This point is valid for *all* ancient and medieval history, for rams continued throughout to be the principal weapon employed against walls. Very few catapults ever built were capable of knocking down a well-built heavy-duty city wall. Their function was usually to clear a section of the wall of defenders so that those operating the ram, or digging to undermine the wall, or building a ramp up to the wall, could do so in relative safety. Big catapults could do real damage to the single-skin battlements *atop* the wall, behind which the defenders would shelter, but the walls proper, constructed (usually) of tightly fitting stone on the outer faces with a rubble (or better) fill, were essentially impervious to catapult shot.[25] See, for example, figures 7.5 and 8.4 below, which show the impact craters created by a variety of caliber weapons, none of which were remotely close to that required to dislodge a stone in such walls.

Ineffectiveness against well-built walls might explain why, despite the number of battles and sieges that occurred while Alexander's marshals, generals, and assorted other subordinates fought among themselves and with others for control of parts of his empire, the next explicit reference to stone-throwers concerns events some twelve years later. This is the siege of Mounykhia (Athens' harbor) by Demetrios the Besieger in 307 B.C. However, some of the many sieges mentioned in Diodoros book 19 were apparently successful very quickly, and he barely mentioned them. We only hear about any catapults in the Mounykhia siege in paragraph 7 of his relatively long discussion of it (DS 20.45.7). If what we were told about this siege resembled what we are told of others, something to the effect that the attacking commander had received "land and sea forces, and a suitable supply of missiles and the other things required for carrying on a siege" (20.45.1), then we would probably not have heard about catapults here, either.

Alexander used catapults more extensively than his father did, and used them in the field as well as in sieges. He used them once, in retreat, from a bank of a river. His "artillery" troops having crossed first, they shot back over the heads of his retreating forces "all the types of missile that are hurled from machines" to the other bank, to deter enemy troops pursuing the last of his men. His archers, whose range was not so great, were ordered into the river to shoot from midstream. The result was an unqualified success, with no man being lost in the withdrawal across the river.[26] It is inconceivable that a military man of Alexander's talents would not have used such a successful tactic again should the need and circumstance have arisen, whether or not our surviving sources tell us so. And indeed we hear of one other such occasion, this time with catapults being used to clear the bank *to* which Alexander's forces were moving, rather than from which they were fleeing.[27]

> Having dismissed the [Skythian] envoys, he embarked his army on the rafts which he had prepared beforehand. On the prows he had stationed those who carried shields, with orders to sink upon their knees, in order that they might be safer against the shots of arrows. Behind these stood those who worked the hurling-engines (tormenta), surrounded both on each side and in front by armed men. The rest, who were placed behind the artillery, armed with shields in testudo-formation, defended the rowers, who were not protected by corselets. . . . Although making every effort, the soldiers could

> not even hurl their javelins, since they thought rather of keeping their footing [on the rafts] without danger, than of attacking the enemy. Their safety was the hurling engines, from which bolts were hurled with effect against the enemy, who were crowded together and recklessly exposed themselves. . . . And now the rafts were being brought to land, when those who were armed with shields rose in a body and with sure aim, since they had firm footing, hurled their spears . . . and began to leap to land. (Curtius 7.9.2–9 Loeb)

The employment of catapults to clear riverbanks, exploiting their greater range, continued as required through the centuries to late antiquity and beyond. So, for example, we hear of Vitellius' troops in A.D. 69 building a tower at the end of a partially constructed bridge to hold "catapults and machines" with which to repulse Otho's forces on the other side of the Po, and he meanwhile built a tower on the opposite bank and "kept shooting stones and torches at the Vitellians." Cassius Dio's description of the regular method of bridging rivers when on campaign in his time (about A.D. 164–230) includes the datum that "the ship that is nearest the enemy's bank carries towers upon it and a little gate and archers and catapults."[28]

LET US THEN CONCLUDE OUR QUEST to find the origins of the two-armed torsion catapult.

There is no good evidence that Philip II (who died in 336) had torsion engines. The "sharp-casters" that he used at Perinthos may equally well have been tension (bow) catapults, and only sharp-casters are certainly attested (via their ammunition) at Olynthos; if the argument of Chapter 5 is accepted that *glandes* are small caliber catapult shot, then small caliber stone-throwers are attested too, but these may be tension catapults again, what are later called stonebows. However, Philip is known to have been responsible for many major and minor innovations in the military sphere, in equipment and in organization, and he was largely responsible for creating the army with which his son, Alexander the Great, created an empire (and let us not overlook the fact that he also created his son!). No other army is known to have been engaged in a program of investment in their armed forces on a scale remotely similar to the Macedonians during his lifetime. Named engineers who are said to have been important figures in catapult design and development are said to have worked for him or his son. The two-armed torsion catapult was almost certainly in existence by 326 B.C. It was present in Athens about then, and the poor state of repair of the equipment suggests that it was new to the Athenians, if not absolutely new, at that time. Alexander, who died three years later in 323, had stone-throwing catapults that could shake walls, and that, probably but not certainly, means relatively large torsion, not bow, engines. Together, this circumstantial evidence suggests that two-armed torsion catapults were invented in Macedon during Philip's reign and were developed into functional war machines during Alexander's.

Other catapults mentioned among Alexander's forces may also have been torsion devices, but bow catapults were probably still the norm at this time, in Macedonia's arsenal as well as everyone else's. The sources imply that his war machines were not noticeably superior to or even different from his opponents', and therefore Macedon's ordnance makers cannot have been producing anything that performed significantly better than their opponents' (nontorsion) machines. Although Alexander occasionally used catapults to great effect, they appear to have had little, if any, effect on his tactics in field or siege. That is a proxy for proof that catapults, even torsion ones, were not yet revolutionary weapons; they augmented the arsenals of their users, they did not transform them. That revolution was still to come.

FIVE

# Little Catapults

*God is on the side not of the heavy battalions, but of the best shots.*
—Voltaire, *The Piccini Notebooks*, 547

THE VERY FIRST CATAPULT was a hand weapon, a sort of crossbow. The development of such weapons led to monster machines that could launch stones weighing 75 kg or more over hundreds of meters, and reconstructions of such weapons attract a great deal of attention from scholars and public alike. However, the continuation and development of catapults as hand weapons has been overlooked in the literature. Modern use of the term "artillery" for ancient catapults plays a role here, perhaps, apparently overriding the fact that the name for the crossbow in almost every medieval European language was derived from the Latin term ballista, which might have suggested that ballistas came in small as well as large sizes.[1] There is a variety of direct and indirect evidence for hand-held catapults of the same sort of size as a modern rifle, powered by bow and by spring, and for ammunition for such weapons, throughout antiquity. Little torsion catapults would have been more complicated and expensive to make than little tension catapults, but they also would have been more powerful, potentially at least. However, because the existence of little catapults is not widely recognized at present, it seems wise to introduce the evidence and argument here.

In 1999 Dietwulf Baatz argued for the existence of what he called torsion crossbows (Torsionsarmbrust), small personal weapons, on the basis of the archaeological evidence of small washers.[2] The washer is the key component of a torsion catapult, from which the rest of the machine can be reconstructed with confidence. Basically, little washers come from little machines; the precise relationships are spelled out in Appendix 1, "The Calibration Formulae." Baatz was absolutely right about the washers; but they are not the only evidence for little catapults. There is actually a lot more evidence to support this idea. One is surviving ammunition. The Dura bolt, as a type of missile shot by catapult, was built for small caliber catapults. In addition, it may be that *glandes*, lead slingshots, were invented for catapults, and were only afterward used for hand slinging too. Further, direct literary evidence for little catapults has been overlooked, and indirect literary evidence for them exists too. We review these types of evidence in turn.

## The Archaeological Evidence for Little Catapults

We begin with the evidence identified by Baatz, and since he discussed it, we can be brief. The smallest washer found to date is 34 mm in diameter (Ephyra no. 6).[3] This small caliber weapon was one of seven catapults whose badly burned remains were found at Ephyra in Epirus (see Figure 6.5). A washer of this size, with an internal diameter of 34 mm, corresponds to a sharp-caster built for a missile about a foot long. Alternatively, it could have belonged to a stone-thrower which, if built to the Philonian formula for palintones, would have been intended to discharge four-drachmai shot (about 18 g weight, the smallest known size for slingshot), and which would have had a stock length of about 645 mm (about 26 inches). The Elginhaugh washer is fractionally bigger, with an internal diameter of 35 mm, and would have been of the same nominal caliber.[4] Another five known catapult washers have an internal diameter of about 40 mm, which corresponds to a stone-thrower with a stock length of about 760 mm (31 inches), or a sharp-caster with a stock length of about 640 mm (25 1/2 inches).[5]

Another class of archaeological evidence for little catapults is a type of missile known as a Dura bolt, after the discovery of such projectiles at Dura-Europos, which has long been recognized as catapult ammunition (Figure 5.1). The head is a common type; the significance of Dura is that the wooden shafts and flights, as well as the metal heads, survived in the dry conditions there. One bolt was found complete; a few shafts survive complete; many shafts are more or less complete. The shape is fantastically aerodynamic.[6] Most shafts would have made bolts about 500 mm (20 inches) long when complete with their heads, but some are considerably shorter. Consider Catalogue no. 823, which is a complete shaft but for the loss of two vanes.[7] It is 318 mm long, and would have carried a socketed head, which would have added perhaps another 15 mm, making the whole bolt about 333 mm (a smidgen over 13 inches, or an ancient foot) long. Different ancient states had different measures of a foot, varying between about 295 and 350 mm, but the Aiginetan foot was 333 mm, just like this bolt, while the Pergamene and the pes Drusianus of Gaul and Germany in the first century B.C. were 330 mm (13 inches).[8] A catapult built according to Philon's formula to shoot this bolt would have had a spring diameter of about 1 1/2 inches, or 36 mm (just 2 mm wider than the smallest known catapult washer, 1 mm more than the Elginhaugh catapult), a stock length of about 585 mm, or 24 inches, and arms about 256 mm, or 10 inches, long. A euthytone built to formula to shoot

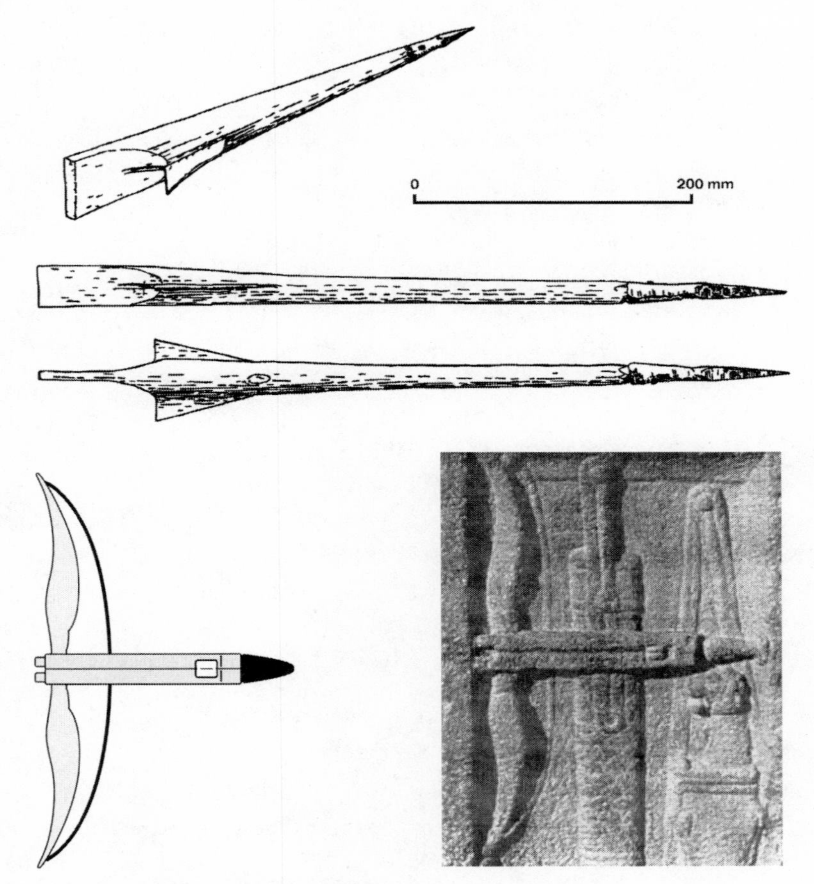


FIGURE 5.1 *(top) The Dura Bolt. This bolt is very aerodynamic; the tapering shape and short wooden vanes make it very stable in flight. The narrow rear section fits between the fingers of the claw of a catapult. Lengths varied considerably; some were scarcely more than 330 mm long, or a little over a foot. (After James 2004)* FIGURE 5.2 *(bottom) The Solignac crossbow. (Drawing by T. D. Dungan after Baatz 1991)*

*kestros* or Dura-type bolts 1 cubit long (18 inches or 460 mm), would have a spring diameter of 2 inches (50 mm) and a stock just 32 inches (800 mm) long. If I assume an archer's position, that is the distance between my outstretched left hand and my right ear. It is not difficult to imagine being able to hold and sight along the "barrel" of a weapon of such a size. The evidence of the washers and the bolts is consistent. They both testify to the existence of personal catapults, hand-held weapons.

Little catapults are depicted on stone in two second-century A.D. reliefs from France, one of them reproduced in Figure 5.2. This is a descendant of the *gastraphetes*. It is a crossbow, for the nut is clearly visible. We will discuss the development and survival of the gastraphetes fully in chapter 7, so I will not elaborate on it here.

There may be another class of archaeological evidence for crossbows in antiquity. There was a type of Roman brooch called the crossbow brooch (Figure 5.3). It has been so named for the obvious reason that it looks like a crossbow. This could be coincidental, but I suspect that it is not. In fact, the type strongly resembles the stonebow (Figure 5.4), rather than the sharp-casting crossbow. The stonebow, which was a crossbow that shot stones not bolts, features in nearly a dozen old illustrations in Payne-Gallwey's *The Book of the Crossbow*, and he devotes six chapters to it.[9]

Crossbow brooches are of course dress items, and they show variations, some of which were introduced for aesthetic or functional reasons, but some of which may also mirror variations in military or hunting crossbows. Some, for example, have straight arms with "bulbs" at the ends, which is oddly reminiscent of the representations of catapults on Trajan's column (see Chapter 9). Was the crossbow brooch modeled on little tension and torsion catapults, crossbows and stonebows and torsion catapults with springs on the outside (as the *kheiroballistra*, on which see Chapter 9)? If not, then what was the inspiration behind this design?

### The Literary Evidence for Little Catapults

There are two direct pieces of literary evidence for little catapults.

The first is in Philon's *Belopoiika* (the most thorough and detailed of all the technical manuals). At 56.3, when telling the reader how to scale machines up and down from a paradigm in order to reproduce a successful machine at a different caliber, Philon talks about the possibility of being required to make "a half-span (ἡμισπίθαμον), or any other size, even an irregular calibre." A half-span is about 4 1/2 inches, or 115 mm, so Philon here specifically entertains the possibility of making a catapult small enough to shoot sharps significantly less than six inches long. Applying his formula for sharp-casters (55.1-11), if we made a machine to shoot such a dart, it would have a washer diameter of two-thirds of an inch, or 13 mm, a case length of 9 1/2 inches, or 210 mm, and arms about 3 1/2 inches, or 90 mm, long. Philon is thought to have been writing about 200 B.C., so this is not only the smallest catapult attested, but it is also the earliest evidence for little catapults, preceding the destruction of Ephyra (167 B.C.) by about a generation.

The second direct piece of evidence is from Vitruvius. He implies that the scorpion, a standard catapult type in Roman sources, is a one-man device. He says explicitly that a single person was enough to turn its windlass (10.1.3). This for him was part of the definition of the *organon* (usually translated as

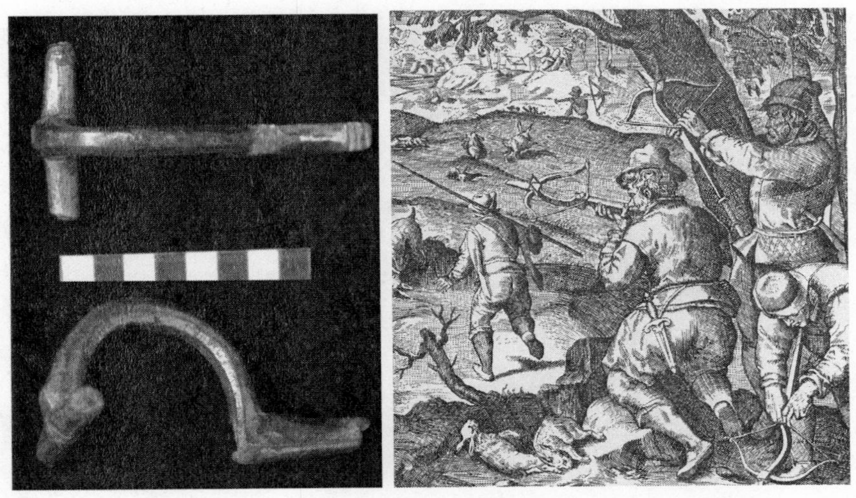

FIGURE 5.3 *(left) A crossbow brooch found in Norfolk (UK). (NMS-BC1567; courtesy of the Portable Antiquities Scheme.)* FIGURE 5.4 *(right) Stonebows; small hunting models depicted in 1578. (Payne-Gallwey fig. 12.)*

"engine") as a subset of the "machine" (*mekhane*): an engine was so designed that a single person could set it in motion. He cites the scorpion as an example of an engine (a one-man device) and the ballista as an example of a machine (a two-or-more man device).

Further, Vitruvius' handy table of calibers and their associated spring diameter sizes begins with two-libra stones (about 1 1/2 lb modern, or 650 g, *de Architectura* 10.11.3). Now this machine would not have been a hand-weapon, being too large and heavy for one person to support, but a two-libra is a relatively small catapult. Spherical stones weighing about two libra found in ancient contexts, such as Arbeia, could have been ammunition for hand-throwing or for two-libra catapults.[10] Those from Burnswark are almost certainly ammunition for small ballistae.

The direct literary and archaeological evidence is clear: long before the development of the *kheiroballistra*, the all-metal-frame weapon of Heron's treatise (on which see Chapters 7 and 9 below), there were small and very small catapults, of sizes that could be considered analogues to the modern two man gun, and personal weapons, analogues to the modern rifle.

ONE MIGHT THEN REASONABLY ASK: why have we not heard about them before? Are they not mentioned in the surviving historical sources? Well, usually the sources do not say anything about the size of the catapults they mention, and we have simply assumed that they were large machines. Sometimes they are clearly small by implication, even if not by direct testimony. For example, when Caesar besieged the hilltop site of Uxellodunum, he built a ramp, topped by a ten-storey turret housing *tormenta*, in order to hinder the defenders' access to water (*Gallic War* 8.41). The only type of "artillery" that one could possibly get up into and use from the ladder-accessed floors of a ten-storey wooden turret, which was itself sitting atop a sixty-foot ramp, is a very small hand-held variety. A similar, albeit less clear-cut instance relates to one of the stories told of A.D. 69. Tacitus tells that, when a group of Vitellians who had no arms except swords were being pelted with stones and tiles from the roofs of old porticos on the Capitoline, they thought it would take too long to send for tormenta and missiles to dislodge the defenders, so threw firebrands onto a projecting colonnade instead (*Histories* 3.71). Tacitus surely was envisioning small arms for such a function in such an environment: if hand-thrown firebrands could do the job, then the tormenta for which they substituted were probably small caliber. The Vespasians went on to oust the Vitellians from Rome, and the fighting in the streets culminated in a battle for the Praetorian Guard camp, against which "the tortoise, tormenta, mounds and torches" were employed, until the last Praetorian defender lay dead in the dust. Since only the gates are said to have been broken down, and the attackers intended to occupy the camp rather than destroy it, one imagines the tormenta in question were small calibre anti-personnel weapons, easily and rapidly carried through the streets and thoroughfares of the eternal city.

In 219 B.C. Hannibal attacked Saguntum in Spain. He mounted catapults and ballistas on every level of a siege tower, and with them drove off the defenders a section of the wall that had been hastily rebuilt after a previous battering. Thereupon he sent in 500 men to attack the wall in that area, and they soon opened up new breaches. Through those breaches armed men passed, and seized a high point, where they set up catapults and ballistas. Now whatever Hannibal's catapults and ballistas looked like, this rapid movement over tumbled masonry and rubble implies that they were very mobile. They were presumably small caliber, easily dismantled and reassembled weapons, if not hand-held varieties. There are no reasonable grounds for rejecting Livy's testimony about the use of catapults at Saguntum, though he could of course be anachronistic in his choice of terms. Ballistas and scorpions were also key elements in the Karthaginians' defense of Capua in 211, keeping the Roman forces at bay at the city gates, and perhaps being responsible for the (fatal?) wounding of Appius Claudius, commanding from the front.[11] These too sound like small caliber weapons.

This example raises the issue of catapults as snipers' weapons—sniping is typically an activity performed with hand-weapons, not big guns. Catapults were set up commanding specific strategic points, such as directly above the part of the wall being assaulted or dominating access to water or other essential supplies. An example of the catapult's accuracy, which also highlights its consistency being exploited to the full, occurs in the *Gallic War*.[12] At Avaricum the Romans had built a massive ramp, which the defenders had set on fire via a mine. The Romans were attempting to extinguish the fire, while a defender was adding fuel to it from the top of the walls. He was hit by a bolt from a scorpion and dropped dead. The man next to him took his place and carried on feeding the fire. He was killed in the same way, so a third man took up the role. Then he was killed in the same way, and a fourth came up, and so on. Caesar comments that that spot was not without a defender until the enemy was cleared on every side, the ramp extinguished, and the fighting brought to a halt.

There are several famous examples of leaders being targeted by catapults, and sometimes hit when they thought they were out of range. The earliest known to me refers to an event that took place about 360 B.C. The Athenian general Timotheos claimed that he tried prudently to stay out of range of the enemy's long-range weapons (unlike another general who was hit and was showing off his war wounds), and when he underestimated the safe distance and suffered a near miss, he was ashamed of his error of judgment.[13] The fact that the missile in question was specifically said to have been shot from a catapult, which had longer range than manual weapons, cleverly mitigates the self-criticism (he was a politician as well as a general).[14] But this story exemplifies the point that catapults, with longer range and greater impact, were used to try to pick off commanders and other notable targets.

Examples can be found from all periods, down to Ammianus and Prokopios.[15] For instance, in the winter of 47–46 B.C., one of Pompey's best generals, Labienus, attacked Caesar's forces in Leptis (Map 6), using cavalry daily to skirmish and harass the defenders. A squadron once chanced to halt in front of the gate, whereupon its captain was hit and pinned to his horse by a scorpion bolt. The squadron fled, never to return. The practice of targeting leaders is perhaps why the "princes and nobles" of so many of Rome's enemies adopted heavy iron or hard hide armor, which apparently was impenetrable to blows, at the cost of making the wearer almost immobile when forced to go on foot.[16] When they expected to engage in the fighting, emperors sometimes chose to wear iron breastplates too. Trajan dressed down when besieging Hatra in order to try to avoid being targeted, but the Hatreni anyway suspected who he was and shot at him; a cavalryman escorting him was killed before they got out of range.[17]

There is evidence of little catapults from peace-time activities too. In the third century B.C. on Keos, three-monthly competitions in "javelin-throwing, archery, and catapult-shooting" were being held.[18] The three events are listed on the stone as if they were kindred types of competition, and the prize for the winner of the catapult-shooting competition was a helmet and spear; there was just a spear for second place. These seem inappropriate prizes for a crew, were catapult-shooting a team activity.

But these are unusual cases because it is pretty clear from the circumstances that they must refer to personal weapons. The general silence on little catapults is, I think, more easily explained as a phenomenon that is apparent rather than real.

### Indirect Evidence for Little Catapults

According to Apollodoros, who flourished about A.D. 100, "some people call one-armed stone-throwers slings."[19] All catapults that shoot stones have a sling for a bowstring. Livy, talking of a particular kind of sling pouch, says that the lead slingshot (*glans*), held firmly while being whirled, "may be sent as if from a bowstring."[20] These point the way to a solution.

What sort of *glans* is shot by a bowstring? Catapult shot. No one seems to have considered lead "slingshots" as catapult "bullets," despite Livy's clue, perhaps (again) because we think of catapults as "artillery" and thus large caliber weapons. Little catapults would have discharged small shot. Catapults would have encouraged the standardization of shot, since they are built to caliber. They may have been the catalyst for the development of cast lead shot, which is not a feature of other cultures' slinging equipment. Indeed, lead slingshot seems to have been invented at the turn of the fourth century B.C.—no certain dated example precedes the orthodox date of the invention of the catapult in about 399 B.C.[21] Hand slingers are not under the same pressure as catapultists to find ammunition of a given size and shape. Moreover, slingers were usually self-supplying, gathering stones as needed, whereas one clear feature that emerges from the inscribed and decorated classical *glandes* is that a central authority issued them, person or state. The workmanship involved in engraving the mold is in some cases extremely fine indeed; see, for example, Figure 8.2 (scorpion *glans*). I find it hard to imagine any ancient Greek state investing this sort of energy and expertise in the production of shot for slingers, the lowest ranking military unit. It seems entirely appropriate, however, for expensive bits of gear such as scorpion catapults. Glandes as little catapult munitions would also explain why we have some bearing legends referring to specific legions, e.g., LEG XV, just as the Cremona battleplate (see Figure 9.1) boasts its belonging to the Fourth.[22] We will develop this further below.

Just as all stone-thrower catapults have slings for bowstrings, so all tension catapults have bows. All projectiles can pass under the label βέλη (*bele*) or tela. *Bele* is usually conceptualized and translated as "arrows," though it liter-

FIGURE 5.5 *Selection of slingshot in the Greco-Roman collection at the British Museum, all to the same scale. The largest shown weighs some 500 g, the smallest shown just 29 g; the largest was found in Attike, the smallest in Rhodes. (British Museum, author photographs.)*

ally means (any) things thrown, and includes things that are neither sharp nor pointed. *Belos*, said Ktesibios (ed. Heron, *Belopoiika* 75.1), is the name given to everything (πᾶν) projected by engines (ὄργανα) or any other sort of force (δύναμις), be it bow, sling, or anything else (τόξου, σφενδόνης, ἢ ἄλλον τινός). For Paul of Aigina, a doctor writing in (probably) the seventh century A.D. and here advising his reader on the extraction of missiles, *bele* were made of iron, bronze, tin, lead, glass, horn, bone, stone, reed, or wood (6.88.1). They were round, triangular, square, or flat. The first of these, "round" projectiles, are clearly what we call shot, though we tend to call them (more accurately) almond or acorn or oval shape rather than round. Recovered lead slingshot (glandes) varies hugely in length and weight, from 2 to 7 cm in length and from 18 g in weight upward.[23] This weight range corresponds to a caliber range from 8-drachmai upward as shot for catapults built to Philonian "stone-thrower" formulae. The big ones were presumably what Appian had in mind when he said that Sulla used catapults that discharged twenty of the heaviest *molubdainas*, lead bullets, in a single volley (*Mithridatic Wars* 5.34)—which killed a large number of the enemy and rendered a wooden tower insecure, he adds. Ignoring for the moment the interesting health and safety issues for catapult crews discharging multiple bullets in order to keep focused on the shot, Griffiths pointed out that the length of cast shot could be regulated easily, so we must assume size (and shape) variations in *glandes* were deliberate and intentional. Figure 5.5 demonstrates the point graphically.

Talking of sharps, Paul says that barbs could be directed backward, making extraction difficult, or forward, making expulsion difficult, or both, complicating the task of moving the head at all. Barbs could even be hinged, says

Paul, so that they were not apparent until one tried to extract the head, at which point they would unfold.[24] They might be as wide as three fingers, or as small as one, or any size in between.[25] Some heads were firmly attached to their shafts, others separated easily. Some were tanged, some socketed. An instrument, curiously called the διώστηρ (which is the same term used of the catapult's slider), which was designed to push the head through when that seemed the best method of removal, seems to have had each end shaped to deal with each type: the "female" end taking a tang and the "male" end fitting into a socket.[26] Catapult bolt heads are surely included in this discussion of sharps that a surgeon might have to extract, though they are not distinguished as such.

Let us turn to Paul's description of wounds caused by "slings" (6.88.9): "One has to recognise [stones, trumpet-shells, glandes or similar objects shot with a sling] by the swelling being jagged and uneven to the touch, by the fact that the wound is not all in a straight line but [becomes?] larger, that the flesh appears to be crushed and livid, and that there is pain accompanied by heaviness. One has to prise those missiles up and lift them with levers or with the spoon-shaped end of the wound probe, if the wound allows for them, and extract them with a tooth forceps or a root forceps."[27] Salazar observed that the effects described in detail by Paul are similar to long-range gunshot wounds.[28] Onasander described the condition as perceived by a common (nonmedical) man: 'the missile . . . penetrates the flesh very deeply, so that it even becomes invisible and the swelling closes over it."[29] So too, apparently, did Livy. In 189 B.C., in Anatolia, in response to Roman advances into their territory, some Galatian tribes had fled to a mighty natural fortress called Mount Olympus. They entrenched and fortified the summit, but had no missile weapons—they had apparently assumed that they would find plenty of suitable stones in situ. Gnaeus Manlius Volso, who led the Roman forces, decided that this was an occasion for missile warfare. He had prepared a vast quantity of "*pila*, velites' javelins, arrows and *glandes* and stones of moderate size that could be thrown by sling."[30] Livy tells how the Galatians were not bothered by open wounds and even sported them with pride, but under a barrage of missiles "coming from unseen and distant sources" were driven to distraction, rage, and fear by "the barb of an arrow or a *glans* that had hidden itself in the flesh through a wound that appeared small" and which they could not find to extract (38.21).

Slingers are not capable of inflicting penetrating injuries of this order; they simply cannot achieve the velocities necessary. A certain amount of energy per unit area is needed to break skin. This minimum energy density or threshold can be expressed as a mathematical formula.[31] To penetrate human skin, a blunt projectile must hit at 49 meters per second or more (111 mph, or 163 fps). To break bone, it must hit at 65 m/s or more (145 mph, 213 fps).[32] The most striking, if gruesome, demonstration of these physical facts is a photograph in a recent forensic science journal which shows what is called "tenting" of the skin at the exit side of an arrow wound (see Figure 5.6). The conically tapered field-point arrow tip (the type used for example for target shooting at an archery club) penetrated the skin of the victim's arm on the side it entered. It then passed through the middle of the arm without touching bone, and stretched the skin several centimeters before coming to rest, without breaking the skin on the other side of the arm.[33] It no longer had enough energy to penetrate the skin to exit the arm.

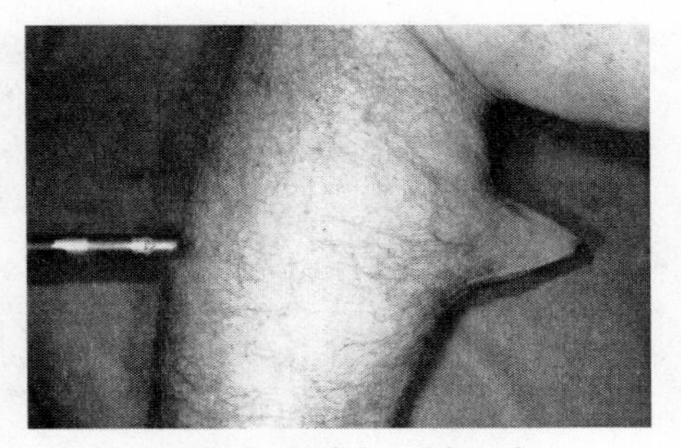

FIGURE 5.6 *Tenting of skin. This photograph demonstrates the resilience of human skin and its resistance to penetration by missiles.*(Eriksson)

Thus, even an arrow, which is sharper than slingshot, will "bounce" off the skin when its velocity has fallen below the level needed to break the epidermis. With cast lead slingshot, which performs very slightly better than any other type of slingshot, a good slinger can consistently reach a velocity of 30-31 m/s (about 68 mph or 100 fps), with best performance of 32 m/s, and worst of 29 m/s.[34] The sort of wounds Livy is talking about can only have been caused by shot from catapults, which had a probable discharge velocity twice that of slingshot, at about 60-70 m/s, which is enough to enable the shot to penetrate skin and bury itself in the soft tissues.

Journée found experimentally that spherical lead shot weighing about the same as two-drachmai shot (viz. 8 1/2 g) caused skin contusions at velocities above 46 m/s, and penetrated muscle at 70 m/s.[35] Ancient cast lead shot, being generally ovoid, would probably have penetrated the skin at slightly lower velocities than spherical does, but still not normally at the sort of speeds achievable with delivery by sling. Even if they did occasionally break the skin, hand thrown shot would not have had enough speed to penetrate very far into the body, still less to lodge in bone. Likewise, hand-thrown slingshot cannot create the sort of impact craters one can find on ancient walls, for example on the curtain adjacent to the Herculaneum Gate at Pompeii (see Figure 8.3). These particular craters probably date from Sulla's attack on the city in 89 B.C. Small caliber craters almost certainly reflect shot that fell short of its intended target, namely, the defenders on the walls.

Extracted from patients, lead shot that had been squashed or flattened on impact with their flesh probably led to the widely reported idea that slingshot could "melt" in flight. Lead's softness, rather than its relatively low melting point, may explain this ancient misconception.[36] Alternatively, glandes may have been shot shortly after casting, when the outside was cool enough to handle but the inside still hot and soft, as appears to have happened at Velsen in Holland, for example. [37]

In these cases the argument is clear. Elsewhere authors probably used the terms "bow" and "sling," "thing thrown," "glans," and "stone" without discriminating among different types of launcher, at least at the small scale with which we are here concerned. When Caesar remarks that Pompey's forces, having brought ladders up to the outer rampart of a camp, were terrifying the defenders of the inner rampart with *tormenta* and missiles of every kind (*Civil Wars* 3.63), what sort of catapults are we to imagine, and how used? Big catapults a long way back lobbing missiles in a high arc onto the rampart would not explain the role of the ladders in the story. Could he mean men standing on ladders outside of and protected by the outer rampart, shooting with crossbow-size weapons at the men defending the inner rampart? Livy, in his description of the missiles brought by Manlius to attack the Galatians on Mount Olympus, used these same terms "arrows" and "slingshot" (*sagittarum glandesque*), yet his description of the wounds caused—shot buried in the flesh—demonstrates the use of small catapults. The Ephyra catapults date from about twenty years after Manlius' battle with the Galatians, and Ephyra 6 was built to shoot four-drachmai shot, weighing 18 g or thereabouts—which is the lightest glans calibre known—or an odd 15-daktyls caliber sharp. In this battle there was a clear distinction between the legionary forces who were marching up behind and the light troops who engaged using range weapons, and the battle was won by the light troops alone (Livy 38.21.14; 22.5). There was a repeat performance, with light-armed troops well supplied with "every kind of weapon" shooting at another tribe of Galatians on another mountain in Anatolia, with the same effect. The Galatians were described as "carrying missiles fixed in their bodies" after the attack by the "light-armed" troops. In the debate between himself and the legates L. Furius Purpurio and L. Aemilius Paulus, his political rivals, who tried and failed to stop him being awarded a triumph, Manlius gives the impression of having been a very pragmatic military commander. His tactics were perhaps novel; that may explain why Hannibal saw fit to write a book for the Rhodians, in Greek, about Manlius' actions in this campaign.[38] The apparent conclusion is that little catapults, personal weapons shooting bullets and darts, were here, at least as weapons borne by light-armed troops. Some scholars have argued strongly for years, on the basis of other evidence, that the use of catapults was not confined to legionaries or phalangites.[39]

Very small caliber shot would, I imagine, be indistinguishable from slingshot. The same laws of aerodynamics operate on both, and ancient apprecia-

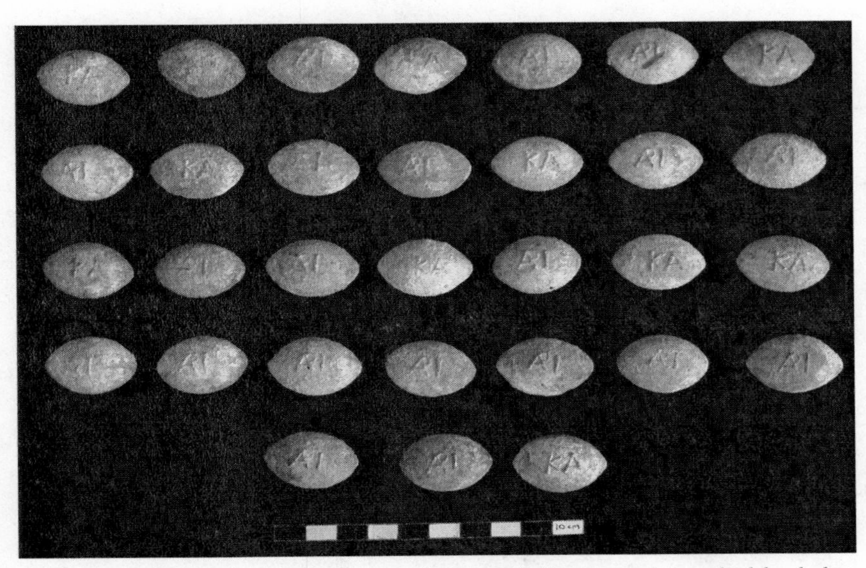

FIGURE 5.7 *Glandes from Stymphalos. These carefully cast and inscribed lead shot were found with 130 catapult sharps in the ruins of the acropolis tower that probably fell around the end of the third century BC. (Ben Gourley)*

tion of the benefits of an ovoid shape, discovered early for slinging, would presumably be transferred quickly to catapult shot. Slingshot found in the ruins of forts is invariably, and naturally, interpreted as slingshot. However, wall-walks usually lack room for slingers to operate, as has been confirmed empirically.[40] Their use in towers is extremely dubious. Even the most experienced and accurate slinger would balk, I imagine, at being asked to launch their shot through an aperture, never mind the loop holes with which towers are usually furnished. The slightest mistake would send the shot ricocheting round the room. Perhaps, if the tower had a flat roof, they stood there, but evidence about the top of a tower rarely survives, and then it sometimes clearly indicates a sloping, tiled roof, such as at Aigosthena or Paestum.[41] Slingshot found in the debris of collapsed artillery towers, such as the 31 lead bullets inscribed KA (for Kassandros, probably, given the date and location) and ARI (?), found together with over 130 iron catapult sharps in the ruins of the acropolis tower at Stymphalos, which collapsed probably in the mid-third century B.C., are surely, therefore, good candidates for little catapult ammunition (Figure 5.7).[42]

For those on the receiving end it would be difficult if not impossible to know exactly what had shot an incoming missile, unless it flew too fast or was obviously, or at least probably, too large or too heavy to have been launched by hand. Indeed, the latter remains the rule used by archaeologists now to determine whether a given sharp was an arrowhead or some other kind of sharp (and we still cannot distinguish between large catapult bolt heads and spear butt spikes).[43] Similarly ad hoc decisions may be needed to differentiate slingshot from small catapult shot. The same problem must have beset ancient writers, assuming that it occurred to them to draw such distinctions in the first place. Paul's and Celsus' patients were probably in no condition to reflect on what had penetrated them and might yet kill them if the doctor could not find and remove it, and patch up the damage that it had caused to their soft tissues. This is perhaps why Philon's advice on preparing for a siege included ensuring the presence of doctors experienced in catapult injuries.[44] Little catapults did not need to kill instantly to be effective; indeed, if one's aim is to remove from the fighting as many of the enemy as possible, wounding is much more effective than killing, for it takes out not only the target but also all the others who help him over time and space, from the front line to civvy street, via whatever passes for an ambulance service and hospitals in the society in question. The philosophy behind some modern munitions is to wound, not kill, precisely for that reason.

The terms that ancient sources use for range missiles, and therefore perhaps also for the troops and weapons that shot them, are more encompassing in their semantic range than are our habitual translations of them—bows and slings, archers, and arrows. But small caliber catapults and catapult-men appear certainly in the historians in their descriptions of the injuries that they cause; if we are told that someone suffered a penetrating injury from a slingshot-like missile, then we can confidently deduce the presence of at least one little stone-thrower catapult, because that is the only ancient weapon that could have been responsible for causing it. Since any stone-thrower employs a sling to project the missile (whether a one-armed catapult, a two-armed catapult, an ordinary (human) slinger or a staff slinger), Apollodoros, Paul, and other people who called a stone-thrower a sling, were, technically, correct in their usage. Note too that the stonebow's slingstring was called a sling in contemporary literature.[45]

## Conclusion

From their invention as personal weapons at the beginning of the fourth century, catapults were first and foremost antipersonnel weapons. It stands to reason that a version that could be held easily in the hands and aimed along the stock, just like an arrow on a bow or a barrel on a rifle, would have been developed by one or more of the many engineers who designed catapults in antiquity, some of whose names are immortalized in the surviving literature. Indeed, the point is so obvious, that little catapults' absence would stand in greater need of explanation than their presence. But we do not have to rely on common sense and historical plausibility; there is good and varied evidence, literary and archaeological, direct and indirect, for the existence of little catapults from at least 200 B.C.

Little catapults' roles in classical times probably included that of sniper, discharging darts, or small shot. Compared with hand-bows, catapults gave a higher velocity and therefore more accurate or deadly fire, at the cost of a slower rate of fire; it is the familiar trade-off of quality versus quantity. Each would be advantageous in different circumstances, so the little catapult supplemented, not replaced, its manual forebears and alternatives, the sling and the bow. Ancient armies had continually to adapt technologies and techniques: the Greeks to fight each other and the Karthaginians, and then the Romans; the Romans to fight the Greeks and the Karthaginians, and then the Gauls, Parthians, Sarmatians, and whoever else either got in the way of their empire or wanted a piece of it. The little tension catapult ultimately became the medieval crossbow. What became of the little torsion catapult remains to be seen.

SIX

# The Age of Experiment: The Hellenistic Period

*The arts and crafts don't reach their fine heyday unless a generous patron comes their way.*—Menander *fr.* 473 Edmonds

CATAPULT DEVELOPMENT CONTINUED APACE under Alexander's successors. The Ptolemies in Egypt, the Seleukids in Asia, the Antigonids in Macedonia, and the Attalids in Pergamon fought over his inheritance for two centuries or so, investing heavily in their arsenals and their fortifications as they did so. Meanwhile, in the western Mediterranean, Karthage invaded and then extended her area of control in the Iberian Peninsula and Sicily, with similar attention to the development of arms, armor, and artillery. It was a period of incessant warfare the length and breadth of the Mediterranean and in that part of Asia conquered by Alexander. The Greek cities in Sicily found themselves challenged first by Karthage, and then by a new power emerging between the Macedonians in the East and the Karthaginians in the West: Rome. Catapult technology, and other war machines, then came under the scrutiny of one of antiquity's greatest intellects: Arkhimedes of Syrakuse. There is an overlap chronologically with the Roman Republic, and I have generally divided the material on the basis of how it relates to the history of the catapult: if the focus is Greek I tell it here, if Roman, then in Chapter 8.

In this chapter, we will explore the history of the catapult in the Hellenistic period, a time of experimentation and then stabilization, when fundamentally different designs competed, and gradually a few well-performing types emerged and became standardized as they were repeatedly preferred for service and modified to try to reach a state of perfection. This is not to say that development stopped once the designs stabilized in the mid-Hellenistic period; it *changed* from the radical development that characterizes immature technologies to the incremental development that characterises mature technologies. Fundamental performance-related advances could still emerge, and did, a couple of centuries later, as we shall see in Chapter 9. But this, the

Hellenistic period, was the period of maximum diversity and variety in the history of the catapult,[1] including the production of niche weapons for specific short-term purposes. Polybios' comment on the siege of Abydos in 200 B.C.—that it was not so remarkable for the variety of the devices employed in the works, but for the bravery displayed by the besieged (16.30.2–3)—inadvertently demonstrates that a wide variety of siege and anti-siege equipment existed then, and that it caught people's attention. Over time, it was the adaptable general purpose catapults that were found to be most useful, and a few of them emerged as "standard" weapons. The period also witnessed a number of responses to the new, powerful catapults, notably in fortifications.

This was a time of huge instability, when ambitious men commanded troops drawn from more or less anywhere between Iberia in the west and India in the east, amalgamated them into cosmopolitan armies, and fought over vast swathes of land, old and new cities, and enormous fortunes in treasure and plunder. Alliances between strong men were made and broken and made again as individuals jockeyed for position or survival. Troops on the losing side in a battle might be killed, or incorporated into the winner's army, or anything in between. Diodoros observed that conditions throughout Greece (and we may imagine it true of many other areas, too) had become unstable and mean because of the continuous wars and the mutual rivalries of the dynasts.

At first, sieges dragged on in the usual fashion until the defenders were relieved by reinforcements from elsewhere or they capitulated for one reason or another or the enemy successfully stormed the walls. For example, Kleonymos, Agiad king of Sparta in the 320s, used *oxybeleis*, sharp-casters, to send messages into Troizen to tell the citizens he was then besieging that he had come to liberate them from their Macedonian garrison, and when an uprising started inside the city, he took the walls by ladders. The catapults were exploited for their greater range, to get messages into the city. At the siege of Pydna in 317, Kasandros requisitioned "missiles of all sorts and machines" from his would-be allies, but did not assault the city "because of the winter storms"—which did not, however, prevent him maintaining a blockade of the port throughout.[2] That winter storms did not prevent a blockade but did prevent an assault is curious, but perhaps indicates a significant number of torsion catapults in his arsenal intended for wall-clearance duties, because the torsion catapult's one big drawback was that the springs did not respond well to a moist atmosphere, becoming slack and prone to decay. Meanwhile the governor of Amphipolis (not very far away from Pydna, and having the same sort of climate) took in days a place occupied by 2,000 of Kasandros' troops. Sieges of headquarters of opposing leaders were noticeably avoided during this period. For example, in 313 Telesphoros, a general of Antigonos, marched against and "freed" all the cities in the Peloponnese that had garrisons installed by Polyperkhon, except Sikyon and Korinth, because Polyperkhon had headquarters there.[3] The success of a siege depended on a variety of factors, only one of which was artillery provision, but it was prudent to avoid rival headquarters, where the best equipment available to that warlord would be ready to repel all comers. Of course, the existence and performance of artillery depended on a number of factors, not just on the general's desire to utilize it, as Onasander pointed out in the first century A.D.

> The general will make use of the many and various siege machines in so far as he can. It is not for me to say that he must use battering rams or city-takers or sambucas or wheeled towers or tortoises or catapults. These particulars depend on the luck, wealth, and power of the combatants involved, and on the skill of the engineers who accompany the army to build machines. The general's personal task, if he wants to use machines in offence, is to employ them on one stretch, for he would not have enough machines to encircle the entire wall unless the city was very small. (*Strategikos* 42.3–4)

It seems to me that in these circumstances, each of the major warlords would have been pursuing his own program of weapons development.

Ktesibios and Philon both comment explicitly that successful designs were discovered by experiment, by trial and error, and that the calibration formulae are, in effect, algorithms for copying designs that produce good, rather than indifferent, catapults.[4] "I understand that you are fully aware that the [war-machine-maker's] trade contains something unintelligible and baffling to many people. At any rate, many who have undertaken the building of engines of the same size, using the same construction, similar wood, and identical metal, without even changing its weight, have made some with long range and powerful impact and others which fall short. Asked why this happened, they had no reason to give."[5] In their introductions, Ktesibios and Philon refer to earlier manuals, and Philon adds that those manuals give different figures for the sizes of parts of catapults, and even use a different parameter ("the first guiding element") for the proportions. We will examine these texts more closely in Chapter 7. For now, it suffices to say that from Ktesibios' time, the spring-hole diameter had been identified as a key parameter, but evidently it took time to discover this, and some earlier engineers used something(s) else or had no principal unit at all (no doubt some later engineers used something else or nothing, too). I think that the designs that we know about are a tiny subset of those produced. These survived because they were fit, because these catapults won battles in practice. The earliest of the surviving manuals, Ktesibios', dates from about 270 B.C. The Wars of Alexander's Successors were, I suggest, a harsh and prolonged trial of catapults in which error could be fatal, not just for the mechanic and artilleryman, but for his employer as well.

The Ptolemies in Egypt, the Attalids in Pergamon, and the Rhodians all emerged as survivors from the Wars, along with the Seleukids in Asia and the Antigonids in Macedon. Not coincidentally, the Ptolemies, Attalids, and Rhodians are known to have all spent much time and money on weapons' development. Nor coincidentally were the catapults that feature in the surviving treatises designed in, for, or by people from Ptolemaic Alexandria, Attalid Pergamon, or Rhodes. The variety of techniques and terminology used in the different artillery manuals is probably a consequence of their authors' associations with different engineering "schools," as Marsden[6] suggested: Ktesibios with Alexandrian engineers, getting commissions from the Ptolemies; Biton with Pergamine, getting commissions from the Attalids; and Philon with Ptolemaic and Rhodian. Rhodes was an island community forever trying to remain independent of its bigger neighbors. As explained below (this chapter), Rhodian traditions after 224 incorporate Syrakusan traditions, as a result of a donation of catapults from Syrakuse to Rhodes. Shortly afterward, then-current Rhodian traditions would have spread to the southern coast of the Black Sea, when Rhodes donated four stone-throwers—called *lithophoroi* by Polybios, rather than *lithoboloi*—plus operators, 300 talents of prepared hair, and 100 talents of prepared sinew, to the Sinopeans to help them resist Mithridates II of Pontus.[7]

Ktesibios' comment, in the opening chapter of his manual, that "the construction of war machines has taught mankind how to live a peaceful life," could only have been written after the balance of power between Alexander's successors had been achieved (at least in Egypt, where he lived). He also says, "after peace has continued for a long time, one would expect more to follow when men concern themselves with the artillery section [of mechanics]; they will remain tranquil in their consciousness of security, while potential aggressors, observing their study of the subject, will not attack. But every act of aggression, even the most feeble, will overwhelm the neglectful since artillery preparations in their cities will be non-existent."[8] This suggests that the process by which the balance of power had been achieved was over but not yet forgotten. And I think that what Ktesibios says about the militarily unprepared is, in fact, a reflection of the history of the Wars of the Successors, with the contrast between sieges which were over in a jiffy and those that either were avoided altogether or were prolonged.

When the breakthrough with torsion catapult design was made, that is, when the formula was discovered and reliable reproduction became possible, I think we should expect it, or its consequences, to show up in the literary sources. Unlike most other ancient technologies, war machines were involved in the military and political events that constituted the bread and butter of ancient histories. A dramatic change in siege warfare occurs during the period 322–303, when offensive siege warfare became increasingly successful, as

McNicoll showed.[9] The data he used was taken from Diodoros Siculus, whose work is the only surviving continuous record of these years. Diodoros refers to one hundred sieges during this twenty-year period, which will not be a complete tally for those very bellicose times, but it does form a statistically valid group. McNicoll divided the data into four-year intervals.

| *Period* | *No. sieges* | *Winning side* | |
|---|---|---|---|
| | | *Attack* | *Defense* |
| 322–318 | 13 | 6 (46.2%) | 5 (38.5%)* |
| 317–313 | 29 | 18 (62.1%) | 9 (31%) |
| 312–308 | 29 | 21 (72.4%) | 8 (27.6%) |
| 307–303 | 21 | 20 (95.2%) | 1 (4.8%) |

*Two sieges (15.3 %) were drawn.

The change is most obvious when the data is revealed as a graph (Figure 6.1)

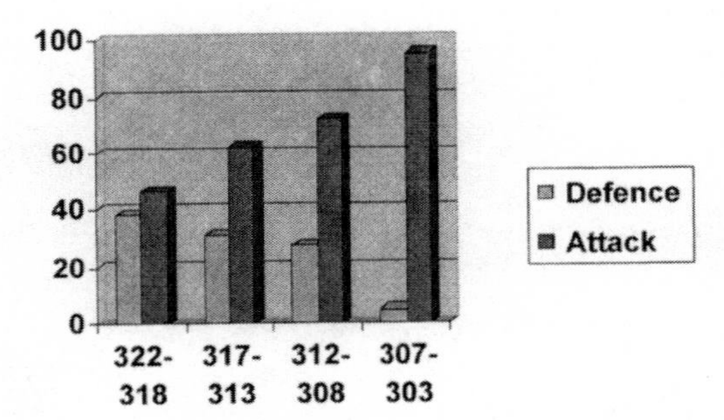


FIGURE 6.1

McNicoll thought that the *increased used* of artillery may explain why siege warfare came to be so decisively in the attackers' favor. I suggest instead that it reflects a great leap forward in catapult design and performance—the emergence of the "classic" two-armed torsion weapon from the mire of (necessarily) uncertain designs that had evolved since the catapult's invention. I think that what we see in these figures is the arrival of honed versions of this technology, machines that began to offer a *significant* advantage over previous models and all other weapons. The first catapults had *not* made a decisive difference to the outcome of battles in which their users were engaged. By 320 there had been perhaps one to three decades of unspectacular development of two-armed torsion machines during which time they had moved from being a new technology with potential to a hallmark technology, at least among the best-equipped armies of the time.[10] The real breakthrough came in the penultimate decade of the fourth century.

This breakthrough was, I suggest, the discovery of the formula and scaling law, so that people understood why some catapults worked well and others did not; they realized that good experimental models could be scaled up to produce good weapons, and good weapons could be replicated again and again, or scaled up or down, without loss of performance. The random element that had plagued catapult technology to this point—not knowing whether a newly built machine would work well or not, despite having been built to *resemble* another one that did work well—was squeezed out by the introduction of mathematics: a key parameter and regular ratios. It was discovered that what mattered was not that the machine *looked* like one known to work well, but that it had sufficient power in the springs for the missile it was meant to launch, and that its structure was proportional to the springs.

Because of the bellicosity of the times, there was a chronic catalyst for innovations in the military sphere. It is very noticeable in the history of military technologies that development is far more rapid and radical in wartime, or when war is thought imminent, when necessity is immediate and unyielding, and state sponsors are more ready than usual to plow manpower and money into untried technologies.[11] The real process of testing by which ancient engineers discovered how much spring-cord was sufficient, and what were the ideal ratios between parts of a catapult, was probably undertaken on the battlefield rather than in the cornfield of some ancient Ministry of Defence. I do not mean that *changes* were undertaken on the battlefield; I mean that *the tests that mattered* took place there, because there failure, from whatever cause, usually meant not "back to the drawing board" but death, injury or enslavement. In the life or death situation of battlefield or siege, survival depended not on expense, nor appearance, nor efficiency, nor intentions, nor potential, nor technical superiority of some sort or another *on paper*. The only thing that mattered was superiority in practice, that is, military effectiveness in that particular environment. Of course, luck also had a part to play, but even cats have only nine lives. And lest (mistakenly) I be thought to be promoting technological determinism, let us not forget the crucially important role of the commander. Lynn White put it well: "a novel technique merely offers opportunity; it does not command."[12] As Polybios pointed out very forcefully, Hannibal beat the Romans because he was a better general, not because he had better equipment (he even swapped his arms for theirs when he acquired them as spoils), and the Romans only got the upper hand when they produced a general of equal caliber (18.28.6–11).

Many different groups of engineers, not just one, would have pursued such trials, and the best of them seem to have arrived independently at approximately similar results. Thus our ancient authors between them describe four different methods for designing the hole-carrier, but all four methods lead to the same shape.[13] When this happens in nature it is called convergence, and it shows that the feature in question (here, the shape of the hole-carrier) is very important to the success of the whole design.

"Technologies mature into stable configurations," said Hall, having pointed out that more than one thousand different designs of motor car had appeared by 1915 but that by 1920 the default was something made of metal with four wheels and rubber tires, a gasoline engine, and a multispeed transmission.[14] Our surviving treatises preserve the catapult designs upon which ancient engineers converged; undoubtedly, these do not exhaust the types of catapults built in antiquity.

The statistics direct our attention to the years 317–313, when the offense began to enjoy a distinct advantage. When we search the sources, the end of that quadrennial stands out. Now the evidence I am about to discuss could be explained by Diodoros starting, at this point in his history, to use a new source that was particularly prone to melodrama. But in conjunction with the statistical evidence, which led us to focus on this part of his history in the first place, I think such a purely literary interpretation is unsatisfying and unconvincing.

In 313 B.C. we hear of a city capitulating apparently from shock and awe (the key Greek term is καταπλήσσειν, to terrify, strike down with amazement). The Kallatians (SW Black Sea coast, Map 5) had rebelled and thrown out Lysimakhos' garrison. En route to them he marched on Odessos, and "he quickly terrified those inside" into submission, then recovered the Istrians "in a similar way." He then faced large forces of Thracians and Skythians who had come to assist their allies, and "he terrified the Thracians and induced them to change sides." The Thracians did not have a reputation for timidity; anything but. It seems that either Lysimakhos had become an astonishingly fierce warlord in a world accustomed to violent men, or he had some new means of terrifying people. In the same year, one of Kasandros' generals launched an attack on the Aitolians and Epirotes. "By gaining such victories in a few days, Philip so terrified (κατεπλήξατο) many of the Aitolians that they abandoned their unfortified cities and fled to the most inaccessible of their mountains with the children and their women." We can do no more than guess about what precisely might have been used against these people to frighten them so, but war was not unusual in either of these locations at this time, and I suspect that what we have here is the faint signature of powerful new catapults. People began to capitulate without a fight, and that, usually, is a response to being challenged by a new and overwhelming weapon to which, at the time, they had no answer. So, at a later period, one elephant panicked the Britons into flight. When the Gauls were first faced with Roman siege machinery, "which they had not seen or heard before," they surrendered without a fight. Back in 313, we have one of Antigonos' generals "terrifying the garrison" on Khalkis into submission. Nine years later, Demetrios "freed" the same Khalkis from a Boiotian garrison, and "frightened" the Boiotians into renouncing their friendship with Kasandros.[15] Kasandros was quite capable of frightening people, too, so this must have been an unpleasant choice for the Boiotians. What we see in these years is the contrast between the Successor armies on the one hand, whose powers were roughly balanced, and everyone else's armies, on the other; they were no longer militarily in the same league, or anything remotely approaching it. While the former could and would stand up to each other and fight it out for supremacy, in this period most other states (and garrisons) were simply no match for any of them, and could be subdued by intimidation, not even requiring a shot to be flung.

Sometimes there is a contrast between the army led by the Successor himself and the armies that his subordinates command. For example, in 309 Ptolemy, "hearing that his generals had lost the cities of Kilikia, sailed with an army to Phaselis and took this city. Then, crossing into Lykia, he took by storm Xanthos, which was garrisoned by Antigonos. Next he sailed to Kaunos and won the city, and by violently attacking the citadels, which were held by garrisons, he stormed the Herakleion, and he gained possession of the Persikon when its soldiers delivered it to him." These sieges appear to be quick, successful, and usually won by storm. Demetrios Poliorketes' siege of Mounykhia in 307 was over in two days. "Bringing up stone-throwers and the other machines and missiles, Demetrios assaulted Mounykhia both by land and by sea. When those within defended themselves robustly from the walls, it turned out that Dionysios [defending commander] had the advantage of the difficult terrain and the greater height of his position, for Mounykhia was strong both by nature and by the fortifications that had been constructed. But Demetrios was many times superior in the number of his soldiers and had a great advantage in his equipment. Finally, after the attack had continued unremittingly for two days, the defenders, severely wounded by the catapults and stone-throwers and not having any men to relieve them, had the worst of it. Those around Demetrios, who were fighting in relays and were continually replaced, after the wall had been cleared by the stone-throwers, broke into Mounykhia, forced the garrison to lay down its arms, and took the commander Dionysios alive."[16]

Thereupon the Athenians "voted to set up golden statues of Antigonos and Demetrios . . . To give them both honorary crowns at a cost of two hundred talents, to consecrate an altar to them and call it the altar of the Saviours, to add to the ten tribes two more, Demetrias and Antigonis, to hold annual games in their honour with a procession and a sacrifice, and to weave their portraits into the peplos of Athena."[17] In other words, the Athenians groveled before them and even worshiped these men as gods. This was not only totally out of keeping with Athenian traditions and history (they refused Alexander much less), it was inappropriate to the situation. The previous ten years, when they were ruled by another Demetrios (he of Phaleron, acting for Kasandros), were among the most peaceful and prosperous in Athenian history, no matter how much Antigonos tried to paint the man a tyrant. One wonders in what sense they needed "saviors." But that was the language of power politics of the day, and I think that what we have in the Athenians' obsequious thank-offerings is another hint of the shock and awe created by dramatically improved catapults, machines which had recently been deployed against them. Diodoros specifically refers to Demetrios' much better equipment.

Cities serving dynasts and their subordinates as headquarters generally continued to be well armed, so when Demetrios moved on later in the year to Kypros to attack Ptolemy's forces there, Ptolemy's commander, based in Salamis, had machines and missiles, which were "carried onto the walls" in preparation for a siege.[18] As a result, Demetrios put more than usual effort into his preparations. "[Demetrios] determined to prepare siege machines of very great size, catapults for shooting sharps and throwing stones of all kinds, and the other terrifying (καταπληκτικός) equipment." Now the stone-throwers were still being used to clear the walls of defenders while the walls were attacked with rams. And siege towers were not yet protected, or not yet adequately protected, from fire, for the *helepolis* Demetrios built for this siege was destroyed completely by fire.[19] However, the lesson was quickly learned, and rawhide, sacks of wool, birdlime, and other materials began to be attached to the faces of the towers, and water stored on the levels to deal quickly with firebrands. Another way of dealing with enemy firebrands was to halt the tower, loaded with catapults, where it could take advantage of the catapults' extra range, that is, before it came within range of bowmen on the walls, so that the shooting was all one way—assuming that the defenders lacked equally powerful catapults, of course.

There were also more unorthodox methods. Agathokles, the tyrant of Syrakuse at the time, who was busy in the western Mediterranean carving out an empire for himself that rivalled those of Alexander's successors in the east, and informed of what those men were doing, decided to attack Utica, near Karthage (Map 1).

> He built a siege machine, hung on it the [Utican] prisoners [he caught earlier] and brought it up to the walls. The Uticans pitied the unfortunate men, yet, holding the liberty of all more important than the safety of those, they assigned posts on the walls to the soldiers, and bravely awaited the assault. Then Agathokles, placing on the machine his sharp-casters and slingers and archers, began the assault fighting from this, applying, as it were, branding-irons to the souls of those within the city. Those standing on the walls at first hesitated to use their missiles, since the target presented to them was their

> fellow countrymen, of whom some, moreover, were the most distinguished of their citizens. But when the enemy pressed on, they were forced to defend themselves against those who manned the machine. As a result, unparalleled suffering and spiteful fortune came to the men of Utica, placed as they were in dire straits from which there was no escape. For because the Greeks had set up before them as human shields the Uticans who had been captured, it was necessary either to spare them and idly watch the fatherland fall into the hands of the enemy or, in protecting the city, to slaughter mercilessly a large number [300] of unfortunate fellow citizens. And the latter, indeed, is what happened. For as they resisted the enemy and employed missiles of every kind, they shot down some of the men who were stationed on the machine, but they also mangled some of their compatriots who were hanging there, and others they nailed to the machine with their sharps at whatever place on the body the missiles changed to strike, so that the wanton violence and the punishment almost amounted to crucifixion.

This same ruthless leader elsewhere used catapults to torture some of his victims: we are told that he shot men bound to catapults. Meanwhile it was said that Antigonos used many of his sharp-casters, together with archers and slingers, against those of his own forces trying to desert to Ptolemy's side during a campaign in Egypt. It is impossible now to know whether this was true, or a case of what we euphemistically call "friendly-fire," or simply the baseless slander of a hostile tradition.[20]

Artillery was still not being used in the field, apparently. For example, catapults are not mentioned when the dispositions of Demetrios' and Ptolemy's armies before battle of Gaza in 312 are spelled out, and Antigonos sent for catapults only once he began a siege, after his expectation of battle was disappointed by the enemy (Lysimakhos) avoiding it and making fortified camp instead.[21] Interestingly in this regard, Polyainos says that when Magas left his capital Kyrene on a campaign, he locked up the engines, missiles, and machines in the acropolis so that they could not be used by the inhabitants—including his friends left on guard—against him when he returned.[22] Large catapults would have been difficult to move along what passed for roads in most of the ancient world, even after the Romans set to work road building, but they were designed to be easily dismantled and reassembled. Consequently, logistics was not remotely the sort of problem faced by the makers and users of large cannon, who normally cast the guns where they were needed. In fact, large shot was perhaps more of a problem to move than the catapult built to shoot it, which is probably why heaps of such shot still lay around ancient sites. Even so, I don't think logistics can explain the divergent use of artillery in siege and battle. One assumes that most catapults were still not very maneuverable, or the rate of discharge was too slow, or they were too valuable to risk losing in a fluid open engagement on the field. Poor maneuverability is suggested by the Karthaginians' abandonment of their machines at the siege of Tunis in 310 when they (mistakenly) thought that a large relieving force was at hand. It may also explain why catapults were not, apparently, deployed against elephants in this period, and reliance was placed instead on triboloi and similar devices in the ground, and on manually discharged javelins and arrows.[23] Polybios explains in detail how the Romans first learned to deal effectively with enemy elephants in his description of the battle of Panormos (Sicily) in 250 B.C. It involved quick moving missile troops, being re-supplied with ammunition and reinforced with fresh troops, enticing the animals within range of the defense works and walls, wherefrom more missiles assailed them too, and they were thereby driven berserk. Polybios does not say so explicitly, but some of these missiles—even those used by the "quick moving" troops—were probably shot by catapult, because just one year later a Sicilian town too insignificant to be named had catapults and stone-throwers, so a large city like Panormos would surely have been equipped with such munitions.[24] That would explain the rather prominent role of resuppliers and reinforcements in the story; someone already carrying a catapult, which is significantly heavier than a bow, cannot carry catapult bolts, which are significantly heavier than arrows, *in quantity*. That is why crossbows were not often employed in fluid field situations in medieval times, but were more commonly confined to static situations such as sieges. It was probably the same in antiquity with both tension and torsion catapults.

The deployment of artillery in the field seems to be a new tactic at the battle of Mantineia of 207 B.C.—and it was implemented by the (stereotypically very conservative) Spartans of all people! "Makhanidas at first looked as if he were about to charge the enemy's right with his phalanx in column. However, on approaching, when he found himself at the proper distance, he wheeled to the right, and deploying into line made his own right wing equal in extent to the Akhaian left, placing his catapults at certain intervals in front of his whole army. . . . Makhanidas' plan was to wound the men and throw the whole enemy force into disorder by shooting at the divisions of the phalanx."[25] These catapults were mounted in wagons. But they were not apparently brought into action, for no sooner had Makhanidas brought them up than his opponent, recognizing the danger, initiated a move that quickly brought the troops of both sides into one confused mass on the field. "Philopoimen [seeing what Makhanidas planned to do] gave him not a moment's leisure but vigorously opened the attack with his Tarentine cavalry." Makhanidas ultimately lost the battle and his life because of his bad generalship when he appeared to have won the battle, as Polybios emphasized. Although the defeat had nothing to do with his novel deployment of catapults, it would have to wait for another commander to have the idea again before artillery would reappear in the field. And if that second appearance was not to be another isolated instance, like Mantineia, he would have to succeed with it. Field artillery was finally developed, of course, but by the Romans not the Greeks and apparently not until the first century B.C. We will consider it in Chapter 8. More than a millennium later, firearms were initially effective only in siege operations too.[26]

The first mention of catapults mounted on ships since the siege of Tyre (excepting Alexander's mounting of sharp-casters on rafts to cross the river Iaxartes and try to clear enemy troops from the far bank) comes in 307, again among Demetrios' forces, this time off Kypros. Both sharp-casters and stone-throwers were put on board, three-span sharp-casters apparently being mounted at the prow of each of Demetrios' 108-plus warships.[27] "[Demetrios] equipped [all his ships] with missiles and stone-throwers and mounted on the prows enough three-span sharp-caster catapults. . . . When the trumpets gave the signal for battle and both forces raised the battle cry, all the ships rushed to the encounter in a terrifying manner. Using their bows and their stone-throwers at first, then their javelins in a shower, the men wounded those who were in range."[28] Here the catapults were used just as we would expect them to be—exploiting their greater range. Note how, in the last sentence excerpted, the simple word "bows" (τόξα) seems to mean the three-span sharp-casters, because as the principal weapon on the prow they must have been used, as well as the stone-throwers, at the beginning of the engagement. They may have been three-span *bow* catapults, not torsion types; this is not the only occasion in the literature when the word "bow" seems to mean bow catapult (*gastraphetes*). Bow catapults might have been preferred for naval duties, as sinew, which powers torsion catapults, becomes slack and prone to decay in wet environments, in which ships exist. As I emphasized in Chapter 3, the gastraphetes was not merely a prototype en route to the torsion catapult; it had a life and a history of its own, and one that continued after the appearance—and the disappearance—of the torsion machine.

The same ships with prow-mounted sharp-casters are mentioned again at the year-long siege of Rhodes in 305–304 (Map 4). Here for the first time we are told extensively about catapult provision and deployment.

> Demetrios, who had an ample supply of everything required for setting up his engines of war, began to prepare two tortoises, one for the stone-throwers, the other for the sharp-casters, each of them firmly mounted on two cargo ships lashed together. . . . He collected the strongest of the light craft, fortified them with planks, provided them with port-holes that could be

> closed, and placed on them the three-palm [1-span, 9 inch] sharp-casters that had the longest range, and the men to work them properly, plus Kretan archers. . . . [The Rhodians] placed two machines on the mole and three on freighters near the boom of the small harbour. In these they mounted a large number of sharp-casters and stone-throwers of all sizes, so that, if the enemy should disembark soldiers on the mole or should advance his machines, he might be thwarted in his plan by this means. They also placed on such cargo ships as were at anchor in the harbour platforms suitable for the catapults that were to be mounted on them. . . . At daybreak [Demetrios] brought his machines into the harbour with the sound of trumpets and shouts. With the lighter sharp-casters that had a long range, he drove back those who were working on the wall alongside the harbour, and with the stone-throwers he damaged some and destroyed other enemy machines, and the wall across the mole, for it was weak and low at the time. . . . When night was closing in, Demetrios withdrew his machines back out of range using tugboats. . . . After carrying on this kind of siege warfare for eight days, Demetrios shattered the machines on the mole with his one-talent stone-throwers, and weakened the curtain of the cross-wall together with the towers themselves. . . . Demetrios withdrew his machines to his own harbour, and repaired the ships and machines that had been damaged, and the Rhodians . . . rebuilt the parts of the wall that had been thrown down by the stone-throwers. (DS 20.83.1, 20.85.1–87.4 Loeb modified)

The fighting began again a week later, after both sides had repaired and regrouped.

> When he was within range, with the fire-arrows (of which he had many) he attacked the Rhodian ships at anchor; with his stone-throwers he shook the walls; and with his sharp-casters he cut down any who showed themselves. Then when the attack had become continuous and terrifying, the Rhodian ship-captains, after a fierce struggle to save the ships, put out the fire-arrows . . . [The magistrates] instructed them to try to sink the ships that carried the enemy's machines. Accordingly, they pushed forward although missiles in large numbers were flying towards them, and at first they broke through the iron-studded boom, and then by delivering repeated blows with their rams, filling the ships with water, they overthrew two of the machines . . . Demetrios constructed another machine (*mekhane*), three times the size of the former in height and width, but while he was bringing it up to the harbour, a violent storm sprang up from the south, which swept over the ships that were anchored, and overthrew the machine. (DS 20.88 Loeb modified)

Note that Demetrios' sharp-casters were specifically utilized for sharp-

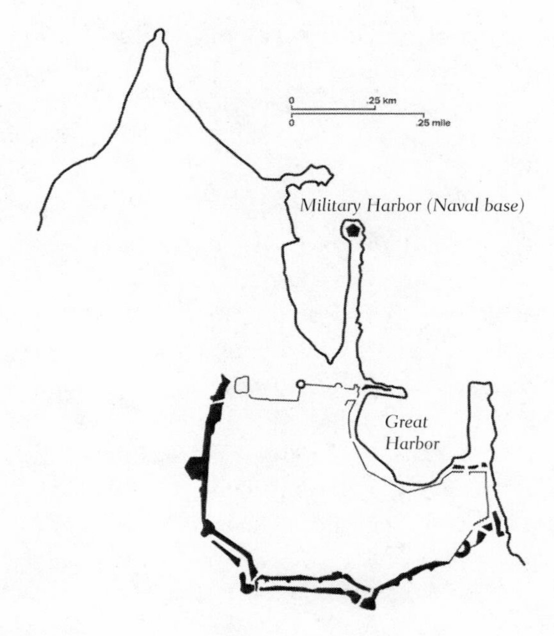


MAP 4. *Rhodes.*

shooting, picking off anyone who put his head above the parapet, so to speak. This seems to be their principal role when the technology settles down; at least, we are not led by the surviving sources to think that massed ranks of catapult-bearers ever stalked ancient battlefields, as crossbowmen sometimes did later.

Demetrios' next attempt on the city was by land, and involved the building of his massive *helepolis*, to be discussed further below. It is clear that new devices, and bigger versions of existing designs, were being produced during the siege to meet the challenges thrown up by these particular opponents in this particular environment on this particular occasion. According to Diodoros, "it was in this man's [Demetrios'] presence that the greatest weapons were perfected, and machines of all kinds far surpassing those that had existed among others." Demetrios' "energy and ingenuity" in siege situations, "being exceedingly ready in invention and devising many things beyond the art of the master builders," earned him the nickname Poliorketes, the Besieger.[29] Some of these innovative ideas would have worked, and some would not. It would seem that the biggest catapults ever made were built for this man.

In time, opponents with ordinary equipment so expected to be beaten by Demetrios that they surrendered strongholds and cities to him without a fight. "After the capture of [Arkadian Orkhomenos], those who commanded the forts in the vicinity, assuming that it was impossible to escape the might of the king [Demetrios], surrendered the strongholds to him. In like fashion those also who guarded the cities withdrew of their own accord, since Kasandros, Prepelaos, and Polyperkhon failed to come to their aid but Demetrios was approaching with a great army and with overwhelming machines."[30]

Demetrios' second *helepolis* was covered with iron plates to protect against fire (always the weak point with these wooden constructions), but the shutters over its artillery ports, variously shaped to suit the variety of missiles being discharged through them, were made of wool-filled hide pads, designed specifically to cushion the blow of shot from the defenders' stone-throwers.[31] Demetrios again used his stone-throwers, as well as his rams, to shake the walls of Rhodes. Since a square stone tower was brought down, and a length of curtain was shaken so badly that it could no longer be patrolled, these must have been powerful large caliber weapons. During this next round of fighting the Rhodians are said to have placed "all" their stone-throwers and sharp-casters "upon the wall" for one particular attack on Demetrios' *helepolis*.[32] Now "the wall" generally means the curtain and towers, that is, the city walls as a circuit. If that is its meaning here, this passage would seem to exclude large stone-throwers from the Rhodians' arsenal at this time. For even a 5-mina palintone, shooting stones about 2 kg in weight, if built to formula, had a stock about 10 1/2 feet long, while a 30-mina stone-thrower hurling 12 kg balls had a stock 19 feet long and an arm-span of 12 1/2 feet. Few city walls—at least in the western portion of Alexander's empire, where they were usually made of stone[33]—were thick enough to enable defenders to operate a catapult with a 10-foot long stock from their parapets; indeed, few *towers* had chambers big enough to hold a 30-mina catapult.[34] However, a one-mina palintone, with a stock about six foot long, or a two-mina, with a 7 1/2-foot stock, would fit easily on top of the wall.[35] Employed in a tower enfilading the curtains, these would also be appropriate sizes to smash scaling ladders brought up against the wall, as Philon recommended, without threatening any wall or tower behind the target. Further evidence of the normally small caliber of stone-throwers on walls is a passage in Philon's *Poliorketika*: three sheets of wicker topped with chaff-filled baskets are said to provide adequate defense from enemy stone-throwers for covered ways.[36] As we saw in Chapter 5, the smallness of some surviving catapult washers also indicates small caliber hand-held weapons launching shot the size of slingshot—stones of a size signified best by the word "bullet" rather than "shot," being propelled by something that resembled a firearm rather than artillery. Some of these washers are much smaller than the smallest caliber mentioned by the literary sources, and the smallest is less than 3.5 centimeters in diameter. This corresponds to a stock length of less than two feet and darts about 12 inches long, if this washer came from a sharp-caster built to formula, and a shot of about 11 grams in weight and a machine with a stock about two feet two inches, if it came from a formulaic stone-thrower.

The Rhodians' equipment probably improved dramatically during the siege, especially after one of their naval captains captured some freighters that were making their way to Demetrios' camp; on board were eleven war-machine technologists, who henceforth would have been employed in Rhodes' arsenal. "[Amyntas] fell in with many freighters carrying to the enemy materials useful for machines; he sank some of these and some he brought to the city [of Rhodes]. On these ships were also captured eleven famous artisans, men of outstanding skill in making missiles and catapults." As Cassius Dio later observed, being an engineer of war machines could be the cause both of being condemned to death, and of being spared. Such was Priskos, builder of war machines for the Byzantines, who gave his compatriots the wherewithal for three years (A.D. 193–196) to keep out Romans wielding "the armaments of practically the whole world." Septimius Severus thus learned to appreciate his skill, prevented his execution on the capitulation of the city, and then employed him on future campaigns such as the siege of Hatra, where Priskos' machines were the only ones to survive the Hatreni's counterattacks.[37]

These stories are sobering testimony to the amount of information about the ancient world that we have lost. Eleven was the number of *additional* and *famous* artificers brought in by one side to work on the siege of Rhodes; I currently know of only thirty-seven names of actual or apparent war-machine builders from all of antiquity, plus three whose names are lost.[38] In about 200 B.C., Philon assumed that any city expecting to be besieged would have at least one artificer and some artillerymen on hand.[39] How many engineers Demetrios had with him from the start of the siege of Rhodes is not known, but must have been considerable as he had already built two *helepoleis* plus their complements of catapults before this additional eleven were captured en route to his camp.

For all his efforts, technology, and expenditure, Demetrios failed to capture Rhodes. The citizens resisted successfully until he found an opportunity to retire without disgrace.[40] Traditionally, the Colossus of Rhodes, a bronze statue of the sun-god Helios (Rhodes' patron deity), 70 cubits high[41] and one of the Seven Wonders of the ancient world, was built from the proceeds of the sale of Demetrios' abandoned siege equipment. One could not wish for a clearer expression of the Rhodians' attitude to these pieces of kit: the world's then-best military technology was sold to the highest bidders.[42] The sale realized 300 talents, according to Pliny.[43] This did not include the value of one of the giant siege towers, for that became the property of the Rhodian engineer who caught it, by designing what appears to have been the first manmade quagmire outside the walls, and he dedicated it to the gods rather than sell it.[44]

Syrakuse, where, as we saw in Chapter 2, catapults were invented, was well equipped with them when the city was caught between the two superpowers of Karthage and Rome. The First Punic War began in 264 and ended in 241, and in the early years of that war (in 263) Syrakuse, under its tyrant Hieron II, became an ally of Rome. That war resulted in the Roman occupation of Sicily, and most of the island became a province, but Syrakuse continued as an independent kingdom. Evidently she took her independence seriously, for when an earthquake hit Rhodes in 224 B.C., bringing down the arsenal and most of the walls as well as the great bronze Colossus that stood on a hill overlooking the harbor, Syrakuse presented the Rhodians with fifty, three-cubit catapults.[45] I deduce from this gift not only that Syrakuse was well armed at the time, but also that she had great confidence in the quality of her catapults, because Rhodes' reputation in artillery construction would have threatened international embarrassment to a donor of inferior equipment. According to Livy, Arkhimedes had been working on Syrakuse's armaments for "many years" before the Romans arrived, ten years later, in 214.[46] When this earthquake hit, in 224, Philon would have been a young man. It is possible that the Rhodian tradition that he continues also preserves, at second hand, Arkhimedes' contribution to catapult design, for the Rhodians would surely have adopted any improvements that the donated Syrakusan models offered, and Philon records Rhodian (and consequently Syrakusan) designs c. 200 B.C.

Hieron's good relations with Rome continued, for during the Celtic War of 225–222, when the Gauls invaded Italy, Syrakuse not only supplied Rome with grain, but also did not demand payment for it until that war was over. Four years later, when Hannibal crossed the Alps in 218, the Second Punic War began. Three years after that, in 215, while Hannibal ravaged Italy, Hieron II's son Gelon died, and a few months later, Hieron himself died. His grandson Hieronymos, "a lad in his teens," succeeded to the throne of Syrakuse.[47] For reasons unknown but probably connected with the Romans' disastrous performance to date against Hannibal, Hieronymos changed Syrakuse's allegiance from Rome to Karthage. This was not his only reform of Syrakusan domestic and foreign arrangements. His rule was so unpopular that he was overthrown after thirteen months, and the Syrakusans executed all remaining members of the family. They apparently did not, however, make amends to or renew their alliance with Rome. A year later, when Rome had raised new legions after the catastrophic defeat at Cannae (in 216), and begun to reassert control in Italy (although Hannibal was still there), Marcus Claudius Marcellus was appointed to the Sicilian province, and went straight to Syrakuse and laid siege to it.

The Romans were optimistic, because they had large numbers of troops. But they did not reckon with the ability of Arkhimedes, or foresee that sometimes the genius of one man accomplishes more than the efforts of any number of men. He was, said Livy, an unrivaled observer of the heavens and the stars (his father Pheidias was an astronomer), but even more remarkable as an inventor and builder of catapults and war machines, by which he frustrated with the least effort whatever the enemy attempted with huge effort. The topographic strength of Syrakuse lay in the fact that the wall extended in a circle along a chain of hills with overhanging brows, which were, except in a limited number of places, not easy to approach even in peacetime. Arkhimedes equipped the walls with every kind of catapult (*genere omni tormentorum*), as seemed suited to each place, and made such extensive preparations that the defenders never had to meet emergencies—every move of the enemy could be responded to instantly with a countermeasure.

Appius, Marcellus' second-in-command, attacked the wall that abuts on the Hexapolis (the greater city, Map 2 inset) to the east with his shields and ladders while Marcellus attacked Akhradina from the sea with sixty quinquiremes, each of which was full of men armed with bows, slings, and a kind of javelin designed to repulse those fighting from the battlements. The Romans intended to attack the towers with various contrivances. One of these was the *sambuca*. Marcellus also had eight quinquiremes lashed together in pairs, and the sambucas were mounted upon them. Pulling with the oars on their outer sides they brought the sambucas up to the wall. Polybios recounts how the machines were contructed.

> A ladder was made four feet wide and as high as the wall, allowing for it being planted at the proper distance. Each side was furnished with a breastwork, and it was covered over at considerable height by a screen. It was then laid horizontal on the inner bulwarks of the ships that were lashed together, and protruding a considerable distance beyond the prows. At the top of the masts are pulleys with ropes, and when they want to use it, they attach the ropes to the front of the ladder, and men standing at the stern pull them up by means of the pulleys, while others stand on the prow and, supporting the machine with props, assure its being raised safely. After this the rowers on both outer sides of the ships bring them close to shore, and they now endeavour to set the machine up against the wall. At the summit of the ladder there is a platform protected on three sides by wicker screens, which four men mount and face the enemy, resisting the efforts of those who, from

> the battlements, try to prevent the sambuca from being set up against the wall. As soon as they have set it up, and are on a higher level than the wall, these men pull down the wicker screens on each side of the platform and mount the battlements or towers, while the rest follow them through the sambuca, which is held firm by the ropes attached to both ships. The construction was called a sambuca appropriately, for when it is raised the shape of the ships and ladder together is just like the musical instrument. (Polybios 8.4.4–11)

Arkhimedes had constructed catapults to shoot at any range. He bombarded the Romans at long range, as they sailed up, with his more powerful stone-throwers and heavier sharps. Plutarch claims that he shot stones weighing up to ten talents,[48] which caused tremendous wave surges and noise even when they failed to hit the ships. As the ships closed and those engines overshot their targets, he used other catapults with shorter range, and so thoroughly shook the Romans' courage that Marcellus decided to withdraw and to try again under cover of darkness. This succeeded to the extent that the fleet got close enough to the walls to be safe from the catapults on the fortifications (they cannot shoot downward, for the missile will fall out under gravity), but Arkhimedes had prepared another defensive measure for this eventuality, which attacked the soldiers on the decks. He had pierced the wall with an extensive series of loopholes that Plutarch described as small and Livy described as almost invisible; they were about a cubit high and a palm's breadth wide on the outside, and a cubit wide on the inside. Loopholes thereafter became common, and we even find Caesar's men incorporating loopholes specifically for use by catapults into a single-use siege-work; a six-story brick tower built to assail Massilia. Arkhimedes stationed archers and "modest" (*modici*) scorpions—which here probably means small caliber torsion catapults, perhaps hand weapons—by these loopholes, and they shot the soldiers through them. Plutarch described these catapults specifically as "short-spring scorpions" (σκορπίοι βραχύτονοι), adapted to strike targets close at hand, he said. I am reminded of Polybios' comment that the length of the sarissa (the Macedonian pike) was, according to the original design, 16 cubits, but as adapted to actual needs, 14 cubits. Arkhimedes apparently reduced the spring height, which would result in high impact, short-range weapons, according to Philon.[49]

When the Romans tried to raise the sambucas, Arkhimedes had engines ready all along the wall, which appeared over the wall as required. Their beams projected far beyond the battlements, some of them carrying stones weighing as much as ten talents (c. 250 kg), others large lumps of lead. Whenever the sambucas approached, these beams were swung round on their universal joint (καρχήσιον), and dropped their load on the sambucas,[50] smashing the devices and damaging the ships carrying them, threatening all those on board.

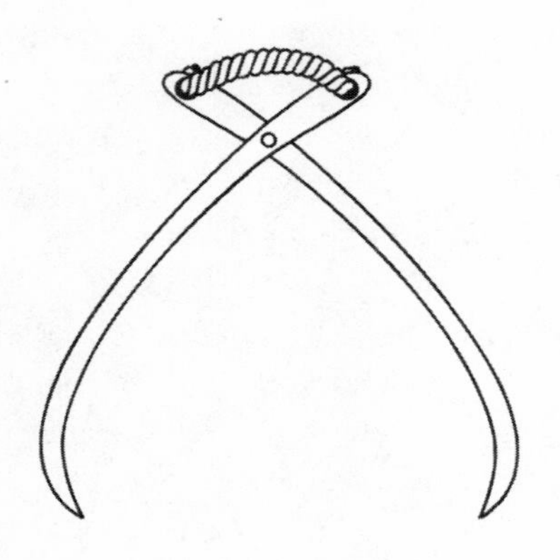

FIGURE 6.2 *The iron hand. (T. D. Dungan after author's sketch)*

Other machines targeted marines on the prows of ships, advancing under the cover of shields, and thus protected from missiles shot by bows and scorpions through the loopholes. These machines dropped stones large enough or numerous enough to frighten people off the prow of a ship, and simultaneously let down an "iron hand" on a chain (Figure 6.2). This "hand" was probably a variant of a device long since used in the building trade for lifting stone.

The rope on a building site that was pulled to make the claws close was replaced by a chain on the battlefield to prevent the enemy from cutting it. The man who piloted the beam would clutch at the ship with this iron hand, and when he had hold of it by the prow, would depress the opposite end of the beam, which was inside the wall. Thus he lifted the ship's prow and made it stand upright on its stern. After securing the beam, he suddenly dropped the ship by pulling a trigger that released the iron hand. Some such ships capsized, wholly or partially, but most submerged and filled.[51] Marcellus withdrew.

On land Appius found himself in similar difficulties and abandoned his attempt too. His men were mowed down while still at a distance from the fortifications by shots from the stone-thrower and sharp-caster catapults. Nothing could stop the stones, said Plutarch, which cut swathes through the ranks of legionaries. The supply of artillery and ammunition was apparently awe-inspiring as regards both quantity and power, as was to be expected, thought Polybios, in a situation in which Hiero had furnished the means and Arkhimedes had designed and constructed the methods. And when Appius' men did get near the wall they were so severely punished by the continuous volleys of sharps from the loopholes that their advance was checked. If they attacked under cover of shields, they were destroyed by the stones and lead dropped upon their heads. The Syrakusans also inflicted damage on Roman troops by the iron hands hanging from cranes, for they lifted up men, armor and all, and then let them drop. Appius eventually retired to his camp and called a council of his military tribunes, at which it was unanimously decided to attempt to take Syrakuse by any means other than storm. And to this resolution they adhered.[52] Ultimately, in 211, Syrakuse fell from within, by treachery, and Arkhimedes died with many of his compatriots when Roman troops poured into the city.

Some of his machines no doubt would have been destroyed, intentionally or accidentally, during the sack of the city. But his genius was recognized within his lifetime and the Romans were never slow to adopt, adapt, and improve other people's technologies if they offered advantages over the Romans' own. It is more than likely that at least some of Arkhimedes' machines were removed from Syrakuse and copied in some Roman workshops, as his planetarium (a device to show the positions of the celestial bodies, possibly a technological ancestor of the surviving Antikythera mechanism) and hodometer (a device to measure distance traveled) apparently were.

MEANWHILE, THE SURVIVING FRAGMENTS of Diodoros' history hint at the relentless struggle for power still being fought out in the eastern Mediterranean by Alexander's successors over a hundred and twenty years after his death (Map 5). In 201 B.C. Philip V of Macedon attacked the kingdom of Attalos I, "even to the gates of Pergamon." In 196 B.C. Antiokhis III (the Great) was in Thrace, refounding Lysimakheia. In 192 Antiokhis' forces were as far south as Delion, near Oropos, in the borderlands between Athens and Thebes, and on Euboia. He himself overwintered in Thessaly, at Demetrias. The Romans intervened and in 191 Antiokhis was defeated at Thermopylai. In 190 he suffered a naval defeat at Myonnesos, as a consequence of which he pulled out of Europe, ordering his supporters in Lysimakheia to move to Asian cities. In the settlement of this war, Pergamon got control (in theory, anyway) of the territory up to the Taurus mountains, and of Antiokhis' elephants.[53]

For new powers were arising in the interstices of the successor states, taking advantage of the chaos and anarchy created by the endemic violence and insecurity to establish control over first a little patch and then, if successful, a bigger one, and sometimes even going on to challenge the once-mighty kingdoms themselves. That was precisely how the kings of Pontos arose (mostly called Mithridates, from 302), and they were followed by the kings of Bithynia (mostly called Prusias and Nikomedes, from 298 B.C.), the Attalids of Pergamon (from 283 B.C.), the kings of the Indo-Greek region (starting with one Diodotos in 256 B.C.), the kings of Kappadokia (mostly called Ariarathes, from 255 B.C.), the kings of Armenia (Tigranes and others, from 189 or thereabouts), and the kings of Commagene (mostly called Mithridates and Antiokhis, from 170 B.C.). It is not coincidental that most of these king-

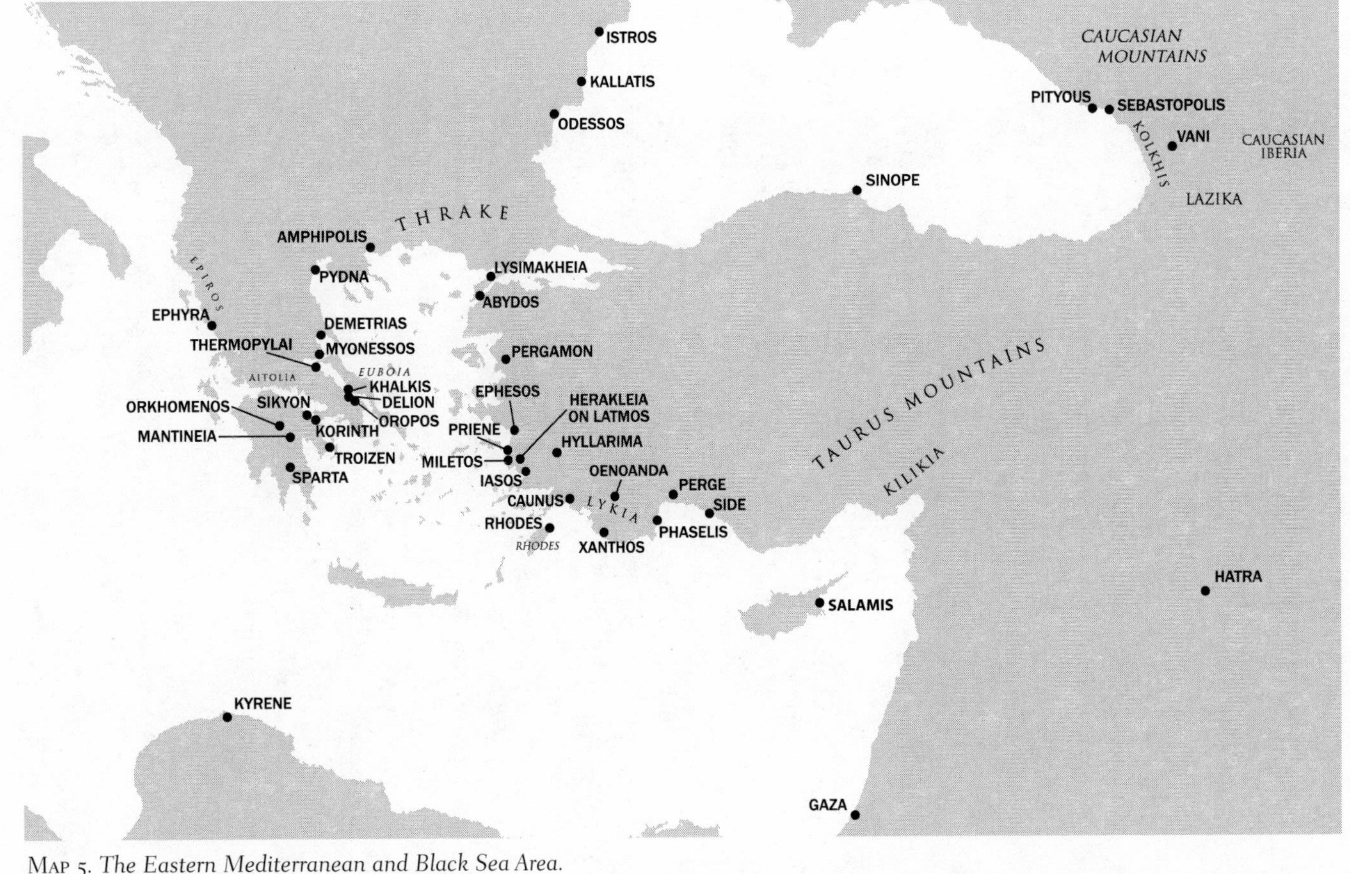


MAP 5. *The Eastern Mediterranean and Black Sea Area.*

doms are in Asia Minor, the territory between the Antigonids in Macedonia and the Seleukids in Mesopotamia. The areas where these kingdoms arose either were left to their own devices or were periodically invaded by the mercenary armies of overextended kingdoms that apparently cared little for the lives, liberty, and possessions of their inhabitants. It was perhaps inevitable that such inhabitants would wish for independence from such overlords, would eventually produce a leader up to the task, and would prepare to keep their own order and fight their own wars instead of other people's orders and other people's wars. Arms production must have flourished in these areas and others like them. Such decentralized arms production probably fostered the development of new weapons, as well as renewed production of old-fashioned but simpler and cheaper weapons.

The Pergamon relief is one of the few surviving depictions of catapults on stone (Figure 6.3), and it was carved for Eumenes II, who reigned there 197–159 B.C.

The cuirass, shields, greaves, sword, and spears all appear to have been depicted at the same scale. There is no reason to think that the catapult is depicted at a different scale, so it represents a catapult with a spring length slightly shorter than a greave. To get a feel for the size, suppose the greave is two feet long, then the catapult, if a palintone, would have a spring diameter of approximately 61 mm which, it so happens, is the diameter of the Pergamon washer (see Appendix 1). This machine, if built to Philon's formula or something close to it, would have been about 775 mm (2´6˝) wide, and have a stock 1170 mm (3´10") long. If it was a euthytone, then we are looking at a slightly *anatonic* (high-sprung) two-span sharp-caster, which would have been about 735 mm (2´5˝) wide and 800 mm (2´7˝) long.

Before 200 B.C., there were missiles with solid iron heads over a meter long, presumably made for appropriately sized sharp-casters to hurl them. From Vani in Kolkhis (inland from ancient Phasis, Poti in modern Georgia, on the eastern coast of the Black Sea between Russia to the north and Turkey and Armenia to the south) comes an astounding sharp: 1040 mm in length (3´ 5˝), 70 kg (154 pounds) in weight, and so large and heavy that the excavator considered it the head of a battering ram (Figure 6.4).[54] However, this looks like an enormous bodkin point rather a ram, and the non-equilateral shape of the socket (which is some 30 cm (1 foot) across) would allow it to lie flat on the stock, in the same way that a regular bolt's three flights are perpendicular to one another, leaving one side unflighted for laying in the groove of the catapult's slider.

Assuming that this giant head, more than 2 cubits long on its own, was mounted on a shaft at least as long as itself, this could perhaps be the only known remains of a 5-cubit sharp, the sort thrown by the largest standard

FIGURE 6.3 *The Pergamon relief.This relief depicts (from left to right) greaves, spears, shields, torsion catapult, and breastplate with sharps behind and sword in scabbard above. If they are all depicted at the same approximate scale, as they appear to be, and the catapult was built to formula, then it was a two-span euthytone or something similar. The (lost) Pergamon washer was the same sort of size. (T. D. Dungan)*

sharp-caster. If so, however, the euthytone formula's shortcoming (being based on length instead of weight of missile) would quickly become apparent, for this head weighs close to three talents, for which a 14-daktyl spring diameter would be hopelessly inadequate. One would need a 3-talent palintone to launch this properly. Could that explain why the head was buried in the filling below the floor level of what is interpreted as a temple complex near the saddle of the hill, dated to the second century B.C.? The fact that this mass of iron was discarded, apparently deliberately (in which case "deposited" is the right term), is in itself significant. Why did they not recycle it? Was it a trophy, shot in from outside by enemies that nevertheless lost to the Vanians, or was it brought in as booty? Or was it of the Vanians' own manufacture, discarded (albeit with religious fanfare and froth) because it was impractical? Was this another experiment that failed? The massive "censer" or smoldering-matter container (firepot?) found nearby perhaps suggests as much. Such a missile would surely have been attention-getting (a "shock and awe" weapon of the second or third century B.C.), but bigger is not necessarily better in catapult technology. Very big stone-throwers may have been built in the Hellenistic period, but they soon fell out of use and were not the preferred weapon of the Roman legions. Perhaps the same was true of sharp-casters, and that too was something that could only be found by experience. The discovery in the same area of a 620-mm-long solid iron "arrow" with iron vanes and nock also suggests strongly that powerful sharp-casters were in the local arsenal. The 620 mm "arrow" appears to be a slightly short version of the popular 1 1/2 cubit, or 3-span caliber.[55]

Pityous (Pitsunda), somewhat further north round the eastern end of the Black Sea, is the findspot of one of the few remaining catapult parts, a washer of 84 mm diameter, but dating to the fourth century A.D. Pityous was a fortress in Lazika, near Caucasian Iberia (ancient Kolkhis area); this and Sebastopolis were the only two foundations maintained as garrisons along this stretch of the Euxine coast up to Prokopios' own day.[56] The Alans, against whom Arrian campaigned and recorded his experiences in the second century A.D., were just the other side of the Caucasus mountains (on which see further Chapter 9).

THE HISTORY OF THE CATAPULT in the Hellenistic period is not only a story of massive machines and outstanding engineers. Less spectacular but in some ways more significant developments were made then too.

The Ephyra catapults were in operation during the Third Macedonian War (171–168 B.C.). About 600 m to the south of the ancient settlement of Ephyra in Thesprotia, Epiros, stands a large building, some 72 feet square, with massive polygonal stone walls some eleven feet thick. It is located on the top of a low hill that commands the surrounding countryside in every direction. It was destroyed in 167 B.C., when the Romans sacked the whole region to try to bring order to the area following the end of that war (Polybios characterized the Epirotes and their leader at this time as lawless and ungodly, and only slightly better behaved than the Aitolians). This building, despite its obvious defensive characteristics, was identified by its initial excavator as a religious site, the Nekromanteion, where people wishing to consult the dead could do so; the river Acheron, where Odysseus went to speak to the spirits of the dead (Homer, *Odyssey,* book 11), flows in the plain below. The twenty-one metal catapult washers, unrecognized as such, were initially interpreted as parts of some mechanism by which the priests could maneuver models and images of spirits of the dead up out of the ground (or rather, from the basement) to hoodwink their superstitious visitors. Now it is possible that the site was the Nekromanteion, since that ought to be near the river Acheron, and a religious site might well need defending in this period, but the mechanical pieces were definitely not from some device of priestly trickery; they were parts of catapults. More recent investigations of this site and others in the area (the Nikopolis Project, 1991–1995) suggest that the building was a fortified farmhouse.

At least seven catapults, which had evidently been located on the ground floor or upper floor(s) of the building, fell into a subterranean cellar when the building collapsed, on fire. It is not surprising that the remains of these

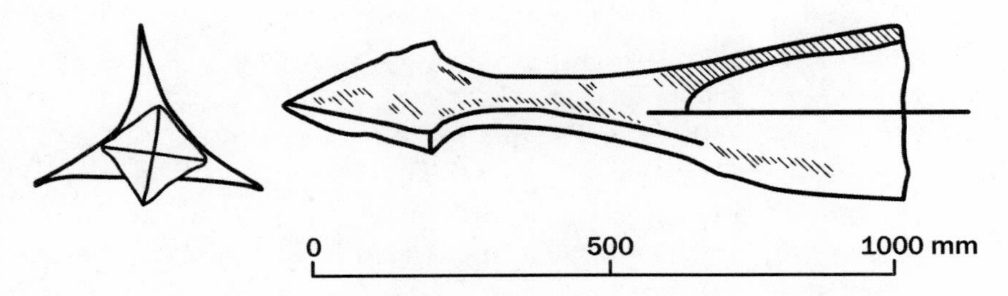


FIGURE 6.4 *Vani megabolt. This massive iron socketed head could have tipped a five-cubit sharp. A solid iron "arrow" some 620 mm (about 1 1/2 cubits) long was also found in the same area. (T.D. Dungan after Lordkipanidze)*

weapons were not recovered and thus entered the archaeological record, for not only had the building burned and collapsed (destroying the wooden and sinew catapult components and leaving even the metal remains in very poor condition), but the whole region was devastated by the Romans. Some of its political leaders were killed and others who aroused suspicion were taken to Rome, and many of its people were marched off into slavery (150,000 according to Polybios), leaving Epiros in a state of chaos and its remaining inhabitants terrorized by a new and ruthless pro-Roman leader. Plutarch described it as the crushing of an entire nation. In the region more widely, scores of cities were destroyed, Macedonia and Illyria were divided into protectorates, and other areas of Greece gave up hostages—this was when the historian Polybios was taken to Rome, as one of hundreds to guarantee the good behaviour of the Akhaians. After three wars with the natives, the Roman authorities would simply not have allowed those who remained to rebuild a strong defensive site such as Ephyra, and Epiros was still "unsettled and disturbed" in 157 B.C., according to Polybios.[57] The ruins remained ruins.

This was a tragedy for the Ephyrans, but is a blessing for us, because no other site has yielded as many catapults from one place and time. The Mahdia shipwreck held three; that too was evidently an irrecoverable loss. Cremona was an important military base destroyed by Vespasian in 69 A.D., and later restored; its ruins hid two catapults about which more in Chapter 9. Azaila suffered a similar fate with similar consequences, though about 150 years earlier. Elsewhere we are usually dealing with isolated finds, sometimes just a single washer. It is likely that metal waste was recovered and recycled unless the parts were really lost or of such low quantity or quality that the recovery cost outweighed the value. This is probably also why almost all the known washers come from small or very small caliber weapons. The Hatra ballista—with the largest surviving washers, at 160 mm or 6 1/2 inches—fell or was pulled from a tower in A.D. 241 when the Sasanid king Sapor I destroyed the city and its kingdom. Hatra, always remote, was thereafter deserted, and the ballista vanished under the sands of this semi-arid land some 50 miles (80 km) south of Mosul. (More on this in Chapter 9.)

The Ephyra washers (Figure 6.5) represent a variety of calibers, from the smallest known surviving, at 34 mm (catapult no. 6), to the second-largest surviving, at 136 mm (catapult no. 3). These catapults were in action about a generation after Philon wrote, and may therefore be expected to have been built to designs and standards very similar to, if not identical with, those described by him. Converting washers into weapons, the 34 mm torsion catapult, if built to formula, and if a stone-thrower, would have shot 4-drachmai ammunition, weighing around 17 grams or 1/2 ounce, and would have had a stock length of 64 1/2 cm or 2´2˝. We could call this a torsion gun, shooting what were essentially bullets (made of cast lead, probably). Alternatively, if this little catapult was a sharp-caster, it would have launched a 306 mm (12˝) long sharp, and could have had an even shorter stock length of 544 mm (1´ 9˝). We could call this (following Baatz) a torsion crossbow. At the other end of the range, the 136 mm weapon, if built to the Philonian palintone formulae, and if a stone-thrower, would have shot 3-mina ammunition, weighing around 1 1/4 kg or 2 1/2 lb, and would have had a stock length of about 2.7 m or 8´9˝. If it was a euthytone, it would have shot a five-span bolt, about 1.2 m or 4´ long, and had a stock length of 2.2 m or 7´2˝. If it was a palintone and was equipped with alternative slider and bowstring, it could have shot both 3-mina stones and 5-span sharps, though not simultaneously of course.

Between these two calibers, one very small and the other alarmingly large (as a sharp-caster at least), were five other torsion catapults, one 56 mm (Ephyra 7), one 60 mm,[58] one 75 mm (Ephyra 5; this is the popular three-span sharp-caster model, with a stock about 1,235 mm or 4´ long), and two at 83–84 mm (Ephyra 1 and 2). The strongly fortified stone square building popularly known as the Nekromanteion was apparently equipped with a range of anti-personnel weapons that between them could discharge ammunition that ranged in size from about one-foot bolts to four-foot stakes, as sharps, and from acorn- to orange-size shot. None of these catapults were big enough to smash a one-talent catapult of the sort Philon recommends for offensive sieges, but the largest machine could have inflicted serious damage on an unplated siege tower and would easily have destroyed the average siege shed or shelter and punched a hole in any *testudo* (tortoise) formed of legionaries' shields. The largest four of the Ephyra catapults would have needed a stand, and perhaps two men, to operate, while the smallest three would have been hand weapons, each wielded by one man. Ephyra 4 would have had a stock length of 3´10" maximum, if built to formula. There is, of course, no reason

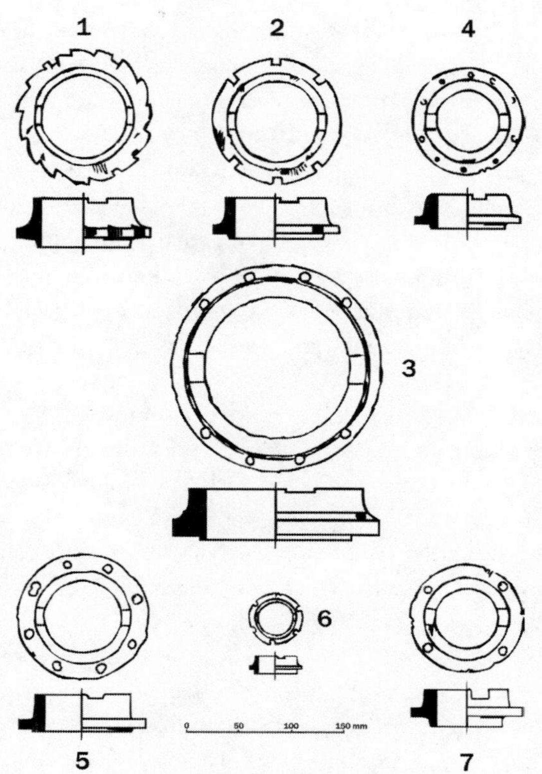


FIGURE 6.5 *The Ephyra washers. Twenty-one washers from the site represent at least seven different machines. Note the variety of features, as well as size, such as the fixing mechanisms: we have ratchet, circular pin holes, squared pin holes, and holes inset or along the edge. Note also the variation in the number of pin holes, the angle of the flange, and the relative depth of the recess for the lever. (T.D. Dungan after Baatz).*

to think that all Ephyra's catapults fell down into this cellar when the building burned and collapsed, and the seven that did may or may not be a representative sample of the whole armament at the site when it fell.

In the triumphal parade in Rome, to celebrate his victory over Perseus of Macedon and the end of the Third Macedonian War, Aemilius laid on a display of spoils he took from the defeated. Now there are divergent stories about this parade, so, for example, according to Diodoros, who was writing about 130 years after the event, the captured arms and armor were exhibited on the first day, while according to Plutarch, writing nearly three centuries after it happened, on the first day was displayed more than 250 wagon-loads of art works—statues, statuettes, and paintings—and the arms were displayed on the second day.[59] However, on the contents of the parade, if not the order in which they appeared, the sources seem to complement one another rather than contradict. Two thousand four hundred wagons full of shields passed by spectators, and 800 panoplies (Greek suits of armor, consisting of helmet, breastplate, and greaves), and then there were three hundred wagons laden with λόγχας and σάρισας and τόξα and ἄκοντια; "spears and pikes and bows and javelins" in the standard translation of these terms. We saw in Chapter 5 how "slings" can mean stone-thrower catapults. So, it appears, a variety of terms for sharps can hide sharp-caster catapults. *Logkhas* means spear- or lance-heads; the *sarisa* is the long pike that distinguished the Macedonian phalanx from the Greek; the *toxon* is the bow, though the plural *toxa* can mean bows and arrows; and *akontion* is the diminutive of ἄκων, so means small javelins, or darts. These wagons were presumably carrying all the sharps in the Macedonian arsenal, so these terms must cover not just infantry, cavalry, archer, and light-armed pointed weapons, but also sharp-caster catapult munitions. Ammianus refers explicitly to *tormenta* shooting "arrows" (*sagittae*, e.g., 19.7.4), and there is no reason to suppose that this usage was confined to late antiquity: even modern scholars frequently call sharp-casters "*arrow*-shooters." Catapults may have been much more common in antiquity than they now appear to have been, because ancient authors do not distinguish their missiles from other range missiles, and they are rarely distinguishable in the archaeological record, either. If a relatively insignificant little building in Epiros held at least seven, how many were contained in King Perseus' headquarters, never mind in the whole of his dominions?

### New Developments in Fortifications: Supporting Technologies and Counter-technologies

The formulae that took the guesswork out of catapult construction and enabled the Hellenistic Greeks to produce powerful machines reliably is the subject of the next chapter. But before that, we turn to consider the issue of fortifications, for developments in catapult technology ultimately demanded developments in city defenses, both to house them and to resist them.

Recall that the first engagement in which catapults were used was in 397 B.C. at the siege of Motya, a little islet on the western tip of Sicily, whence Dionysios had driven the Karthaginians across the island. The survivors from Motya settled a place called Lilybaion, a promontory about five miles south along the Sicilian coast (Map 1 inset). Through the next century, the Karthaginians slowly reestablished themselves in western Sicily, but in 277 B.C. history was repeating itself and another strong Greek military commander, Pyrrhus of Epiros, had driven them back to Lilybaion. The city was reinforced from Karthage to such an extent that "there were so many catapults, both sharp-casters and stone-throwers, that there was not enough room on the walls for all the equipment." Pyrrhus' army consisted of 30,000 infantry,

1,500 cavalry, elephants (unnumbered) and "siege engines and many missiles." This time, the Karthaginians' armament was superior to the Greek, and although Pyrrhus constructed on site engines "more powerful than those he had brought from Syrakuse," he still could not take Lilybaion, and he lifted the siege after two months, during which time many of his men were killed and many more injured. About fifteen years later the Romans besieged another Karthaginian city, Myttistraton, about five miles inland from the central north coast of Sicily (Map 1 inset), and after a seven-month siege employing "many siege engines" but nevertheless losing many men, they left empty-handed.[60] Evidently neither the Greeks nor the Romans had stone-throwers capable of swinging a siege against the Karthaginians in their favor in the first half of the third century. When the Romans finally took Myttistraton on the third attempt in 258, they razed it to the ground.

Lilybaion's defenses consisted of walls with close-set towers and an outwork ditch. Engineers and catapult builders like Poluidos and Philon, among others, were thinking about and working on the defensive aspects of catapult technology too. The emergence of powerful stone-throwers in the third century B.C. led to changes in fortifications. We are first told explicitly that the Romans threw down a city wall with machines in 254 B.C. (ταῖς μηχαναῖς κατέβαλον τὸ τεῖχος), though the machines in question might not have been catapults, of course. However, just four years later, in 250, it was the Romans' turn to besiege Lilybaion, and when they "constructed a stone-throwing engine, the Karthaginians built another wall inside the first." Here is the first indication of the existence in Roman hands of catapults that could rock walls—but yet again we see that new technology did not immediately revolutionize warfare for those who were using it. The defenders set fire to the Roman war engines and other siege machines, and according to Diodoros "the Romans were rendered helpless by the burning of their engines."[61] They did not leave, but both sides were periodically reinforced, and this siege went on for ten years, ending in a peace treaty.

Hitherto, simple single-skin walls had been quite adequate for most purposes. Carefully cut, close fitting, largish stones were laid on top of one another, without mortar, and were usually smooth faced to hinder attempts to scale them. Such fortifications, designed to stop infantry with spears, swords, and slings, can look remarkably flimsy to the modern eye. In remote areas where ready-cut stone has not since been required and taken for other purposes, ancient curtain walls may still rise meters high yet be only one stone thick, the wallwalk that once accompanied them having been made of wood and supported by nothing more durable than timber scaffolding. With the advent of stone-throwing catapults that were capable of shaking such walls—of hitting with such impact that they could dislodge a stone and bring a section tumbling down—walls became thicker. Parallel single-thickness stone walls were filled with rubble and bonded with huge stones laid crosswise, to make walls that are meters wide rather than centimeters, where there is room to stand on top of the wall, instead of having to walk on planks beside it. Towers, which had long since been constructed to strengthen weak points, guard gates, give a further height advantage to the defenders, and to protrude beyond the walls allowing defenders to rain missiles down on those who attacked the walls to either side of the tower, remained single-walled for as long as catapults were not sufficiently powerful to threaten them. First-generation artillery towers have thin walls and, on their top floors, open "windows" that served as embrasures for catapults (see Figure 3.1, Aigosthena). With the advent of powerful, formulaic, large caliber torsion stone-throwers, towers become progressively more squat, trading height for solidity, because those either side of an intended breach point became the principal targets of offensive artillery bombardment. Sometimes towers are not even tied in to the masonry of the curtains, so that if the tower falls, it does not pull the curtain down with it (and vice versa). This is not to suggest that artillery bombardment could bring down a well-made thick wall or tower. But there was no room anymore for flimsy fortifications, which could have been rendered unstable and unsafe, if not actually brought down, by large caliber stone-throwers built to formula.

Nonrectangular designs of towers and bossed masonry became increasingly more common from the third century B.C., reflecting or reflected by the theories of engineers like Philon to minimize the effects of bombardment. Besides exposure to enemy bombardment, other matters Philon considered important in the design and location of towers are the desired field of fire from its embrasures, the ability to enfilade the curtain wall, command of forward ground or access routes, and liability to attack by sapping (undermining), siege towers, or rams. Undoubtedly, some sites that were deemed indefensible against the new weapons were abandoned.[62]

Fitting catapults to towers can resemble a scholarly equivalent of designing a new kitchen with scaled cut-outs to see what might fit where, while keeping the sink by the window and still leaving space to shut the door. The possibilities are numerous. For example, Kirschen suggested that Herakleia-on-Latmos' best-preserved tower might have housed lots of small catapults on each floor;[63] Ober that it housed two modest (10-mina) stone-throwers; and McNicoll that the top floor housed one large ballista shooting large caliber shot. Sharp-casters seem to serve the function of space-fillers in this game, for no obvious reason. Given that all literary sources always report many more sharp-casters than stone-throwers, especially in the significant cases when the total armament of a city was revealed after it fell, this game, if played at all, should perhaps be played with rules that reflect a ratio of sharp-casters to stone-throwers of at least 4 to 1, if not 10 to 1. There is also a noticeable desire by moderns to emphasize size, and scholars who play this game seem to be interested principally in discovering what is the largest possible stone-thrower that would fit in the space available.[64] Overenthusiasm, constrained by floor size alone, sometimes puts quixotically large caliber weapons into such buildings,[65] neglecting other significant factors such as ceiling height and access. For example, if the room is a floor or more higher than the wall-walk(s), and the personnel could only have got in by climbing an internal staircase or ladder (such as at Oenoanda, Perge, and Side), it is more likely that they carried in hand weapons than that they scaled a ladder in order to operate a large stone-thrower that could only have been constructed inside the room.[66] Moreover, some consideration should be given to the ammunition. The obvious has a tendency to be overlooked: a large stone-thrower needs large stones. Gunners can't simply pop up and down ladders with 25 kg stone balls tucked under their arms and load them into catapults as if they were tennis balls. Heavy ammo needs lifting equipment or relays of gunners with the arms of Garth operating on very short shifts, which is not an ideal situation in a crisis, which is precisely when they would be needed. Open-backed towers (as at Herakleia-on-Latmos and possibly at Perge) would in theory allow machines to be installed and removed at will, but they raise architectural problems regarding the flooring of the top chamber—the floor on which said catapult is supposed by some to be sitting.[67] Such a supposition also makes two big and very questionable assumptions: (i) that strong enough cranes were available to lift large caliber stone-throwers in and out (they would have been very heavy), and (ii) that there was no roof or a removable roof to the chamber in which said catapults were installed (so that the crane could operate). From the defenders' perspective, the most important function of a large stone-thrower was to destroy enemy siege engines.[68] If the machines were built-in to tower rooms, as some scholars suggest, their effectiveness would depend entirely on the enemy bringing their machines up in range of a tower that was large enough, and with big enough windows, to hold a stone-thrower that could destroy them. Relying on the enemy's stupidity is not a strong tactic for the defense, and there is no evidence that any ancient ever did this. There is instead evidence that large catapults were not sited in towers.

Philon explicitly recommended that large stone-throwers should be movable, not built-in to a tower. He says further that the largest catapults should be mounted on emplacements at the *foot* of the wall and outworks, where they have plenty of room. He says, "it is necessary to oppose each [enemy stone-thrower] with two 10-mina stone-throwers, *which must be moved to whatever point* the enemy move one of their stone-throwers, so that you may shoot at it, break it and smash the engine."[69] Moreover, stone-throwers can be set to high trajectories, and their missiles are ruthlessly destructive to siege engines (but not walls) when falling in a high arc, so from ground level they can lob stones over the wall from inside to out as effectively as enemy stone-throwers can lob them from outside in. They do not need the height offered by towers. Indeed, the range is far greater if they are laid at 42° or thereabouts than if the entire machine is raised to the top of the highest tower and then is shot horizontally—though they have less force on impact as a result, losing about 30 percent of their energy to wind resistance. In some towers either the window is too small or low or there is not enough clearance from floor to ceiling to set a decent elevation for a large catapult, a problem that grows with the caliber of machine and the consequent increase in length of the stock.[70]

Further, if something went wrong with a large stone-thrower discharged within a tower—if, for example, an arm broke, or a spring, as they often did, according to Philon—the misfired ordnance might demolish part of the tower—not a desirable scenario at any time, but especially not when an enemy was trying to achieve the same from the outside. Schramm apparently nearly killed Kaiser Wilhelm II in 1904 when demonstrating one of his large stone-throwers (in the open), when the sling slipped under the stone and cast it heavenward instead of forward. An artilleryman of Ammianus' acquaintance was disintegrated in a similar situation. Imagine the chaos caused by even small shot discharged within a room if it misfired and did not exit the room through the embrasure.

There are thus many intrinsic reasons why stone-throwers, especially large ones, would have been better placed on the ground and in the open than in towers. Further, large stone-throwers are not useful as anti-personnel weapons, therefore they are inappropriate for enfilading fire along a wall, which is the function of most towers—the apertures to the flanks, overlooking the walls, being as important as those to the front in most towers. Small sharp-casters had another advantage over stone-throwers in towers, especially when the enemy was close; the nock of an arrow could hold it on the bowstring even when the catapult was aimed downward, whereas there is nothing to prevent shot falling out if a stone-thrower is depressed. Marcellus supposedly thought that by getting close to Syrakuse's walls he would get "under" the range of Arkhimedes' catapults, only to find that at least some of Arkhimedes' machines could shoot down there, too, and now he was in range of smaller machines shooting through loopholes in the lower courses of the walls as well.

The offense needed to get their stone-throwers close enough to shoot horizontally if they hoped to have an impact on a stone wall, but the defenders did not need the force of a horizontal shot to destroy wooden siege engines. Stones simply dropped from above will do the job, if only one can get a sufficiently big stone to the target. A stone-thrower catapult, if compared with, say, Arkhimedes' stone-dropping crane, trades the size of stone (smaller) for force on impact (greater), and delivers it at a distance from the wall, instead of immediately beneath it. Any machine that is at the base of the wall is already doing damage to that wall, so it is better to destroy enemy engines at a distance, if possible.

Philon's catapult emplacements in front of the walls are the most obvious place to put stone-throwers on the ground. The platforms that were once identified as "engine stands" *within* curtain walls are no longer thought to be such. McNicoll first pointed to the problems.[71] These platforms, known from a handful of sites (Ephesos, Herakleia-on-Latmos, Miletos, and Seleukia Pieria), have embrasures in the wall that are the wrong size and height for the machines that were imagined to have sat on them to shoot through. Consequently, the platforms are now interpreted as the floors of guardhouses, dungeons, or suchlike. Much work remains to be done on matching catapults to so-called artillery emplacements.

SEVEN

# The Technical Treatises

*It is universally agreed that in the technical arts (among which we include the invention of weapons) progress is due not to those of the highest birth or immense wealth or public office or eloquence derived from literary studies but solely to men of intellectual power.*
—Anonymous *De rebus bellicis* Preface 4

We owe most of what we know about ancient catapults to four men: Ktesibios, Philon, Biton, and Vitruvius. Ktesibios probably lived about 270 B.C., Philon about 200 B.C., Biton anywhere between about 250 and 133 B.C., and Vitruvius about 50–25 B.C. (Heron, who preserved Ktesibios' work, lived about A.D. 60.) Here we examine the Greek works; Vitruvius will be discussed in Chapter 8, and Heron's short text on the *Kheiroballistra* in Chapter 9, because their brief accounts are our chief sources of evidence for Roman developments, and it does not do to conflate their evidence with that of the Greek authors who are writing about different machines in a different world. Likewise, the anonymous author of the *De rebus bellicis*, Ammianus Marcellinus, and Vegetius are reserved for Chapter 10, Late Antiquity. Here, in this chapter, I will quote extensively from the Greek authors, because what they say is the foundation of all histories of the catapult, and the history of ancient technology more generally. Modern interpretations of the texts will change; the texts are the constants. No other branch of ancient technology is blessed with texts like these explaining the structures, functions, and purposes of ancient machines.

My purpose in this chapter is to discuss the technical treatises as evidence in the history of a technology; it is not to reproduce them. Nevertheless, it is in these works, which were written precisely to enable readers to build good catapults, that the necessary information is given, albeit in a form rather indigestible to modern tastes. Parts, materials, and methods are all discussed in more or less detail, and demand more or less commentary. Moreover, to get any real feel for the size of the machines in question, one has to compute their overall dimensions using the given ratios. These highly technical matters I have reserved to an Appendix, where I have also given tables of calibers for the principal "standard" machines, the palintone and the euthytone, based on Philon's treatise, so that one can see at a glance what size machine might have shot what size and type of ammunition. Here, we look at the men who wrote these treatises, at their historical contexts, and at their own views about what they were doing, and why.

Our story begins in Alexandria in Egypt. Ktesibios was a Greek from this Alexandria, Alexander the Great's most famous and successful eponymous city (he founded seventeen Alexandrias altogether). When Ktesibios walked its streets, the city had been in existence for a mere sixty years or thereabouts, but it was expanding rapidly. It had a cosmopolitan mix of Greek, Egyptian, and Jewish inhabitants, and much of it must have resembled a building site. After Alexander's death, Ptolemy, a friend since his youth, one of his right-hand men and a highly successful military commander in his own right, seized this part of Alexander's empire. The new Macedonian ruler of Egypt adopted the structure and trappings of what was already, at that time, a very ancient monarchy. Perhaps showing a lack of confidence that is characteristic of the nouveau riche, this immigrant family followed Egyptian custom in numerous ways, including in their portrayal in art and literature. More unusually, they barely acknowledged any discontinuity through the generations, each successive Macedonian pharaoh bearing the same name as the first in the dynasty. Pharaoh was Ptolemy, and Ptolemy was pharaoh, always, as the identity of the individual was subordinated to the interests of the dynasty. Outsiders gave them additional or surnames to distinguish between them.

The first Ptolemy, surnamed Soter (Savior), self-styled King from 305 until his death in 282 B.C., built up Alexandria, and made it the capital of his domain, which extended beyond Egypt, variously, as wars with others of Alexander's successors allowed. The second Ptolemy, surnamed Philadelphos (Sister-lover),[1] who reigned jointly with his father from 285, and solely from 282 until his death on 29 January 246, built up the Museum and Library, both of which were inaugurated by his father but grew rapidly under the son. This man, Ptolemy II, was probably King when Ktesibios was in his prime and wrote his *Notes* (Ὑπομνήματα).[2] Ktesibios worked in a number of engineering areas, mechanical, pneumatic, and hydraulic, and the *Notes* probably covered all major aspects of his work. They have not survived the two-thousand-year interval between his time and our own, but Heron's edition of that part of the *Notes* that was concerned with war machines, *Ktesibios' Belopoiika*, does survive.

Like Degering and Marsden,[3] I believe that Heron's edition corresponds very closely indeed to Ktesibios' original; in fact, I think that fewer anachronisms are liable to arise if we call this work Ktesibios', rather than Heron's, and that is what I have done throughout this book. The most important reasons for this are three. First, all machines mentioned in it are earlier than the machines that Philon describes as standard for his own time, c. 200 B.C., so the work belongs, historically, in the third century B.C. Second, Heron appears to add nothing to the work. He does not attempt to bring it up to date by adding any descriptions of artillery in his own time, about A.D. 60, which would have shown advances on those described by Vitruvius, for example. Nor does he supply the sort of detail that he gives in works of his own authorship: neither technical details such as we find in his *Kheiroballistra*, nor mathematical precision such as we find in his *Metrika*. He does not even bring technical terms up to date, so we find, for example that "trigger" is σχαστηρία in Ktesibios' *Belopoiika* 78.2-3 (as it is in Philon's *Belopoiika*, and in a fourth-century B.C. inscription) whereas it is δρακόντιον in Heron's *Kheiroballistra* 126.2–3). Third, Heron subtracts nothing from the work. *He* could not really claim, as is stated in the introduction to this work, that no one before him had written on artillery construction for the public, though Cuomo (2002) makes a valiant attempt to justify it in the context of Heron's time. Thus, I think that this is Ktesibios' voice, not Heron's.

Since Heron's role with this text was therefore essentially as publisher, we have to ask, "why did he republish Ktesibios' centuries-old *Notes* on war-machines?" Many reasons have been advanced in answer to this question. For example, Marsden argued that he did it because it was still the best account of the subject for the layman; Cuomo argued that he wanted to show how geometry, retrospectively, provided solid foundations for solutions that were initially found empirically.[4] Both of these may be right. In any case, Heron must have expected it to be either sufficiently interesting or sufficiently useful to many to justify his efforts. Another reason may be because the zenith of war machine construction, in some ancient scholars' estimation, came in the early Hellenistic period, the era of Demetrios Poliorketes, when Ktesibios was alive, rather than later, and no one (then or now) would locate it in Heron's time. Perhaps Heron thought so, too. Certainly, the biggest machines were built in the late third and early second centuries, not later. Ktesibios not only lived at that time but was recognized then, as he is now, as one of the most brilliant and original engineers of antiquity. So intellectuals like Heron may have considered his work, though old, as at least as good as, if not better than, then-current treatises.

Ktesibios (for thus shall I style its author, with Heron given *after* the title, as is conventional with a modern edition of an old work) says that people before him had written many documents on making war machines in which they gave measurements and the arrangement of components (73.6–7). He continues that these writings assumed expert knowledge of the subject by the reader and did not bother, for example, to describe the construction process from start to finish. Ktesibios sells his own work as a supplement to these documents, explaining the basics such as nomenclature, and the original motivation for each engine he describes.

> Writers before us have composed many documents on the construction of war machines dealing with measurements and designs, but none of them describes the construction of engines by type, or their uses, but they apparently wrote for those who understand it all. We consider it beneficial to supplement their work, and to clarify things with respect to war engines, perhaps even those out-of-date, so that the exposition will be easy for everyone to follow. Therefore we shall speak about the construction of complete engines, and the individual parts thereof, about their names, and about their assembly and cord-fitting, and further about the uses and measurements of each, speaking first about the differences between the engines and the first development of each of them. (Ktesibios *Belopoiika* 73.6–74.4 Heron)

Five points arise from these introductory remarks.

1. The earlier writings were documents that did not share Ktesibios' pedagogical aim—to make the subject intelligible to everyone. So what was the intended audience for those works? Why were they written at all? The obvious answer to this question, it seems to me, is that these writings were essentially the calculations and notes (and possibly drawings too) that master craftsmen would have made prior to making the components and assembling a new machine. There would be good reason to keep and copy them, lest the whole machine or one or more of its components needed rebuilding at some time in the future. Arms, springs, and stanchions were prone to breakage. Fire was the principal weapon against siege engines, and after a successful sally by the defenders, they sometimes needed to be rebuilt, in whole or in part. So notes, figures, and drawings could serve as reference for repairs, or inspiration for future builds. Another reason might be if the designer did not actually build the machine, in which case such documents were more like blueprints, to be handed over to the artisans who did build it. A final possibility for their existence could be to enable the engineer to analyze the design on paper.

2. The documents were apparently not secrets. Ktesibios writes as if such documents were common, if not commonplace. However, they are meaningless to most people, and comprehensible only to those who already understand the principles of catapult construction.

3. Πλείστας ἀναγραφὰς, many documents, implies many different designs. This is in keeping with his opening remarks, suggesting that those documents were specifications for individual machines, not abstract formulae to apply as required.

4. This claims to be the first attempt at synthesis. As such, Ktesibios was probably the first author on machines to abstract from specific details on materials and dimensions, which would only ever be appropriate to specific machines. Some apparent omissions and tendency to incompleteness may arise from this.

5. Ktesibios recognized that knowing the history of their designs—"the causes of their development" or why certain features were introduced—helped people gain a proper understanding of the catapults of his own day, and as such I think he should be considered the father of the history of technology.

Ktesibios explains the creation of artillery thus.

> Originally, the construction of these engines developed from hand-bows. As men were forced to use them to project a somewhat larger missile and at greater range, they made the bows bigger, and the springs in them—by "springs" I mean the curvy bits [to the handle] from the tips, from the stiff parts made of horns. Consequently, these bows were very difficult to bend, and required more force than was possible pulling with the hand. To this end, they devised a machine like this [the *gastraphetes* follows]. (Ktesibios 75.3–9 Heron)

After giving a description of the construction of the gastraphetes, he moves on to the next development: the torsion engine.

> By means of the [*gastraphetes*] a larger missile was projected, and to a greater distance. Wanting to increase further both the missile and the discharge, they investigated making the arms of the bow more powerful, but they were not able to achieve their intention through the horns [bows]. So they did the other things as described above, but manufactured the arms from very strong wood and made them bigger than those in a bow. They knocked up a frame of four strong beams . . . [description of simple spring frames follows]. (Ktesibios 81.3–9 Heron)

Ktesibios glosses over the radical change here, offering no explanation at all for the origin of the torsion spring. He does, however, show how snags discovered in the process of manufacture, or use, led to modifications and refinements:

> When it turned out in the construction mentioned that the twisting and stretching of the cords could not be carried very far because the cross-pieces AD and BG could not take the spring-cord, they put levers over the holes, and made them the same as those mentioned. Then the twisting of the lever was difficult again because it rested on the cross-piece and touched every-

> where. Therefore, they were compelled to add washers, as we call them. (Ktesibios 83.5–11 Heron)

Ktesibios explains clearly how and why one thing led to another.

> Since the force of the arms has become powerful, it is necessary to make the pull-back powerful as well, because the same force is needed to pull back the arms. Therefore, instead of the so-called belly-bar (καταγωγίδα, Marsden's "withdrawal-rest"), they added an axle to the end of the case, at a right angle to it, and easily turned. They turn it by square hand-spikes on its ends. As it was turned, the slider and the bowstring it carried were withdrawn. . . . But on the larger engines the pullback of the arms was still difficult. So they withdrew them with pulleys. (Ktesibios 84.1–10 Heron)

Some requirements would have become obvious during the construction process. Others became apparent with usage.

> It is necessary at certain places, I mean those parts that endure some stress, to put small iron plates and fasten them with nails, and use sturdy wood, and secure these parts by all possible means. Make the parts not suffering thus of light and small wood, taking into consideration the whole shape, size, and weight of the engines. Of all made, not many are constructed for immediate use, and consequently, it is necessary for them to be easily disassembled for transportation, and light, and not expensive. (Ktesibios 102.3–10 Heron)

Key discoveries with the sinew for cordage and bowstring were made empirically too:

> It is necessary to use the sinew from the shoulders and backs of all animals except pigs, for they are unusable. It is necessary to understand that the back and shoulder sinews of other creatures are very useful. It has been discovered that the more exercised sinews of an animal are better toned, such as those from the feet of a deer, and those from the neck of a bull. Regarding other animals, think accordingly. (Ktesibios 110.4–8 Heron)

Besides necessary modifications, there were also useful ones.

> So that the slider may be pulled forward without difficulty in the larger engines, or rather, so that it may be drawn forward when the axle is turned the opposite way, let it be as follows . . . (Ktesibios 85.7–8 Heron)

Some features appear to have been optional, and included by some catapult-builders but not by others. Ktesibios' descriptions are consistent with the variation in washer design that we find in the archaeological record—the washers sometimes being made of metal, and thus surviving better than the wooden parts of a catapult. He says (96.6–98.1) washers may be made of metal or wood, tenoned (and thus the washer is fixed to the hole-carrier and does not rotate with the lever) or rimmed (and thus the washer rotates with the lever), and if rimmed, they may be furnished, or not, with a counterplate. Each choice has consequences for the rest of the construction. For example, if tenoned, one must not make recesses in the washer for the lever, because on this model the lever (not the washer) turns (and the lever is held in place by friction). A rimmed washer, however, will need recesses in the top to hold the lever firmly, and there will need to be either ratchet teeth round the washer collar and a pawl on the cross-piece, or additional small holes in the rim and in the frame, and counterplate, if used, to take retaining pins, to stop the washer slipping back after tightening. One would expect the increase in complexity and cost of a rimmed washer to correspond with an increase in performance or durability, but that was not necessarily so, and the additional cost may have been perceived to outweigh the additional benefit.

> We have sidestepped all the securing of the hole-carriers and washers, which costs, for a two-cubit [euthytone], not less than 80 drachmai. (Philon *Bel.* 62.17–19)

In any case, we see that there were simultaneously, for essentially the same design of catapult and the same problems, different solutions that would have suited different clients with different budgets. It is for this reason that I resist the attempt to order the Ephyra catapult washers into a sequential, supposedly chronological, order. I think that from the circumstances of their deposition in the archaeological record, they more likely coexisted.

One passage of Ktesibios' treatise reveals very clearly that he thought trial and error played a major role in the development of the catapult.

> It is necessary to understand that the record of measurements [what are now loosely called the calibration formulae] was obtained from trials alone. The older generation, thinking about shape and arrangement only, did not particularly shine in the projection of the missile, because they did not use harmonious proportions. Those who came later, subtracting here and adding there, created the well-proportioned and effective engines described. The said engines, specifically, all their parts, are based on the diameter of the hole that receives the spring. For the spring is the chief parameter. (Ktesibios 112.9–113.4 Heron)

History seems to confirm his view (Chapters 2–6 above): early catapults did not "shine" in performance.

The calibration formulae follow. The principal one, to establish the spring diameter in the first place, is essentially the same as that specified by Philon (which is discussed in Appendix 1), except that Ktesibios is less precise with respect to the cubic roots when (as is usual) the result is not a whole number. Ktesibios simply tells the reader, "if the product does not have a cube root, it is necessary to take the nearest and add a tenth part."[5] Philon recommends that the tenth be adjusted up or down depending on whether the nearest whole number is more or less, to compensate for the excess or deficit arising from the rounding of the root. Ktesibios then goes on to explain how to find the cubic root, in order to scale an effective catapult up or down, by the method of two mean proportionals. Curiously, he does not give any of the methods now known to have been discovered by his time, but a different one, and the method he uses is essentially the one used by later catapult-builders, suggesting that solutions to applied mathematical problems were handed down within crafts.[6]

The next pharaoh in the dynasty, Ptolemy III Euergetes (Benefactor), continued to support technical and intellectual work in Alexandria. He was an aggressive king who restarted a campaign against the Seleukids in the 240s, taking Phoenicia and a considerable amount of Syria from them. It was during his reign (246–February 221 B.C.), in 238 in fact, that the Canopus Decree was passed, which ordered a calendar of 365 days plus an extra day every four years—a leap year.[7] It was also he who invited Eratosthenes of Kyrene to move to Alexandria, to become tutor to the royal children and to succeed Apollonius of Rhodes (author of the story of Jason and the Argonauts) as Head of the Library. Eratosthenes corresponded with Arkhimedes, and famously calculated correctly (give or take a percent or two) the circumference of the earth. He also devised a new, and a mechanical, solution to the problem of duplicating the cube, and this relates directly to catapults—although catapults were not the only obvious application for this bit of mathematical wizardry. Eratosthenes' method involved a sort of three-dimensional slide rule, by which one could find, relatively easily and mechanically, the solution to the problem of changing the size of a solid without changing its shape. To the nonmathematically minded, this probably appeared, as Eratosthenes himself claimed it was, much more amenable than Arkhytas' cylinder method, Eudoxos' curve method, or Menaikhimos' cone-cutting method. What is particularly interesting is that he set this instrument up in a public place, and dedicated it to the public.[8] It suggests that he expected "the public" to make use of this invention.

The calibration formula for stone-throwers involves cubic roots. To find cubic roots requires scaling in the opposite direction from duplicating cubes. Obviously, a method for duplicating cubes can easily be manipulated to find cubic roots. Assuming it remained in service, Eratosthenes' method would have simplified the business of making different caliber stone-throwers by scaling from an existing model. The use of cubic roots in the calibration formula was probably encouraged by the discovery of simple methods to find them, as it would have been discouraged by methods too difficult for the average artillery-builder to employ confidently. Methods of manufacture in the real world are usually a compromise between the theoretically desirable and the practically achievable. Eratosthenes' discovery of a simple method to find cubic roots, and his construction of the relevant instrument in a public place, could have facilitated the use of cubic root formulae by artillery-builders in Alexandria, and perhaps even encouraged experimentation with new calibers.

PHILON WAS PROBABLY WRITING about thirty years later, when Ptolemy V Epiphanes (Conspicuous) was on the throne (205–180 B.C.), though he may not have written the *Belopoiika* in Egypt. Ptolemy V inherited the throne aged five.[9] Not surprisingly in this circumstance, large parts of the Ptolemaic empire were lost during the early part of his reign: territories beyond Egypt's boundaries, such as Syria, to other Successor Kings, and Upper Egypt itself, to natives in revolt following rival pharaohs, who severed the old ties between Upper and Lower Egypt. This was one of the lowest points in the history of Ptolemaic Egypt. It was a time when the Egyptians fought many sieges, both to retain and to acquire cities, and was a time when Philon's *Poliorketika* would perhaps have been most useful.[10]

Philon came from Byzantion, which probably had a reasonable arsenal of its own. We know it supplied bow-catapults to neighbors in trouble from the fourth century onward, and he seems to have learned more about catapult-construction in both Rhodes and Alexandria.

> We shall communicate to you exactly what we discovered in Alexandria through extensive association with the artisans (*tekhnitai*) engaged in such matters, and through acquaintance with not a few chief builders (*arkhitektonoi*) in Rhodes, from whom we learned that the most highly regarded engines more or less conformed to the design we are about to describe. (Philon *Bel.* 51.10–14)

Rhodes had a significant arsenal at the time. Recall that in 220 B.C. they sent four unusually named torsion catapults, plus personnel, plus prepared hair, plus prepared sinew, to Sinope, which was then being threatened by Mithridates II of Pontus (see Chapter 6).

Philon could apparently move freely between arsenals, even though they were sometimes at or close to war at the time that we think he was active, around 200 B.C. His hometown of Byzantion, as the most easterly city in

Europe and in a strategic position commanding the straits into the Black Sea, was fought over occasionally by various successors, and sometimes maintained a more or less precarious independence. The Byzantines enjoyed good relations with Egypt, some of the time; there was, for example, a temple dedicated to Ptolemy II in Byzantion.[11] They also maintained good relations with Rhodes, some of the time. Rhodes succeeded in maintaining her independence and good relations with both the Ptolemies and Seleukids (and, later, the Romans too), who at times sought to incorporate her into their respective empires. But this world was very unstable, and both enmities and alliances could be short-lived. For example, Rhodes declared war on Byzantion in 220, when the Byzantines were said to be fleecing traders operating in and out of the Black Sea and the Rhodians rattled their sabers as policeman of the seas.[12] Thirteen years later, in 207, Rhodes and Byzantion were working together with Ptolemy IV to try to mediate a peace deal between the Aitolians and Philip V of Macedon (and thus deny the Romans an excuse to invade Greece). In 202, after some provocation—including an act of sabotage by an "architect" from Tarentum[13] who was acting for Philip V, that destroyed by fire thirteen of their ships and ship-sheds—Rhodes declared war on Philip, who by now controlled Byzantion. In the next year at the battle of Khios, Byzantine contingents were fighting with Rhodes and Attalos I of Pergamon against Philip.[14] Either Philon's ability to move between and study catapult-making at Byzantion, Rhodes, and Alexandria was unaffected by these events, or we have his dates wrong.

Philon's treatise on artillery construction (*Belopoiika*) is by far the longest and most precise of those remaining on the topic. It was part of his compendium on *Mechanics*, which included, for example, books on pneumatics and automata and what we now call mechanical engineering, and harbor construction and city defenses and what we now call civil engineering. The work on city defenses, the *Poliorketika*, offers further information of use to us; we will have cause to cite it later. Philon's account of the catapult development process is fuller than Ktesibios'.

> In the old days, some engineers discovered that the fundamental element, principle, and measure for the construction of engines was the diameter of the hole. It was necessary to acquire this not by chance or at random, but by a certain method that made it possible to make analogous engines in all sizes. This was found by nothing other than increasing and diminishing the perimeter of the hole. The old timers did not take it to the limit, as I say, nor establish the parameter, since their experience was not based on much work; however, they sought the key feature. Those who came later, carefully considering earlier mistakes, and looking hard for a fundamental factor from the

> experiments thereafter, introduced the principle (*arkhe*) and science of construction, which I say is the diameter of the circle that holds the spring. Alexandrian artisans came upon this first, having great wealth, because the kings they had loved glory and technology. (Philon *Belopoiika* 50.14–26)

I read this to mean that Alexandrian engineers were well paid and encouraged by the Ptolemies (recall Dionysios I's treatment of artisans in Syrakuse), rather than that the Ptolemies employed them in some kind of state-owned artillery research establishment. The image conjured up by the latter reading is a modern one that has no place in the ancient Greek and Hellenistic world. There is no evidence, literary or archaeological, for such an institution in this time; in fact, there is almost evidence of absence. For example, when Strabo describes the Seleukid military base of Apamaea, he talks of the 500 elephants, the larger part of the army, the war office, the royal stud of more than 30,000 mares and 300 stallions, the horse breakers, drill-sergeants, and all paid military instructors (16.2.10); he makes no mention of artillerymen, engineers, or still less of weapons research establishments.

A natural-born engineer would want to experiment with different designs, as a natural-born poet would want to play with words and a natural-born mathematician would want to play with shapes or numbers. Combine those inclinations with a phenomenally wealthy patron who was prepared to pay generously for new weapons, as he was for new poems, new syntheses, new pictures, new animals, new machines, and new medicines (among other things), and you have ancient Alexandria at its height. The Library and Museum was not a university with laboratories and faculties employing Greek geniuses; they were part of the Palace—Pharoah's library, his temple of the Muses (= Museum), his zoo, his garden. A few people were employed to work in these facilities; most no doubt were slaves. The Librarian's was the only post of distinction. It may have involved a salary. Intellectuals were in this city, selling their wares just as doctors and potters and cobblers sold their wares; the essential difference between the men we speak of and other craftsmen was the specialized nature of their product and thus their market.

Elsewhere in the ancient world a few individuals with special skills are known to have received annual payment from the government of various cities at various times; we know most about this system with regard to healers. When and where we find evidence of it, it is usually for a single individual to be paid an annual fee. "Doctors" in such posts were sometimes required to train apprentices or act as coroners as part of the contract, but they were not required to provide a free service to the citizens, who still had to pay for medical care, which was everywhere private except in the Roman army. The state payment was a retainer to induce the doctor to come to and stay in the city so that he or she was available to the citizens of that state; it was not a living wage for his or her labor. This system seems to have developed to attract and retain especially skilled individuals, not just healers, who were welcome everywhere and were highly mobile geographically, and the Rhodians even developed it to provide schooling at state expense (Polybios 31.31). Vitruvius tells a story about a Rhodian "architect" (chief builder) called Diognetos who lost such an honorarium to a foreign competitor.

> Diognetos was a Rhodian architect to whom, as an honour, a fixed annual payment, commensurate with the dignity of his art, was granted out of the public treasury. Then an architect from Arados called Kallias came to Rhodes, gave a public lecture, and showed a model of a wall, over which he set a machine on a revolving crane. With this he seized a model helepolis (giant siege tower) as it approached the curtains, and brought it inside the walls. The Rhodians, when they had seen this model, filled with admiration, transferred the yearly grant and honour from Diognetos to Kallias. (Vitruvius 10.16.3 Dover modified)

This was shortly before Demetrios Poliorketes turned up outside Rhodes and Kallias' design was revealed as impossible to implement. After a certain amount of groveling, the Rhodians persuaded Diognetos to accept the honor back again, with the bonus of being able to keep whatever siege machines he could catch, and he more than proved his worth by keeping Demetrios "the Besieger" out of Rhodes for over a year, until the latter gave up trying to get in. Diognetos was not "employed" by the Rhodians in our sense, and he worked from home. The relationship between the Alexandrian engineers and the Ptolemies was probably similar.

Some intellectuals were probably drawn to Alexandria (as to other princely capitals) because Ptolemy (as other Successor kings) cultivated the arts and commissioned works of all types, and could pay handsomely for things that pleased him. The description of the procession of Ptolemy II, which gives us the word "pomp" as in "pomp and ceremony," is staggering in its scope and depth, and must have been a truly awesome spectacle. Ptolemy displayed his power, his resources, and the quantity and quality of labor at his disposal through exhibitions such as this. Paeans of praise written by poets such as Theokritos were another sort of display, and one that could take the message further afield.[15] Alexandria may have been more attractive to intellectuals than most other cities because spare copies of texts from the palace library were placed in a "daughter" library that was open to the public, which was an unusual benefaction of the time. Nor should we overlook the possibility that competition between royal centers may have led to very generous rewards for high-flyers. In such an environment, it is easy to imagine the best catapult engineers earning enough rewards to be able to experiment extensively with their own models and machines in their own time and their own space when designing the next catapult with which to impress the king.

Philon goes on to stress that empiricism had a key part to play in technological development, citing building as a different but analogous case. It was discovered that e.g. columns which were, in fact, equally thick and straight, *seemed* not to be so, and that by adding here, subtracting there, tapering, "and conducting all possible tests" the builders made them *appear* to be straight and symmetrical (although in fact they were not now, 50.30–51.7). As tests were conducted to discover the right proportions that would make buildings look right, so tests were undertaken to discover the right proportions that would make catapults discharge missiles that would strike with lightning-like impact at a distance (51.8–10). Philon later indicates that these tests included varying the length of the spring, the length of the arm, the angle of the arms relative to the stock,[16] and whether sinew or hair worked best.

> We do not investigate the minor causes—such as extending or shrinking the height of the spring, or lengthening the arms or shortening them or making them to stand noticeably forward or back, or whether sinew or hair is best—just as we have explained above. For these things have been investigated already, as I said before, and being in the public domain, common to all, they have been tried many times already and in all sorts of ways. (Philon *Bel.* 68.10–15)

Despite that, I am not sure that Philon or at least some of his contemporaries were really secure in their understanding of the mechanics of the catapult or the principles of scientific construction. For, astonishingly, he says that people who wanted to maximize the amount of spring-cord *enlarged the hole but did not enlarge the hole-carrier* (leaving the edges extremely thin and weak),[17] because "it is not possible to make the hole-carrier bigger, for that would deviate from the dimensional scheme" (57.7–8). This implies that he thought that the hole-carrier was, and the spring-hole was *not*, part of the dimensional scheme.

Now an oversized spring-frame would require that the stand be moved forward (or the stock weighted) in order to balance the machine and make targeting easier, but that is not the point. The spring-hole diameter is supposed to be the key parameter. Yet here we are told that catapult-builders, having built a machine to scale for a certain hole size, then shaved away at the hole to enable them to increase the amount of spring-cord they could get in. This could be explained away as a superficial slip on Philon's part, but he then criticized this practice because they unduly weakened the structure in the process of enlarging the hole. (He adds that they plate the hole-carrier to try

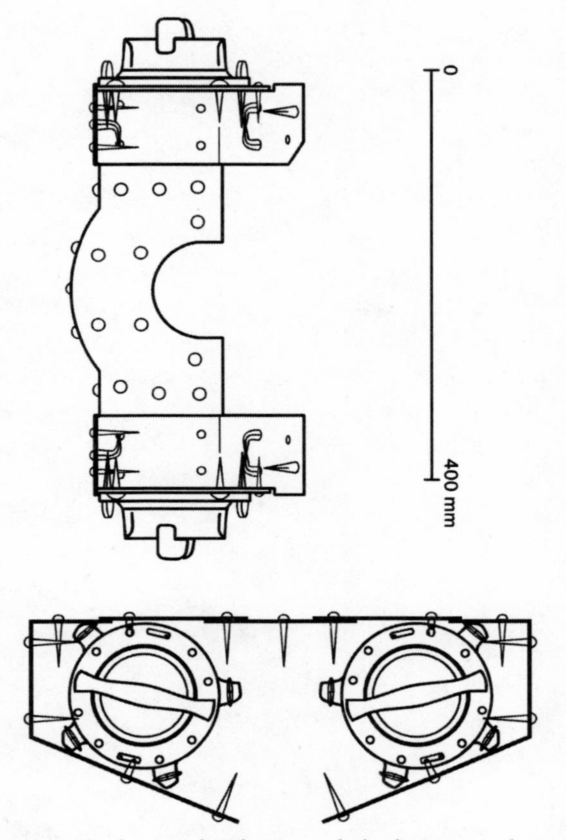


FIGURE 7.1 *The Caminreal catapult. The spring holes here are so large relative to the hole-carrier that the nails that fixed the thin metal plating to the wooden framework penetrated the spring space, and were bent to avoid damaging the springs. (T.D. Dungan after Vincente et. al. p. 179)*

to strengthen it, but sometimes there is so little wood left to fix to that the plates have to be thin and weak too, and they bend in use. The Caminreal catapult, Figure 7.1, demonstrates the point nicely.)

A more scientific mind would have criticized this practice not because it weakens the hole-carrier but because it jettisons the principle of catapult-building by formula. Philon's criticism is, however, consistent with his belief (expressed several times) that theory does not outweigh practice. Further, he himself jettisons the spring-hole altogether in the design of his wedge-catapult (though he tries to retain the algorithms based on spring-hole diameters! *Bel*. 62.29–63.14). In fact, it looks as if the ancients set more store on the relative proportions of the machine (which were expressed in spring-hole diameters) than they did on the spring-hole diameter itself. The importance of preserving the proportions was probably learned by experience too, because trials with a simulation model suggest that any change in proportions of the principal components leads to poorer performance.[18]

Despite his comment that no one before him "has dared to depart from the established method" (58.35), it is apparent that ancient catapult-builders had a rather flexible attitude to "the established method" in general and the calibration formulae in particular, and that although the latter were designed to take the guesswork out of successful catapult-engineering, formal and informal adaptation continued, with mixed results.[19] But surely, this is what we should expect in a vigorous culture. Torsion catapults, even those built *exactly* to formula, were not the last word in military technology. The desire and hope for improved performance continued as long as wars were fought or anticipated, and some engineers continued to seek new and better weapons and war machines. Philon used the discussion of spring-hole enlargement and its drawbacks as an introduction to his account of radically different designs of catapult like the bronze-spring and the air-spring, including his own (not so radically different) wedge engine, and it is likely that he exaggerated the problems with the then-typical two-armed torsion weapon in order to try to sell his own brand of catapult.

He certainly misunderstood the mechanics of the torsion spring action, thinking, against common opinion of the time (*Bel.* 68.29–31), that half of each spring was essentially redundant (68.31–69.23). This was how Philon rationalized Ktesibios' invention of the bronze-spring catapult, which applies its force to one side of the arm only. But it seems to me highly unlikely that Ktesibios had any such "reason," or mechanical misunderstanding, in mind. The arrangement of springs and arms on his bronze-spring sharp-caster catapult is essentially the same as that on his air-spring stone-thrower catapult (78.13–15). Ktesibios' air-spring catapult, described by Philon right at the end of his *Belopoiika* (77.9–78.22), was an adaptation of his cylinder and plunger pump—one of Ktesibios' most famous inventions, and one that carried his name throughout antiquity. Hydraulic Ktesibian pumps were at the heart of Roman fire engines. They were also to be found, for example, at the faces of mines and in domestic environments as pressure-washers; the Ktesibian pump was particularly useful wherever forceful jets of liquid were required. The pneumatic version appears in the air-spring catapult, and in the "bellows" for another of his very popular and famous inventions, the water-organ, beloved by Nero and played by God—according to Tertullian at any rate. We do not know in which order Ktesibios produced his catapults, but I imagine that the air-spring came before the bronze-spring, because the air-spring used the same technology as the water-organ and the water-organ was reputedly one of his first inventions. If so, the arrangement of arms and springs was worked out for the air-spring, and then applied to the bronze-spring, rather than vice versa.

FIGURE 7.2 *Shears. This design relies on springy metal (steel or bronze) in the bend of the U to function; the bronze-spring catapult relied on similarly resilient metal. (Author photo)*

Philon then describes his version of Ktesibios' bronze-spring engine, which, if it coincided with Ktesibios' in detail, was fortuitous because he did not have access to Ktesibios' construction notes for this catapult (*Bel.* 67.29–68.3)—and indeed, they do not survive in those of Ktesibios' *Notes* republished by Heron. Although the Greek word *khalkos* means variously copper, bronze, iron, or metal, Philon describes the ingredients for the springs as *red* khalkos (i.e., copper) and tin, so it is probably rightly known as the bronze-spring engine. This section of his treatise includes a discussion of research by Greeks on foreign steel swords to discover the cause of their extraordinary resilience. Three factors were identified: (i) superior iron for the raw material, which was (ii) well wrought when hot to remove impurities and (iii) subjected to prolonged but gentle hammering on both sides when cold (71.17–27). It is not clear from the discussion whether this knowledge was gained by talk or by test, but Wells inferred from it that Philon did not see these blades made, because cold hammering works for bronze but the effect is usually achieved with steel by tempering. Philon seems to have appreciated this, even if he did not express it as such, for his explanation of these swords' resilience is that heating and working the metal produces a kind of sandwich of hard layers either side of a soft middle. This is in contrast to the typical Greek blade that was soft wrought iron with a hard steel edge.[20]

Marsden thought that this passage spoke loudly against the idea that steel bows were in existence in Philon's time. Why, he asked, would Philon have expected his readers to "deny the possibility" of a springy-bronze-powered catapult if springy steel bow catapults were familiar? Why would Philon have chosen the relatively obscure analogue of far-western barbarian sword making if he could have cited the relatively familiar analogue of bow catapults, such as Biton knew? Now, while this argument has force, the fact is there were other sorts of springy bronze that Philon could have but did not cite that would definitely have been familiar to his audience. Greaves were put on, and held in place, by the springiness of their bronze, and nothing else. Early Greek greaves were laced up, but they soon learned to make bronze sufficiently resilient that they could be shaped to the owner's legs, almost meeting at the back of the calf, holding sufficiently tightly not to slip, and yet elastic enough to put on. Helmets may have been the same but I know of no discussion of the point by archaeologists. Meanwhile cosmetic and surgical kits typically contained several sets of tweezers, which again were commonplace applications of springy bronze. In the domestic (and cultic) environment, springy bronze pins had been used to hold fabric since Mykenaian times. And from about 300 B.C., U-shaped shears, of the type held in one hand and now used for topiary work, relied on springy bronze or steel in the bend of the U to function (Figure 7.2). Most ancient shears of this type were made of one piece of metal, usually iron, but a bronze pair is among the Roman material from Caerleon in Wales. In Pompeii, shears were commonly made of iron cutting blades riveted to a bronze spring curve. Flinders Petrie discussed a selection of such shears in the large collection of University College London in 1917.[21] The earliest are dated to about 300 B.C., and are Italic, from La Tène, and the Gallic culture of Bologna. This is consistent with the Celtic home of springy metal swords known to Philon, and precedes by about a generation the probable date of Ktesibios' application of springy bronze to the catapult. There are a number of such shears from various locations around Greece on display in the Archaeological Museum in Athens, but they are only vaguely dated (Hellenistic or Roman).

Further, there are precedents for metal bows. Two bows made of bronze were found at Susa.[22] One is flat, the other slightly stacked. No scientific techniques have been applied to the Susa bows, as far as I know, to analyze their chemistry or structure. There are references to "steel" bows in 2 Samuel 22.35 and Job 20.24,[23] but, as in Greek, perhaps the Hebrew word translated here as steel (King James Version) could also mean copper, bronze, or even simply metal. If so, then the Syrians and Jews used such bows, according to these Biblical sources. Those found at Susa were 24 inches (60 cm) long, which is at least six inches shorter than common composite bows, but they are made of different material so that is not very significant. Bows depicted on a fifth-century B.C. Skythian bowl in the Hermitage collection and on Marcus Aurelius' column of the second century A.D. are very short (just the length of the archer's torso), but we do not know the degree of artistic license employed in instances like this. For comparison, Blackfoot bows in the nineteenth century were 30–36 inches long, and a typical Skythian bow was about 30–40 inches (75–100 cm) long.[24] De Morgan thought that the Susa bows were votives, while Rausing argued that they were not; the latter pointed out that the staves are solid bronze and of considerable section, which used a lot more metal than was usual for votive offerings at Susa. He also observed that, insofar as the "copper bow" excludes wood from the stave, and used an alternative, normally rigid, material, it is related to the joined bow made exclusive-

FIGURE 7.3 *Detail of the Foundry Cup, Berlin. (Author photo)*

ly of horn.[25] I know of one object that could possibly be evidence for metal bows in Greece. On the wall of the foundry depicted on the Foundry Cup (Figure 7.3), there appears to be a bow. Plaques and heads are hanging below it. Other things on the wall are directly connected with the foundry men's work, and this artist is known for his realism; could this be a metal bow? If so, must it be a decorative one, intended to be put into the hands of a bronze sculpture, or could it be a functional metal bow? In any case, Cavalier's study of the saw depicted elsewhere on the same cup suggests that this is a very accurate representation of a Greek bow in the first half of the fifth century B.C.[26]

The springs in Ktesibios' design were presumably made of bronze—and Philon's were presumably made of the particular bronze that results from the recipe that he gives in *Bel.* 70.2–6. From the perspective of the metallurgist, there is no reason in principle why bronze of this composition, worked in this way, should not have produced a springy bronze.[27] Marsden was particularly dismissive of this machine and it has received relatively little attention from either scholars or reconstructers. There is no other trace of the bronze-spring catapult in the surviving literary or archaeological record, but this does not mean that it was made only twice. Literary sources rarely specify what sorts of catapults were employed in battles and sieges they report; indeed, they sometimes fail to mention catapults where we know they were used because traces of their presence are found in the archaeology. Parts of a mere thirty-six machines have so far been recognized in excavations and collections, yet we know that sometimes hundreds were deployed at just one site.

Philon then considers the repeating catapult, which shot sharps. Interestingly, he says Dionysios of Alexandria, its designer, called it "the little scorpion" (σκορπίδιον). Even more interestingly, Dionysios is said to have been working at Rhodes at the time. This might be further evidence of the free movement of catapult-builders between arsenals if we assume he was working not just *in* Rhodes but *for* the Rhodians, or further evidence of ingenuity shown during a siege, in which case he could have been working for either side. The year-long siege of Rhodes, for which, as we saw earlier, Demetrios drafted in many engineers, losing eleven to the Rhodians en route, would be an obvious occasion for such an invention.

Dionysios of Alexandria's repeating catapult shot sharps that were 25 dactyls (about 19 inches) long, an unspecified number of which were loaded into a magazine. This had a device that automatically dropped one missile into the groove on the slider as the catapult was drawn. Philon does not offer any thoughts or observations on the inspiration or motivation for this invention. The one feature of this repeating catapult that he saw as a problem—the fact that the contents of the whole magazine fall in a tight group[28]—would actually be advantageous in a siege situation, during most of which time the enemy was stationary. This was especially true when both sides needed to concentrate firepower in one area (usually, around the point of attack by tower or ram): the defenders, to attack the siege engine, and the attackers, to keep the defenders' heads down. It was also true for the defense per se, for as [Hyginus] stated explicitly, artillery was best placed in prominent sites overlooking any entry points to the town, around the harbor (if such there be), and covering corners.[29]

The same may have been true in other situations too. Lucan tells a story of a casualty among Caesar's forces in a naval battle off Massilia, wherein one Tyrrhenus was rendered blind when the lead shot of a Balearic slinger hit him in the head. After realizing that he was otherwise all right, and regaining his composure, Tyrrhenus asked his friends to position him "as they would a catapult" on the boat, that is, in the right place to hurl missiles, so he could then keep on fighting effectively.[30] The implication is that even in a naval engagement, when all targets and targeteers were perforce moving in four dimensions, a catapult would be positioned to cover a critical field and would keep shooting along essentially the same trajectory; it was not swung about from side to side or up and down like a Bofors gun tracking an airplane.

Philon's worries about the repeating catapult's wastage rates would have been valid for field use, but of no concern to either the Rhodians or Demetrios, both of whom were famously spendthrift with their missiles (and we should never lose sight of the fact that catapults were used by people who probably perceived themselves to be fighting to the death rather than theorizing in an armchair). For example, when day broke after a ferocious night attack on one of Demetrios' siege machines, he "ordered the camp followers to collect the missiles that had been shot by the Rhodians, since he wanted to estimate from them the armament of the forces within the city. Quickly carrying out his orders, they counted more than 800 incendiary missiles of various sizes, and not less than 1,500 catapult sharps. Since so many missiles had been hurled in a short time at night, he marvelled at the resources possessed by the city, and at their prodigality in the use of these weapons."[31] Although the Chinese equivalent was employed until at least the nineteenth century, Philon observed that Dionysios' repeater had not found noteworthy application (in his experience, anyway) and he seems to have included it because it was complex and ingenious.

The same might be true of the last catapult he describes, Ktesibios' air-spring stone-thrower. Philon imagines that this machine was the result of Ktesibios' recognition that compressed air could supply the necessary power to catapult arms, and consequent application of his pre-existing knowledge of pneumatics to catapult construction (77.16–17). This is a very plausible explanation for this radically different type of design, and in my view the air-spring probably preceded his bronze-spring design, as mentioned above.

One major difficulty with both this and the bronze-spring catapult, and possibly the principal reason why they did not find favor (as it seems they did not), would have been the tuning of the springs to equalize their power. If the springs do not move simultaneously, the bowstring is pulled faster on one side than on the other and pulls the missile off line, with potentially devastating results (a misfired stone could destroy the center stanchions, for example). A torsion catapult could be tuned either aurally, by plucking the cords and listening for the note, or mechanically, by measuring the diameter of the cords with callipers, and adjustments could be made by twisting the levers of one spring until the sound or measurement of both were the same. It is far from obvious how one could adjust the bronze-spring or the air-spring, at least without dismantling them, and then trying again. Schramm found it impossible, with early twentieth-century A.D. technology, to make a pair of identical cylinders and pistons,[32] though his characterization of "identical" was probably a good deal more exacting than the ancients'. Scientific analysis of the composition of the bronze of the four washers in one (the Caminreal) catapult showed amazingly wide variation, with copper content, for example, varying from 32 percent to 76 percent, and lead content varying from 18 percent to 65 percent.[33] Clearly, one cannot easily make four washers from the same smelt *and* the same mold. But even if the bronze springs or cylinders and plungers for a new bronze- or air-spring catapult *were* made from what began as "the same" smelt and cast (sequentially) in the same mold, their performance characteristics would still have been unknowable until tried.

This introduces the issue of safety. All catapults are designed to maim, kill, or crush, at range. Even a small catapult contains huge forces that, should the catapult break, can wreak more havoc on its operators than it would have done on its target had it not broken. Several modern reconstructions have imperiled their builders, and so it was in antiquity. Ammianus Marcellinus records that during the siege of Maiozamalcha in A.D. 363 someone who was standing behind a catapult was dismembered when a stone misfired because one of the artillerymen had not fitted it securely to the sling (24.4.28): "his limbs were so torn asunder that parts of his body could not be identified." In these circumstances, faith in the empirical method is not simply a philosophical choice by a scientist or engineer; it may arise more directly from personal health and safety considerations. Copying common patterns was a less risky strategy than following uncommon ones, and copying uncommon ones was less risky than trying new ones. Innovative designs might offer improved performance, in some respects, over existing designs, but they also introduced risks, and those risks were not just financial. New military technologies were, literally, risks to life and limb for their users, because if they failed when they were needed, then the people who commissioned them, as well as those who operated them, were more likely to be killed, injured, or enslaved. Thus, the risk of technological obsolescence was opposed by the risk of becoming technological guinea pigs.

Those generals who were most daring on the battlefield, and those communities who were most daring in their politics, were probably also those who were most prepared to take the risks of developing and deploying new military hardware. Most people, I expect, followed the lead set by those of the daring whose gamble paid off—whose machines worked, and whose battles were fought and won. The absence of battle meant the absence of real trial and the continuance of the then-traditional. That, I think, is why development is more rapid in wartime. It is not just that during war there is urgent need for new weapons—for it could be fought using the existing ones—it is also because during war there is opportunity for new weapons to show their worth, if they have any. Conversely, retrenchment encourages the continuance of traditional technologies and discourages innovative ones. But we digress.

Ktesibios' compressed-air and bronze-spring catapults were designs radically different from existing bow and spring designs. They represent an attempt to think about powering the machines in totally different ways, using materials that behaved in ways similar to sinew and hair and that were significantly stronger than either. The compressed-air catapult exploited discoveries he had made with other pneumatic devices, and the bronze-spring represents lateral thinking of the highest order. Unfortunately for Ktesibios, but perhaps fortunately for all those caught up in the battles of the time, these designs were more difficult to translate into functioning machines than usual, and they disappear from our view.

We leave Philon with consideration of a couple of machines mentioned in his other war-related treatise. An otherwise unknown machine appears in his *Poliorketika* C 13 (= 91.45 Th.), the *eneter*. In medical contexts, this word means a type of syringe for giving enemas. Lawrence suggested (1979: 90) that here it means flame-thrower, which would be a natural extension of that device in a military context, and would work as a countermeasure against the weapons and devices specified for it in D 48, though this is a very corrupt section of text.

> Against rams, reaping-hooks and crows: eneters, cranes, lassoes (or wheels) and rings (or rams) of other kinds. (Philon Pol. D 48 = 100.17–20 Th)

FIGURE 7.4 *Paestum/Poseidonia, southeast corner tower. The small, close-packed bosses on the face of the top floor around the embrasure are the sort of design recommended by Philon to deal with bombardment by very large caliber stone-throwers. (T.D. Dungan after Adam)*

What seems to be required is something that can catch the shaft of an enemy device, or something to set fire to it. There may be another otherwise unknown machine in C 19 (= 92.28 Th.), a stone-thrower called a *karbation*. However, this word is a *hapax legomenon* (a word that occurs once only), and some scholars have preferred to interpret it as a misspelled version of a word for rawhide shoes, so that here it must mean something like a rawhide shelter for artillery. What is said about this *karbation*—that it is necessary to throw stones as big as possible from it—is so brief that even if it was a stone-thrower, that is all we can say about it.

IT IS NOT CLEAR WHETHER STONE-THROWERS were yet capable of doing serious damage to well-built walls when Philon was writing about 200 B.C. He comments in the *Poliorketika* that if the finial blocks are mortared and clamped then stone-throwers' missiles will be unable to dislodge the battlements and will ricochet off. This implies rather weak stone-throwers. On the other hand, he advises that areas of wall liable to be targeted by one-talent (about 25 kg or 60 lb.) stone-throwers should be faced with protruding blocks of the hardest possible stone, protruding a span (about 9 inches, 23 cm) out from the wall, and sufficiently close together that a talent-ball (about 10 inches, 25 cm diameter) could not touch the wall surface in between.[34] A wall built to this specification would resemble a three dimensional checkerboard, each space between protruding stones being about 9 inches cubed. Something similar survives on the top floor of the southeast corner tower at Paestum (Figure 7.4).

More commonly, the stones for wall faces were simply drafted at the edges and the centers left rough, to leave natural bossing, stones bulging out in their centers, and thereby ready to absorb the force of impact and protect the joints without undue fuss and effort by the masons.[35]

The one-talent stone-thrower was the most powerful in existence in Philon's time.[36] Since three-talenters had been deployed at the siege of Rhodes a hundred years earlier, this should be taken to mean exactly what it says: the most *powerful*, not necessarily the *largest*. Bigger was not necessarily better in terms of velocity or impact. Cannon with bores approaching a meter in diameter were developed in fifteenth-century Europe to throw huge stones,[37] but they were weeded out as guns and gunpowder technology matured and converged around a few good types. Something similar may have happened with catapults; a smaller missile travelling faster can cause as much damage as a larger missile travelling more slowly. Philon recommended using two 10-mina engines (which would have a stock about 13 feet or 4 meters long) as a counter-battery to try to destroy enemy stone-throwers, with or without the addition of a 5-span sharp-caster (which would have a stock about 6 feet 8 inches or 2 m 4 cm long).[38] He recommended 20-mina stone-throwers for harbor defense, to destroy or sink any small craft trying to force their way in,[39] plus a 30-mina (=half-talent) stone-thrower if the harbor had a wide mouth. The 30-mina was evidently recommended for additional range, rather than for any additional weight of stone it could launch. This seems to be confirmed by his next recommendation, where he says that it is essential to use "fire-bearers" (πυροφόρα) and "hurling-spears" (δορυβόλα) against enemy ships.[40] Stone-throwers were always palintones, but palintones could shoot both stones and sharps; they simply required a change of slider and bowstring. I think that what Philon is saying here is that it is necessary to use a 30-mina palintone to cast large sharps and incendiary sharps against enemy ships. The 30-mina, or half-talent, palintone seems to be his preferred general-purpose catapult. For he comments that if they are well made, are suitably placed, and have skilled operators, then the city that owns such machines will suffer no disaster during a siege.[41]

The principle of presenting a tower corner to the front, which was developing at the time Philon was writing, seems to have had as its aim and raison d'être to make catapult stones glance off instead of collide. Impact craters on

FIGURE 7.5 *Impact craters on wall at Ostia. This wall is near the Vigiles' station, inside the city, and the impact damage to the wall was therefore perhaps caused by target practice or civil war rather than warfare proper. A Pompeian style grain mill sits (out of place) in the bottom right corner. (Author photo)*

flat walls at Ostia (Figure 7.5), Pompeii (see Figure 8.4), Eleutherai, Drymaea, Plataea (Lawrence, pl. 10, 28, and 38, respectively), and elsewhere reveal the sort of power with which stones could be shot—but these missiles clearly did not threaten the integrity of the wall.

Another solution to the problem of bombardment was the semi-circular tower; Philon recognized that stones that were shaped like the voussoirs of an arch, here laid flat on the ground so that they were wider outside than inside, could not be pushed inward by stone-thrower impacts.[42]

Philon's personal inexperience of catapult construction and use is evident in his reliance on other people's testimony when he gives his recommendations for the dimensions of the springs and arms of a palintone.

> They said that this size especially [a spring 9D, plus twice the height of the lever, long] was fit in practice and seems to be neither too short nor too overstretched, but to have a medium and standard format. Those with longer springs shoot far and are easier to draw, but their impact is weak and ineffectual, while those with shorter springs are hard to draw and do not shoot very far, and the arms of these engines are often distressed. . . . They said the best length for the arm was six diameters. Arms shorter than these are hard to draw and are not able to throw the stone much distance, while longer arms are easy to draw but, being underpowered, the discharge is as

> weak as the other [with too-long springs]. Consequently, those who have practical experience order the use of the above-mentioned size. (Philon *Bel.* 53.18–30)

Philon offers a mechanical alternative to laborious and error-prone calculations of multiples and fractions of hole diameters which has to be undertaken by anyone wanting to duplicate a euthytone at a different scale. This method would work for any similar scaling problem and he apparently gave it in the first (now lost) book of his *Mechanical Compendium* as well as here in the book on catapult construction. Something similar was probably applied in shipbuilding to produce what Sleeswyk calls a *lineage* of ships with the same basic proportions, albeit at a range of different sizes.[43] Philon tells the reader to take the missile of the pre-existing catapult and use it as the basis for a ruler, say A. The ruler must be the length of the missile, whatever it is; this length will be a pseudo-cubit. The ruler is then divided into six parts (pseudo-palms), the divisions being marked by a line drawn at right angles to the side of the ruler. One part is then divided into four parts (pseudo-dactyls), again marked by perpendicular lines. One of those four parts is divided again into four (pseudo-quarter-dactyls), similarly marked. Now take the missile of the catapult that one wants to build, and make a ruler B by it, following exactly the same procedure. Use ruler A to measure the pre-existing catapult, and ruler B to make the new one. Everything will have its analogue quickly and accurately, even if the missile is an odd caliber.[44]

Philon's recommended method of solving the duplication of the cube problem is similar to Ktesibios', but is more complicated and shows a connection to, if not influence from, Menaikhmos' method.[45] He gives two methods of drawing the template for the hole-carrier of a palintone (*Bel.* 52.20–53.7). This presumably reflects his education in two different "schools."

OUR THIRD AUTHOR OF A TECHNICAL MANUAL describing how to build catapults is Biton. We met him briefly in Chapter 3, because he is our principal source of information on bow catapults, otherwise known as gastraphetai. Biton describes two bow stone-throwers, a *helepolis* (city-taker siege tower), a *sambuka* (covered mechanical scaling ladder), and two bow sharp-casters. None of these four catapults is a torsion engine, though he was apparently writing at least a century after they appeared, and was writing in one the most advanced centers of culture and learning at the time, so his silence on torsion engines cannot stem from his being ignorant of them. Because it is assumed that bow catapults were rendered obsolete by torsion catapults, Biton's treatise seems to present a problem: why was he writing about old technology? Of course, if one drops the assumption that bow catapults were rendered obsolete by torsion catapults, then there is no problem requiring an explanation. Instead, Biton becomes evidence for the continued existence and employment of *gastraphetai*.

What we know of this man is only what we can extract or deduce from his text. He dedicated his treatise on *The Preparation of War Engines and Catapults* to "King Attalos." There were three kings called Attalos, all related, all rulers of the independent kingdom of Pergamon in Asia Minor. The first reigned 238–197 B.C., the second reigned 158–138, and the third reigned 138–133. To the first we owe one of the world's favorite sculptures, the Dying Gaul, commissioned to celebrate his victory over the Galatians around 230 B.C.[46] The second is now perhaps best known for giving his name to the Stoa (portico) in Athens that he funded.[47] The last is now famous for bequeathing his kingdom to Rome, and thereby hastening the Roman conquest of the Hellenized East, and the end of the Roman Republic. Tiberius Sempronius Gracchus' plans for the sudden new wealth that came to Rome as a result of this bequest led directly to his assassination on the Capitol, and thus to the Roman Revolution, that is, the breakdown of republican politics that would lead, over a century and a couple of civil wars, to the emergence of an emperor.

We do not know which Attalos of these three Biton was addressing. The possible chronological range spans 136 years, and different scholars have preferred different kings. Lewis argued that, given the opening lines, wherein Biton talks explicitly of repulsing enemy offensive engines, the most plausible context for his treatise is an attack on Pergamon. He identified the most likely of few such occasions in known Pergamene history as 156–155, when Prusias of Bithynia besieged the city during the reign of Attalos II. At this time, Pergamon's defensive artillery, which had probably been first-rate when installed, had not, Lewis suggested, been needed for over thirty years, and the torsion springs (at least) might well have been inoperative. The last is a reasonable point in principle: maintenance, still less replacement, of apparently redundant machinery over a generation would show an unusual lack of complacency that Ktesibios at least did not expect his readers to have.[48] Lewis suggested that unless stores of prepared fresh sinew or hair cord existed, it would have been impossible to repair broken torsion springs, and in a siege situation, it would be difficult to bring in supplies. Hence, he argued, Biton offered instruction on how to make a variety of bow catapults, because that type of catapult did not need spring-cord. However, his premise is false, because the citizens of Salonae (in Dalmatia), when caught in a similar situation, used the hair of their women to make catapult ropes, so did the Thasians in a similar situation, so too the Roman matrons, so why not the Pergamenes too? Lewis' argument also overlooks potential problems sourcing materials for bow catapults during a siege, especially matured timber of the right kind and size for the limbs of the bow. It also fails to appreciate how long it takes to make a bow: a minimum of a year.[49]

The argument also seems to me to be too categorical. Attalos II was quite capable of launching aggressive wars himself, and while I can imagine a city's artillery being found wanting, I cannot imagine a complete lack of spring-cord as well. In fact, this particular city's attention to the maintenance of its garrison around 200 B.C. was commented on by Polybios, and while the city itself may have been spared military action for a generation (though just because we do not know something happened does not mean it did not happen), the Pergamene army had not. Polybios has Eumenes II (Attalos I's son, between them reigning from 241 to 160) claiming that he and his father contributed more military forces, army and navy, to aid Roman armies in Greece and Asia than did any other of the Romans' allies, and that they personally both fought on the field of battle. Livy has Perseus assert in 172 B.C. that Eumenes is a far worse tyrant over Asia than Antiokhos was and that Rome's allies cannot relax as long as the palace of Pergamon stands because it looms like a citadel over the heads of neighboring states. Eumenes was followed by Attalos II, who used military force in Kappadokia in 156, before Prusias attacked, and who was planning to use it in Galatia in the same year, but was apparently intimidated by the prospect of Roman intervention. Indeed, when Attalos sent an envoy to tell the Romans of Prusias' attack on the city and make a plea for their help, the Romans at first ignored him because they suspected that Attalos was intending to attack Prusias and that this was a false accusation and a pretext for an Attalid first strike.[50] So I expect there was usable spring-cord in Pergamon. Biton may rather have scoured Pergamon's Library—which was second only to Alexandria's—for descriptions of old war machines not because he desperately sought an alternative for torsion weapons but because it was said that the greatest war machines were made in the time of Demetrios Poliorketes. This brings us back to the title of his treatise, which also speaks against a defensive purpose: *war engines* as well as catapults. Biton included instructions for a *helepolis*, a city-taker, a siege tower for offensive, not defensive, purposes (see 52.5–7 W). Further, he emphasizes that a significant range of materials needs to be prepared in advance of the actual construction; this is hardly consistent with composition of the text during an unforeseen siege, and its content as useful substitutes for unforeseen contingencies. Bows are not made overnight. A *sambuka*, a mobile, tilt-able, covered, giant, scaling ladder, described in 57–61 W, is an offensive device too. So I am unpersuaded by Lewis' argument, and continue to seek a plausible historical context and date for Biton's text.

Another clue to the possible date of the text is the designers mentioned in it, particularly Damis of Kolophon and Isidoros of Abydos. Drachmann tentatively identified this Damis, who designed this sambuca, with the Damios who was an admiral under Eumenes II, Attalos II's elder brother, about 168 B.C. He suggested that Isidoros, designer of a large bow stone-thrower of about 40 lbs. caliber, might be the Isidoros who was admiral under Antiokhos III (the Great) in 191 B.C. In either case, the text would then have to have been written in the second century, not the third, for Attalos II or III, not I. Marsden argued that, since the name Isidoros was common and we do not know the hometown of Antiokhos' admiral, the grounds for these identifications were extremely weak. And so they are. However, more arguments can be brought to bear in support of Drachmann's identification. Isidoros was admiral during the short period when Antiokhis occupied Western Asia Minor, Thrace, and Northern Greece, and Isidoros' catapult was specifically designed in Thessalonike (Map 1). Abydos, Isidoros' hometown, had put up heroic resistance to Philip V in 200 B.C., according to Polybios, and he also tells us that in the next decade Antiokhos made extensive preparations for, and engaged in, war, by land and sea, with the Romans and their allies, the Attalids of Pergamon.[51]

Further, it seems not to have been noticed hitherto that both these machines are particularly suitable for *naval* warfare, while both homonyms were specifically *admirals* (*nauarkhoi*), not generals or generals-cum-admirals (*strategoi*). The sambuka was initially designed for assaulting city walls from ships, according to Athenaios Mekhanikos, and featured prominently in this capacity in the siege of Syrakuse in 213–211 B.C. and in Mithridates' siege of Rhodes in 88 B.C. Isidoros' bow catapult, unlike a torsion catapult, would have been relatively unaffected by the salt spray inevitably generated by a battleship under oars during everyday use, never mind in naval combat. Novelty is not claimed for either design; they are presented only as particular versions of pre-existing designs. Therefore, it would not be surprising if their designers had a different day job, so to speak.[52]

Biton's text ends with a paragraph that appears to be addressed to someone who is actually going to try to build the machines he describes, though the omissions in important details are then particularly curious. Could the author have expected his audience (or the copyists their readers) to be able to supply such information from their own experience? Or could the explanation be that this was not, after all, a real manual?

> We have described whatever we thought especially suitable for you. We trust that you will work out similar ideas from these. Do not be distracted by the specific measurements we have used; you will not have to use those same measures. If you want to make something bigger, do it, so too if smaller; just try to preserve the correlation between the designs and the measures given above. (Biton 67.5–68.1 W)

This phraseology may reflect the tendency to ascribe to a king the actions of his minions, or a bit of empty flattery, but it could also reflect the author's belief that his dedicatee, King Attalos, will genuinely get involved in the construction. It is consistent with the preface, which does not appear to be addressed to someone familiar with the niceties of catapult construction, for it carries the advice "try to use common sense. It is necessary to utilize together *both* the parameters *and* the proportions [literally, rhythms] of the things proposed herein" (44.3–5 W). The Attalids were descended from semi-slave stock, maintained an image of being "of the people" throughout their dynasty, and promoted family values, with the result that Pergamon was "a very dull court" by normal Hellenistic standards. Getting personally involved in the production of war machines would not be an unsuitable thing for such a person, and a military dynast, to do. Some Roman emperors (noticeably, those who also did not lead scandalous private lives and were "good" by modern standards) got involved in practical engineering problems. Trajan's chief engineer Apollodoros once made the mistake of criticizing Hadrian's efforts in this department—when Hadrian became emperor, Apollodoros was banished and then killed. One of the things about Hadrian's design for a temple of Venus and Roma that Apollodoros found fault with was that "the machines" could have been accommodated unseen in a basement, had he included one. Evidently, Hadrian designed the necessary plant as well as the building, and saw no reason or need to hide it out of sight. What exactly the machines were remains an open question.[53]

Of our three Kings Attalos, the third seems to me the most suitable recipient for Biton's final piece of advice, for he was reputed a bad king but a good student. Diodoros describes him as becoming cruel and bloodthirsty soon after his accession to the throne (34/5.3.1). He *was* a bad king at least insofar as he was the last of his line and the last King of Pergamon; the last of any dynasty must be deemed a failure. But he had only been king for about five years when he died prematurely in 133, so he had little time to rule and much time to study (he was about thirty-two when he inherited the throne). He wrote a work on agriculture that may or may not have had a strong pharmacological interest. Kosmetatou reasoned that he must have been a competent student in botany and pharmacology to have justified the praise he received on these topics from Galen,[54] and he would have had a vested interest in toxicology, because poisoning was one of the preferred methods of assassination at this time and other kings were investigating toxicology too. But it is worth remembering that national pride and intellectual ambition may have inflated Galen's opinion of Attalos the scholar-king—for Galen too was from Pergamon, and Galen sought the company of Rome's elite. Asserting a king's interest in medical matters would have furthered Galen's agenda to lift medicine out of the category of trades and services and into what we would call professional respectability, and might even persuade the emperor that Galen was worth talking to.

All things considered, I favor a second-century date for Biton's work, and if Attalos III was the dedicatee, his title as King requires this treatise be dated to the period 138–133 B.C. More important than the date, and irrespective of it, I believe that the machines described by Biton were in fact, as he claims them to be, functional war machines for his own age. That is not to say that they were the machines that Marsden or Schramm reconstructed; the text is inadequate in a number of ways and both Marsden and Schramm depart from it sometimes. Marsden does not, for example, even attempt to follow the instructions with regard to positioning the bow on the mountain *gastraphetes*. This is not a minor detail; this fixing has to be able to withstand the full power of the catapult. His interpretation of the "arm-bowl" (which he translated as "cup-bracket") as an early form of the "universal joint" supporting the entire machine could instead be interpreted as the part which supports the arms of the bow either side of the stock. Notwithstanding such problems, there is nothing in the text to support the idea that Biton was presenting these machines as nostalgic replicas. That idea and others like it arise, I believe, from the widespread but false assumption that since bow catapults preceded torsion catapults, any bow catapult described by someone living in the torsion catapult age must be an old catapult, if not an antique.

An apparently unique find from Pergamon could bear on this problem (Figure 7.6). In 1886 on the acropolis, near the munitions store south of the palace, were found "iron bracings" (*Eisenverstrebungen*) or a "clamping frame" (*Spannrahmen*), along with sharps (described as arrow and lance heads) and numerous stone balls. Over 950 ballista balls are known from Pergamon, but just one catapult washer has been recognized among the finds (and it is now lost). Gaitzsch, in consultation with Baatz, considered and rejected the possibility that this metalwork belonged to a torsion catapult.[55] They did not apparently consider the possibility that it might have belonged to a tension catapult. But Biton explicitly talks about fitting iron braces and metal plates to his bow stone-throwers ("and let each [brace] be fastened on with iron plates"), and he was of course writing for one Attalos of *Pergamon*.[56]

I think the following scenario can explain Biton's short treatise: he was asked by a king Attalos to provide guidance on the construction of a stone-thrower. He provided two designs for bow stone-throwers, and then, seeing that this had taken few words (only a couple of pages in modern layout), and wishing to please and impress his king (historical analogies abound), he supplemented it with a description of a giant siege tower built by Poseidonios of Macedon for Alexander the Great. Now his commission did not include this, and so at the beginning he asks the king not to scoff at him for including it—and others that he is going to add. "I have set out to describe the construction of a stone-thrower engine, King Attalos, and do not scoff if some engines happen to be of another type." Biton wants to hold on to his rich and powerful patron. What Successor king and general could fail to be interested in Alexander's war machines? This machine was historic, but few had equaled it since. Its inclusion also gave Biton the opportunity to point out that he knows a thing or two about surveying,[57] and has written a treatise on *Optics*, and, in particular, that he knows a technical method for ensuring that a siege tower is higher than the city walls it is intended to assault (52–53.1 W). Perhaps Polybios' castigation of Philip, Alexander's father, for once approaching a city wall with scaling ladders too short for the job, was recent gossip (Polybios lived c. 200–118 B.C.). That war machine consumed some more words (one more page in Marsden). He could reasonably expect one of Alexander's successors to be interested in another device for storming cities, and to know that great deeds in war are a sure road to immortality (these people live here; we talk of them still!). So now he adds a description of how to build a sambuka, "for this engine provokes great deeds in military contests" (57 W). This consumed a little more text (another page in Marsden). He ended his work with descriptions of two sharp-casters, both of them *gastraphetai*, but both of them unusual for other reasons. Biton explicitly remarked that "often times the local conditions do not suit the same types of engine" (48.3–49.1). One of these machines shot two bolts simultaneously. The other was specifically described as a mountain catapult, which may mean it was very easy to assemble and disassemble, but could refer to another feature ignored by Marsden, the differentially sized tripods and the resulting steep angle at which the stock was laid at rest. With each of these catapults needing about half a page, the document now formed the equivalent of a respectable pamphlet. In Marsden's edition, the whole treatise takes six pages (compare Ktesibios', which takes fractionally more than twelve, and Philon's, which takes a little over twenty-four).

In short, my explanation for the content of Biton's treatise, and which applies to all proposed dates, is that bow catapults were not rendered obsolete by torsion catapults. Marsden almost entertained this idea; he opted for the earliest possible date, at the start of the reign of Attalos I, because he assumed that Biton's content reflected his own personal preference, which Marsden described as "an obstinate predilection for non-torsion artillery." He also pointed out that the calibration formulae and lists of dimensions might not then have been well known outside Alexandria, but that, I think, is irrelevant, for Biton could have built irregularly proportioned torsion engines, just as people had done before the calibration formulae were discovered and

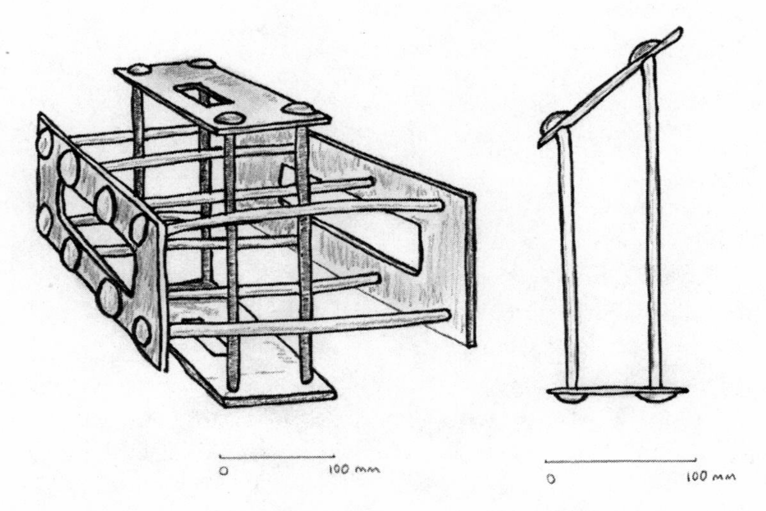


FIGURE 7.6 *Iron bracings from Pergamon. This mysterious artefact was found near the munitions store on the acropolis of the city. It could perhaps have been reinforcing metalwork for a tension catapult, of the kind recommended by Biton, who wrote for Attalos of Pergamon. (Author sketch after Gaitzsch)*

apparently continued to afterward. Marsden also recognised that such weapons might still have been "considered useful in certain circumstances," especially those "where the springs of torsion engines were not so reliable" (1971: 6). The usefulness argument retains its force right down through the ages. Torsion artillery simply is not always superior to non-torsion artillery. As I pointed out in Chapter 1, the most advanced armies of antiquity retained archers and slingers in their ranks, as did armies millennia later, even of those nations pleased to call themselves "developed." New military technology rarely renders old military technology obsolete. This truism is even celebrated in popular culture: Arnold Schwarzenegger, playing one of America's finest soldiers in the finest modern army, ultimately defeated the even more hi-tech-equipped alien Predator by using Stone Age weapons. The simpler the technology, the more likely it is to survive, because more people can make it, in more places, and know how to use it.[58] In essence, a bow catapult is a powerful bow that needs a ratchet and pawl and a trigger, that's all.[59] It is a good deal simpler than a torsion catapult, both to make and to use. This also explains why all four of the principal manuscripts on Greek artillery construction include Biton's text (along with Ktesibios'), unlike Philon's, which only two of them have. Biton's text was more interesting, to more people, over more time, than Philon's was; that is why it survives to us in a larger number of copies than Philon's. It was interesting because it was useful, and it was useful because people wanted to make *these* machines—bow catapults.

The bow catapults that Biton describes are not just oversized bows though: the first two he describes are designed to shoot stones, not sharps. I argued in Chapter 4 that the classic two-armed torsion catapult was the result of the marriage of the power source (the sinew spring) of the one-armed stone-thrower, with the cruciform shape of the mechanical bow with stock. I now want to extend this argument and suggest that the bow stone-thrower, such as those designed by Kharon of Magnesia and Isidoros of Abydos, might have been the result of transferring *back* to the bow catapult the grooved slider and sling-like bowstring (we might call it a slingstring) of the two-armed stone-throwing torsion catapult. If so, then the Magnesia that Kharon came from is most likely to be the one on the Maiandros River (Map 3), in the neighborhood of Miletos and Latmian Herakleia and not, as suggested in Chapter 3, the Magnesia in Thessaly, which is the most likely Magnesia if Kharon was a contemporary of Philip, as is usually supposed simply because he made tension, not torsion, machines.[60]

Kharon—like other mechanical wizards we have met—was apparently working in Rhodes when he designed his stone-thrower. Rhodes went into dramatic decline from 168 B.C., when the Romans promoted Delos as a free port, and Rhodes suddenly lost a huge portion of her regular income from harbor dues. Her exports, at least as reflected in the presence of her wine jars abroad, fell off at about the same time.[61] (Alexandria too suffered from the 160s, when Ptolemy VIII Euergetes II, otherwise known as Physkon, Potbelly, began contending with his older brother Ptolemy VI Philometor and *his* son Ptolemy VII Neos Philometor for the throne, creating a very hostile environment for intellectuals and others for a generation. Alexandria disgusted Polybios when he visited, probably in the 140s or 130s.) Kharon can therefore be assumed to have been active before this time. A date for him in the first third of the second century would be consistent with those for Damis and Isidoros, discussed above.

## Knowledge and Understanding

Did the ancient engineers understand the mechanics of their machines? Did they realize that the range and "power" of the missile shot depended chiefly[62] on (i) the discharge velocity, (ii) the angle of elevation at which it was discharged, and (iii) the air resistance? The answer to this is surely yes, in a superficial sense, though I do not think they fully understood why it was so, or sometimes even how it was so. Let us take them in reverse order.

Despite the fact that the ancient philosophers had a bad to tenuous grasp on the physics of moving bodies, I cannot believe that any practicing catapultist (of a euthytone at least) did not spot that the air somehow played a role in the flight of a projectile. To take the most basic examples, shooting into the wind would have reduced the range he could achieve, shooting with the wind would have increased it, and a cross-wind would have deflected the projectile sideways. This would be common knowledge to archers, javelin throwers, and perhaps slingers, and anyone trying to hit a target in different weather conditions would soon have learned it himself or herself. Even Aristotle (whose theories on motion suggested otherwise) spotted that sometimes air created resistance and reduced speed. Vegetius advised specifically on the issue: headwinds deflect and depress your missiles, while aiding the enemy's.[63]

The real question is: did the ancients recognize that the less energy a projectile loses to air resistance, the more energy it delivers from the catapult to the target? Did they realize that a low drag missile has greater impact on the target than does a high drag missile shot by the same device over the same distance at the same target?

The principal source of drag on a projectile is the vanes, but they are necessary to stop the sharp rotating in flight and veering off target. The same function can be achieved by tapering the shaft of the sharp, which allows smaller vanes (and therefore less drag), because air pressure will bring a tapered shaft back on alignment if it veers. Another source of drag is turbulence, which can be reduced by shortening the shaft.[64] So a short tapering shaft causes less drag than a long non-tapering one. The very low drag of the tapering, relatively short, tiny-vaned Dura bolt suggests that the ancients came to understand most of this—discovered, no doubt, empirically, as Philon repeatedly asserts. The aerodynamic properties of this design are so good that it could not, I believe, be accidental.

Now a longbow arrow needs to be long in order that the head is still supported when the bow is fully drawn, but there is no such mechanical constraint on a catapult bolt, which theoretically can be as short as a couple of inches. In practice, such short missiles may have been more hazardous to their users than longer sharps, because a number of modern reconstructions have nearly killed their makers. Misfired short bolts laid well back on the stock can ricochet off the frame to wreak their havoc in the area of those operating the catapult.[65] The crew, it is suggested, were better protected if the sharp head projected beyond the frame when the catapult was fully drawn, so that even if it misfired, the sharp did not come backward. But what is true of short sharps is more true of shot, so I do not find this argument persuasive. Moreover, if the springs were *very* taut, or the bow *very* stiff, a catapult could have launched a really short sharp whose head was still in the aperture. The distance between brace and draw position on a medieval crossbow was typically just six inches or so; if the springs (of wood or sinew) are very strong, the string does not need to be withdrawn very far. Moreover, sub-formula-weight missiles fly better and further and the catapult performs better if the arms are not withdrawn as far as they would be for a formula-weight missile. If the Qasr Ibrim "foreshaft," which is 133 mm long (about 5 1/2 inches), is the complete tail for the 90 mm long heads found with it (and the absence of any trace of glue on its vanes suggests as much), then the whole sharp would have been about 180 mm long (or 7 inches; the tang of the head fits into the socket of the shortshaft; see Figure 1.4). This would correspond to a 2-palm caliber weapon with a spring-hole diameter of about 20 mm or an inch.[66]

The role of the angle of elevation would have become apparent through practice, even if the ancient engineers did not or could not calculate it on wax tablet. Thus, if they had written it down, they would probably have said that the optimum angle of elevation to achieve maximum range was not 45° (the theoretical optimum in a simplified mathematical world) but about 40° for a sharp with low drag (such as a Dura bolt), and 20° for a sharp with high drag (such as a longbow arrow).[67] Some ancients thought that other factors which do not in fact make much (if any) difference to the performance of a catapult did make a difference, such as the length of the sharp, which, according to our surviving sources, is supposedly the key datum for the euthytone. In fact, all that matters is the weight of the projectile, and sharps of the same length could have very different weights and thus behave very differently.

The role of the velocity of the projectile was perhaps the most difficult thing for the ancients to assess. Range could be measured, and range was a function of discharge velocity, so they could in principle have measured the one via the other, had they possessed the necessary concepts. But there is no evidence that they had such concepts. Velocity, acceleration, and mass are all untranslatable into Greek. There were a variety of ideas about physical processes concerned with motion. The most famous, or notorious, is Aristotelian dynamics as expressed in his *Physics*, which required that the cause of motion be in contact throughout its motion with what is moved (*Physics* 8.10 = 267a). For, he reasoned (and others thought likewise; this idea was not confined to the Peripatetics), the javelin's natural motion was down: if you picked it up and let it go, it would fall to earth. The thrower's (forced) motion caused it to move (unnaturally) forward and up—but what about after it left the thrower's hand? Why did the javelin not drop to the ground then and there? His explanation was that the force was imparted from the thrower's hand into the *air* that supported and continued to push forward the javelin. The same basic idea seems to have been shared by others at the time—Aristotle may have refined the idea but he did not invent it. Some have said that ideas like this on motion prevented the ancients achieving in dynamics the type of outstanding results that they achieved in statics, where, for example, Arkhimedes' law of the lever and similar discoveries stand as fundamental contributions in the history of science. Perhaps so; but there were rival schools with alternative physical theories, so that does not really explain the failure here. Elsewhere Aristotle's contribution was more positive: independently of the ideas expressed in a fully worked out form in *Physics*, he observed that speed, distance, and momentum are related, and that an object was itself struck as hard as it struck its target.[68] In the same context of projectile motion, a thousand years later, at the other end of antiquity, and still trying to understand what happens when a javelin is thrown, Philoponos replaced Aristotelian ideas with the notion of impetus, wherein the force is transferred not the air but to the javelin, which idea then survived to Galileo.[69] The ancient Greeks and Romans discovered in practice what they could not grasp in theory, and some of them, like Philon, recognized and emphasized that fact.

EIGHT

# The Romanization of Catapult Technology

*In a novel kind of warfare, novel methods of waging it were invented by each side.*—Caesar *Civil Wars* 3.50

Although the Romans adopted Greek catapult technology, Roman catapults are not réchauffé Greek models. The same basic technology is involved and the mechanics do not change, of course, but over the centuries that Rome dominated Europe and the Near East, continuing development of a reduced range of types is apparent. For example, there are changes to the height-to-width ratio of the spring-frames, changes to the shape of the arms, and changes in the materials of construction. It is probably not coincidental that as catapults became standard issue for the legions—from some uncertain date, unfortunately, but probably early during the first century A.D.—so they became more standardized. This is not to suggest that there was central production of catapults that were distributed to the troops across the empire, which is not what happened, normally at least. "Architects" and engineers, like Vitruvius, working in and for the legions wherever they were stationed, could and did make as well as maintain catapults, along with any other machines or structures required or desired by the commanding officer. Thus, one of the Vindolanda tablets records an order for 100 pounds of sinew rope to be delivered to the fort from Catterick (Cataractonium, Map 6), suggesting that repairs if not manufacture were to be undertaken at the fort.[1] But there was probably only one such engineer per legion,[2] and most of them would have been trained within the Roman tradition, the most influential of them in the arsenal in Rome or in one of the other principal cities, and diversity would have been reduced compared with the sort of diversity that existed when there were independent ordnance traditions. As the Romans eliminated the military competition, fewer independent traditions of arms manufacturing survived, though their engineers were no doubt absorbed into the Roman stream. Rome acquired Pergamon in 133 B.C., Massilia in 49 B.C.,

Rhodes in 43 B.C., and Alexandria in 31 B.C. As a result of these interconnected factors, fewer types of catapult seem to be in existence, but those that survive continue to be developed.

I have divided the story over the next three chapters, corresponding to the three familiar chronological periods of Roman Republic, Roman Empire, and Late Antiquity. The Roman Republic overlaps chronologically with the Hellenistic period, and I have reserved material for this chapter when it is relevant to the themes of this chapter or has a distinctly Roman focus.

## Vitruvius

We begin with a brief examination of our fourth author of a technical treatise on catapult construction, Vitruvius. He published his book *On Architecture* about 25 B.C. This book is a comprehensive guide to building and furnishing a city, from finding good water sources and initial surveying and laying out of streets and residential blocks (*insulae*) to making a sundial that works for that location and war machines with which to defend it. We are here concerned with the book on mechanics (book 10) and in particular with the chapters on war machines (chapters 10–12).

Vitruvius spoke from experience. He built catapults for Julius Caesar's forces, and may have served on the campaign in Africa in 46 B.C. He went on to serve Octavian as military *architectus*, principally occupied again as catapult engineer. Vitruvius sought work from Octavian, as architect in our sense, after he had won the civil war and become the emperor Augustus, whereupon he began to commission projects to transform Rome from a city of brick to a city of marble, which created huge opportunities for engineers to make their mark on the city and psyche of Rome.

Vitruvius begins his account with the formula for a sharp-caster (called a scorpion) that is unchanged from Philon's time: the spring-hole diameter (here denoted by D) should be one-ninth the length of the sharp the machine is intended to shoot, and this unit is still the key parameter for the rest of the construction. He then gives the stanchion sizes, which result in a square frame that was 6D x 6D. Vitruvius later says (10.6) that the spring height can be adjusted up or down (from 6D), depending on the intended use of the catapult, giving rise to the terms *anatonos* and *katatonos*, long-sprung and short-sprung, respectively.[3] The *anatonos* (stanchions >6D), with relatively long springs for its spring diameter, is less powerful than a machine built exactly to formula. The *katatonos* (stanchions <6D), which has springs that are short and squat relative to their girth, is more powerful than the "standard" version. The *katatonos* would be harder to draw but would strike with more powerful impact, while the *anatonos* would be easier to draw but strike with less vigor. When shooting fire-lances this would be an advantage, since too rapid flight extinguished the flame.[4] Vitruvius goes on to say that the deficiencies in each case can be compensated for by changing the length of the arms: short arms to increase the power of an *anatonos*, long arms to increase the ease of drawing the *katatonos*. This last adaptation (changing the arms' length to compensate for longer or shorter springs), which is based on the lever principle, offers an alternative to making the catapult with an undersized or oversized spring-hole to vary the performance characteristics. As Fleury noticed, Vitruvius repeatedly signals the necessity of adapting the components to the circumstances, especially as the calibers move into the very small or the very large, and this variability is not always recognized in modern scholarship.[5]

Vitruvius gives the length of the case as 19D, whereas Philon had 16D for a sharp-caster. A windlass, necessary to span ("draw" or "cock") the weapon, is attached to or at the end of this case, but precisely how it is attached is unclear from the text as we have it, and Philon did not talk about the spanning mechanism at all, so cannot help. Wilkins reconstructs a box enclosing the windlass *around* the end of the case, retaining a stock length of 19D, and pointed out that Schramm seems to have come to this conclusion too, since he did the same in the drawing in his 1917 article. But in his earlier 1906 and 1910 publications Schramm showed the 3D length windlass *added to the end* of the 19D case, making the whole 22D long, and Marsden followed that (as did Fleury). Wilkins argued that this "results in a far longer stock than is required." Recently, Ober, in a paper that fails to distinguish between tension and torsion, palintone and euthytone catapults, transformed Marsden's exaggerated "working length" into a *formula* for stock length (30D)! [6] This is more than a third as long again as it should be.[7] With Morgan, Wilkins has shown empirically that the machine is balanced and serviceable with a 19D stock. As he notes, no certain solution to the problems is possible unless a well-preserved *scorpio* windlass and housing is found in the ground (2000: 83). It is conceivable that such evidence exists in the remains of the Azaila 2 catapult, but it has yet to be examined with that question in mind. I just note that a 16D case, *apud* Philon, plus a 3D windlass housing, *apud* Vitruvius, adds up to a 19D stock. 19D is also the case length of the palintone, of course.

The arms are the same length as a Philonian euthytone, but they have changed shape from Philon's time; they are now curved. The curve is also based on the spring-hole diameter. Marsden interpreted the text to mean that the appropriate shape can be drawn by multiplying that diameter by eight, and treating it as a radius. This gives a shape that allows the arm to travel slightly further and thus generate more force than could a straight arm in the same machine.[8] This is the sort of refinement we might expect from what was by now a mature technology: a minor change to a major component, which gave a significant improvement in performance. This is besides the sort of

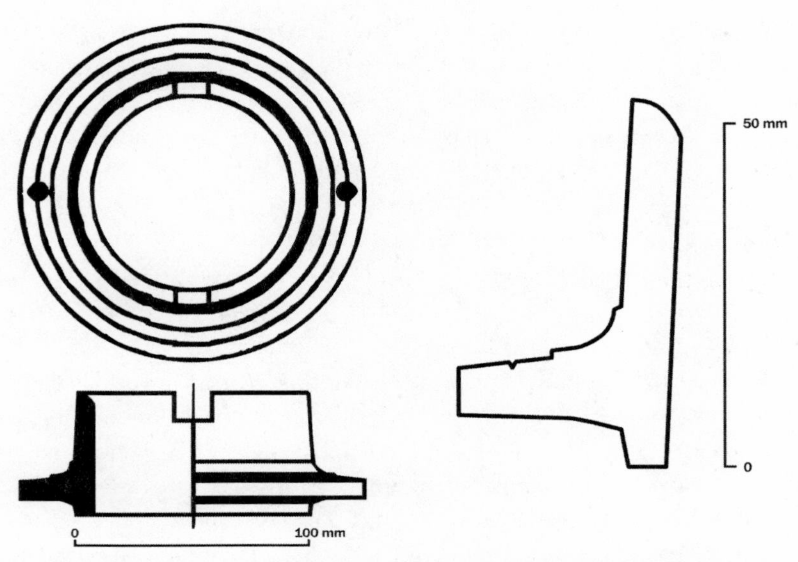


FIGURE 8.1 *Caminreal washer detail. The top inner edge is flared, in the manner (later) recommended by Vitruvius. (T.D. Dungan after Vicente et.al.)*

recurrent minor variations that arose more from personal preference or particular applications and made little difference to performance, such as, for example, alternative methods of drawing the weapon. "The designs of stone-throwers *reveal numerous variations*, though all are produced with one and the same object. Some, you see, are stretched by levers and windlasses, some by pulleys, others by capstans, some even by systems of drums" (Vit. 10.11.1).

In terms of the absolute size of a scorpion, Vitruvius states clearly that a single person was enough to turn its windlass (10.1.3). This, for him, was part of the definition of the *organon* (usually translated as "engine") as a subset of the "machine;" an *organon* was so designed that a single person could set it in motion. He cites the scorpion as an example of an *organon* and the ballista as an example of a machine. So scorpions were small catapults, and although Vitruvius does not say so, we know from Livy that they came in two recognized sizes, smaller and larger (see below).

Let us then turn to the Vitruvian ballista. The chief, in fact, the only real advance in the palintone design of Vitruvius' time that is observable through his description is the shape of the upper section of the washer. This is now elliptical (though the exterior bottom part of the washer must remain round, as it will rotate) and flared at the top. The intention seems to be to compensate for space lost to the iron lever, over which the sinew cord is wound (10.11.4–5). The flaring is seen very clearly in the archaeological material; the washers from Caminreal, dated to about 75 B.C., show a distinct flaring in the top 6 mm of the inner edge (Figure 8.1).

They do not have elliptical cross-section, however, so that was presumably a development of the latter half of the first century B.C., or was something more easily recommended on paper than implemented in the workshop. The flaring would undoubtedly make it easier to load spring-cord into it, but would make no difference to how much spring-cord could be inserted into a washer of a given diameter. The elliptical interior shape of the washer would make a difference to performance (assuming the *minor* axis of the ellipse was the same as the spring diameter of a circular washer, of course), but whether anybody actually made one of this shape has yet to be confirmed in the archaeology. As Marsden observed, the apparent reduction in spring-hole diameters in Vitruvius' "handy table" of ballistae calibers can be explained either by the elliptical washer hypothesis or by a copyist's error in the units of measure (from *unciae*, 12 to a foot, to digits, 16 to a foot), but not by both simultaneously. If they were originally *unciae*, then Vitruvius' figures correspond to Philon's. If they were digits, then Vitruvius' machines apparently were greatly underpowered when compared with Philon's. Marsden concluded that Vitrivius originally said *unciae* and not, as appears to us in the very defective manuscripts, digits, and therefore his calibers and spring sizes corresponded to Philon's. Fleury concurs.[9] In all the key respects, the Vitruvian ballista is unchanged from a Philonian palintone. The ballista stock is 19D, the arm 6D, and the springs are still about 9D long by default but the washer is flared at the top where the lever interferes. The performance was probably very similar, too; if elliptical washers were ever incorporated in reality, they may perhaps have achieved as much as 10 percent more power over regularly equipped versions.[10] Although Vitruvius does not say so, an improved washer design was used on euthytone (sharp-caster, scorpion) catapults, too, as Caminreal has demonstrated, though as mentioned above it demonstrates only the flared top, not the putative elliptical shape.

Thus we see that in the late republican period the standard model of both the palintone and the euthytone follow the formula discovered in the Hellenistic period, indicating that the general purpose design, optimizing range *and* impact, remained the favorite model (see Appendix 1: "The Calibration Formulae"). However, modifications to component shapes that offered improved performance (curved arm, flared and possibly oval washer) had been developed and incorporated into those models. Utilitarian Philonian plating of the wooden frame, spectacularly preserved in the Caminreal catapult (see Figure 7.1), was sometimes implemented in the form of a decorative as well as functional battleshield that completely covered the front of the machine, leaving only one or two small holes—one for the slider and bolt, and the second, if original and deliberate, for the wedge that locked the frame and stock components together (see Figure 9.1).[11] This seems to have occurred before c. 80 B.C. when Azaila, which was possibly native Iberian Sedeisken, Roman Beligio, fell to Pompey's forces.[12] A small aperture for the slider and bolt between two central stanchions had been a regular design feature for euthytones at least since Philon's time (c. 200 B.C.) if not since the euthytone's invention, and he introduced what was effectively a single central stanchion in his wedge-engine. I would imagine that this assisted the catapultist's aim, rather than hindered it, since a small aperture is a device that many sighting instruments employ. Whereas we would naturally imagine laying the bolt and then aiming, as with a gun, I think that with these catapults, the target must have been sighted along the stock and *then* the bolt laid and shot, fairly quickly. The only alternative would have been to look over or around the frame, which would result in very fuzzy targeting. What the single central stanchion prevents is aiming and shooting *simultaneously*. The battleplate would have offered the catapultist great protection, and the catapultist's need to be protected was presumably considered a greater priority than his need to aim and shoot simultaneously. The completeness of the cover suggests that high accuracy could be expected from enemy snipers, which is what we would expect if the enemy snipers were also catapultists. It also implies that catapults were most accurate when shooting stationary targets, or targets moving directly toward the catapult. And that is precisely how we find them used up to this period.

Variation in individual component sizes for specific, nongeneral, purposes continued to be advised in theory and, as we shall see further below, implemented in practice. The scorpion came in two standard sizes, larger and smaller, as did the ballista. But there appears to have been a third kind of catapult too—referred to by the generic term "catapult." For when Scipio captured New Karthage (modern Cartagena) in Spain in 209 B.C., he seized "120 catapults of the largest kind, 281 smaller ones, 23 large ballistas, and 52 smaller ones," and "a vast number of larger and smaller scorpions, armour and weapons" (Livy 26.47). Reordering this to make the information easier to process, there were

> a vast number of smaller and larger scorpions,
> 281 small catapults,
> 120 large catapults,
> 52 small ballistae, and
> 23 large ballistae.

The Karthaginians apparently had approximately twice as many of the smaller version of each type as they did of the larger. Scorpions, in larger and smaller varieties (distinguished as *major* and *minor* in Latin, or *skorpion* and *skorpidion* in Greek), were the most numerous type, and indeed were lumped together with armor and weapons in Livy's reckoning. The prevalence and apparent ordinariness of "scorpions" is consistent with the argument of Chapter 5 that scorpions were one-man catapults (as Vitruvius said): they were little catapults, personal hand weapons. Perhaps this was the regular name for those catapults represented in the archaeological record by the smaller washers, which would have discharged sharps of 1 to 1 1/2 span long, or *glandes* weighing 30 to 35 g. It is perhaps not coincidental that some of the unpublished *glandes* of this weight in the British Museum sport tiny, beautifully carved scorpions, revealing the highest quality engraving of the mold (Figure 8.2).

This one came from Rhodes, famed for its engineering traditions. Another of these glandes is flattened at one end, from the impact of hitting its target one assumes. Another, found in the Nile delta and weighing in at a relatively heavy 46 g (for a "larger" scorpion?), has its scorpion standing very proud of the surface, and displays the thunderbolt, characteristic of the Macedonians, on the other side.[13] The scorpion is not a type listed in any numismatic reference work for Greek or Roman coins, therefore it would not appear to be doing duty on *glandes* as a symbol of the issuing state.[14] It was, for a time, a symbol of the Praetorian Guard, but that doesn't explain its appearance in Hellenistic Greek graves. Perhaps, then, it indicates the caliber of the shot on which it appears: "for a scorpion catapult." More work on "slingshot" needs to be done, as it appears to have been more or less completely neglected to date. Even the British Museum's collection has yet to be catalogued (and some of their *glandes* have been there since the 1800s), so who knows what sits in museums and private collections around the world?

Despite Livy's shortcomings as a historian, it is hard to explain his three types of catapult away, and many scholars have wrestled with the problem of terminology here and elsewhere in Latin sources. If we suppose that by "catapults" Livy meant ballistas and scorpions combined, we then need to explain why he uses the term "vast" for a subset of a specified number, and that makes no sense. Whatever "catapults" were, the Karthaginians had more than five times as many of them as they had ballistas—though not as many, apparently, as they had scorpions. In the surviving historical texts, whenever numbers of catapults are given, there are always fewer stone-thrower catapults than "catapults," which term, when used without qualification, usually seems to mean sharp-casters, as it does, for example, in this passage of Polybios: "a hundred and fifty catapults and twenty-five engines for throwing stones."[15]

Like others, Wilkins (2003: 16) supposes that the Karthaginians' "catapults" were large sharp-casters, discharging two- and four-cubit sharps (small and large, respectively), while the scorpions were the smaller sharp-casters shooting two- and three-span sharps (small and large, respectively). This is an attractive guess, though if these machines were nothing more than the same design in four sizes, I would expect all distinction between them to lie in the size, not the name. Why not have, for example, personal, small, medium, and large scorpions? If the ballista is the palintone, and the scorpion is the euthytone, what type of catapult is left to be called "catapult"? I think that here again the solution may lay in the issue of standard and nonstandard designs. We regularly forget the non-Philonian, non-Vitruvian weapons. In all likelihood, one-arms (*monagkones* or onagers) and bow catapults (*gastraphetai*) were in existence alongside the palintones and euthytones. Alternatively, or additionally, "catapult" might mean any palintone or euthytone that was not a ballista or a scorpion, as described by the designs we associate with those names. In other words, "catapults" could have meant all the nonstandard catapults. Most ancients were probably more concerned with the outward appearance than with the inner workings of these machines. Consequently, significant structural or functional details may have been ignored in their accounts, even when they were impressed by the new devices; such a problem bedevils the history of mechanical clocks in the thirteenth and fourteenth centuries, and it is not helped by the fact that clock terminology remained unchanged throughout.

FIGURE 8.2 *Scorpion glans. GR 68, 1-10, 61, weighs 31g. The 50-pence coin is one inch (27 mm) wide. (Author photo)*

Lienhard pointed out that a machine normally receives a permanent name only after it has achieved a certain level of maturity and settled into people's lives. To demonstrate the point he cited the variety of extraordinary names that once existed for something that still has at least four regular names in English/American, to wit, the plane/ aeroplane/ airplane/ aircraft. He also observed that the terms "machine" and "engine" are commonly applied to new devices, and they may or may not be dropped over time, so that although we still have sewing machines, our search engines run on devices that are no longer called analytical engines but computers.[16] So we may suppose that a device that never made it to maturity or that failed to penetrate the popular imagination—or at least, the written sources' imaginations—remained anonymous, just another "catapult" or "ballista" or whatever was the generic term for a throwing machine favored by that author. Prokopios, talking of topography, avers that names are always appropriate to the things they describe initially, but as the names travel by rumor to people ignorant of the facts, false opinions arise about the origin and meaning of the terms (7.27.17–20). This point is valid for technology as well as topography. We will see a concrete example of it in Chapter 9.

There were more types of catapult than we know about now; the historical sources repeatedly refer to them. For example, Attalos I apparently had "ballistas, catapults, and every sort of *tormenta*," hurling *tela* (the Latin for missiles, equivalent to Greek *bele*) and huge stones at the attack (with Rome) on Oreus in 200 B.C. (Livy 31.46.10). Here "every sort of *tormenta*" seems to mean catapults that were not then classified as "ballistas" or "catapults." While *tormentum* can mean the windlass and any machine that employs it, and therefore noncatapult devices used at the time could be the "other *tormenta*" to which Livy and other sources occasionally refer, that does not seem a very likely interpretation in this instance, as in most instances in which the word occurs.[17] Similarly, when Fulvius came to celebrate his triumph over the Aitolians and Kephallenians in 187 B.C., he displayed (along with twenty-seven captured leaders and generals, vast sums of gold and silver, thousands of Attic and Macedonian coins, nearly 800 bronze statues and 230 marble ones) great quantities of weapons and catapults, ballistae and "every variety of tormenta" (*tormenta omnis generis*), which here too seems to mean catapults.

In 87 B.C., Athens, which at that time was under the tyranny of a philosopher called Aristion, had joined the wrong side in the Third Mithridatic War, and was visited with a siege by Sulla, who was on the brink of civil war with his colleagues back in Rome. After a failed initial attack on the city, Sulla collected equipment and siege materials of all kinds, iron, catapults, "and everything of that kind" from the Athenians' traditional enemy the Thebans. The scale of the operation was phenomenal. According to Plutarch, 10,000 pairs of mules were needed to move his siege engines, some of which broke under their own weight—suggesting that the age of experiment was not over by Sulla's time, at least for the Romans, who had yet to take the Rhodian and Alexandrian arsenals under their ample wing. Having exhausted other timber supplies, he cut down the trees in the Akademy to construct huge siege towers; and he demolished the long walls connecting the city to the harbor, to provide material for his siege mound. A successful sally by the defenders sent about ten days' worth of machines up in smoke. His opponent in Peiraieus (Athens' harbor), Arkhelaos, had many engines (πλεῖστα ὄργανα) too, but these did not have great range because when Arkhelaos led all his troops out on a sally they had to keep under the walls "so that the defenders on the walls could reach the enemy."[18] In due course Sulla deployed some catapults that apparently each discharged twenty of the heaviest caliber lead shot simultaneously, which did enormous damage to a wooden tower, as well as kill lots of men.[19] Appian does not tell us how 20 shot were loaded into one catapult, nor what caliber or type of catapult was being used for this operation. It seems to me that the only simple explanations involve *either* a one-armed design, so that the shot were loose in the sling, *or* a barrel of some sort, so that they were contained until fore of the framework, but neither idea has much to recommend it. One would need great faith in one's onager to load lots of loose, large caliber lead shot in its sling, and a barrel on a catapult lacks any supporting evidence except for Zopyros' double-barreled sharp-caster. In that design—which is for a *gastraphetes*, a bow catapult—a hollowed beam is laid over the slider to form a barrel (in this case, two barrels), but even that would not work here without serious modification. That is not to say that the Romans could not have managed such a modification, but the further the modification has to move from Zopyrus' design in order to launch shot instead of sharps, the more we move into the realm of speculation.

FIGURE 8.3 *Impact craters on the walls of Pompeii. These are mainly small calibre impacts, created by missiles that were anti-personnel and did not threaten the integrity of the wall. Recent excavation of houses behind this section have revealed glandes as well as ballista balls, presumably munitions that were aimed too high (as these were aimed too low) to hit the defenders on the wall. (Author photo)*

Sulla's attack on Pompeii a couple of years earlier (89 B.C.) used stone-throwers of a variety of calibers. Impact craters are visible on the curtain adjacent to the Herculaneum Gate (Figure 8.3), and on the wall beside the spot where the Vesuvian Gate once stood, further along the north wall. Generally small caliber craters, which could have been created only by shot from small caliber catapults, these probably represent shot that fell short of their intended targets, which were almost certainly the defenders on the walls. Other shot flew over the wall to land on the buildings behind, where recent excavations have discovered widespread destruction of residential and commercial properties in the area, and "many" ballista balls (and more lead sling bullets) have been found in the deposits created in the subsequent clear-up.[20]

When they erupted, the Roman civil wars raged over the length and breadth of the Mediterranean, and one urban casualty was Azaila, in Aragon, Spain.[21] Along with other settlements in the Ebro valley such as Caminreal and Calagurris, La Corona, and Valencia (Map 6), Azaila had apparently sided with Sertorius, who had declared independence from Rome when Sulla was dictator, and now Rome, in the person of Pompey, arrived to bring this area back in the fold, which took about four years, 76–72 B.C. On victory in Spain, Pompey refounded a number of settlements, Azaila and Caminreal among them, leading to the deposition of the detritus of the recent battles in their foundations, whence they were found in modern excavations. The remains of at least two and perhaps three catapults turned up at different locations in Azaila, but the finds fell victim to the Spanish Civil War and some of them are lost.

The site of Azaila, which covers about twenty hectares, is in the province of Teruel, Bajo Martin region, about midway between the Basque country and Catalonia. The Romans destroyed a native (Hallstatt) settlement here in the late third century B.C., then the Romano-Iberian city that grew up from the ashes was destroyed by young Pompey when he put down the rebellion of Sertorius in the 70s B.C. The rebuilt Romano-Iberian city was destroyed for the third and final time around 49–45 B.C., following the battle of Ilerda, after which it was deserted. As a result of the desertion, the site is in relatively good condition and the stratigraphy straightforward.

In 1942 a catapult washer was found in the remains of what is described as a chief's house on the Akropolis, and I designate this catapult as Azaila 1 (Figure 8.4). The washer has an internal diameter of 94 mm, exactly the same size as Mahdia 1. The washer has two holes on opposite sides (like the Caminreal catapult) to push pins through to engage with a counterplate below after adjusting the tension of the spring. It also resembles the Caminreal washers in the height : diameter ratio. It was not, however, made from the same melt, because the Azaila washer has a significantly higher copper content. A spring diameter of 94 mm corresponds to a slightly over-sprung 3-span sharpcaster or a slightly over-sprung 1-pounder (Roman) stone-thrower.

Another catapult had been found in 1925 in a 20 cm thick layer of carbonised material and ash on a temple floor, where it had apparently fallen from above.[22] I designate this catapult as Azaila 2. What survived and was identified was the frame, bronze and iron plating, complete in places with charred wood, all of which was burnt and badly preserved, but it is remarkable that either timber or metal plating survive at all—both are usually removed or decayed by the time a site is excavated. The fragments of the metal plating are straight, of variable length, from 30 to 80 mm wide (most are about 50 mm), and 3–10 mm thick. In some cases, two plates survive facing one another with their nails piercing the hollow that remains after the wood of the beam that they reinforced was burned away. Other plates reveal a curve, and evidently came from the stanchions at arm height. Another piece appears to be the battleplate, with central aperture for the slider, like the Caminreal and Cremona examples. Other metal parts may be—uniquely—

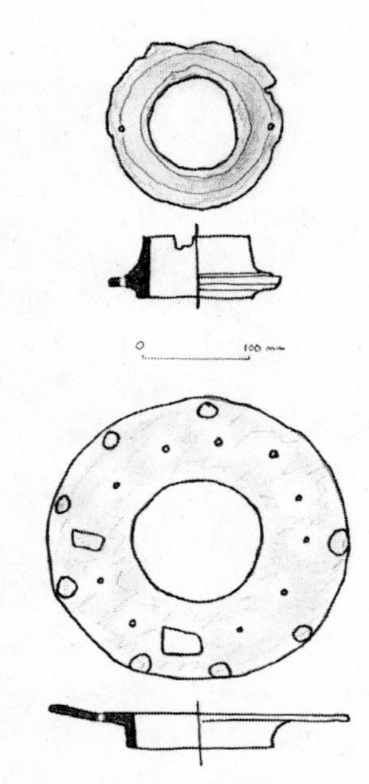

FIGURE 8.4 *The Azaila washer and counterplate. (Author sketch after Díez)*

part of the winch mechanism and the handspikes. The components and dimensions of the "temple catapult" (= Azaila 2), notably the hole-carrier, which is estimated at 60 x 10 cm, and stanchions at 35 cm high, indicate a sharp-caster with a spring diameter in the region of 87.5–100 mm (3–4 inches), that is, a catapult built to shoot missiles of about 3-span to 2-cubit length. So the washer and the frame belonged to similar size catapults, but were apparently found at two different locations. When they entered the National Archaeological Museum in Madrid in the 1940s, all the pieces were recorded as if they belonged to one catapult.[23] It is possible that a third catapult (Azaila 3) was found, in the form of a complete counterplate—the ring of metal that is fixed to the wooden frame, on which the washer sits. The counterplate cannot have come from the same catapult as the washer because, although at first sight the washer appears to fit in the counterplate nicely, the pin holes do not line up. The pin holes on the washer are too close to its projecting bottom edge to have aligned with the pin holes on the counterplate, which are further in from its edge than the washer flange is wide. So I have given the counterplate a separate number for clarity (Azaila 3); future research may cast more light on the issue.

The counterplate has an L-shape profile, a feature shared by the much later Hatra ballista and not by the approximately contemporary Caminreal catapult, whose counterplates are flat. However, like Caminreal, the Azaila counterplate has twelve pin holes to lock the washer, at an average of 30° apart (but individually varying from 23 to 34° apart). The counterplate has an internal diameter of 122 mm. The washer to fit it would have been smaller than the internal diameter of the counterplate, since its lower edge would sit in it, with the washer's flange resting on it. A formulaic 2-mina stone-thrower would have a washer diameter of 120 mm, but that is too tight to fit on here. The distance between the pin holes and the central hole in the counterplate varies from about 30 to 50 mm, suggesting that the washer and its pin holes would have to tolerate lateral adjustment of not 2 mm but up to 20 mm.

A formulaic 4-span (or 2-cubit) sharp-caster would have a diameter of 100 mm, allowing the required 20 mm tolerance. Ancient catapults were sophisticated machines, but this is a salutary reminder that they were not the products of precision engineering in the modern sense. Bolt heads corresponding to missiles of this sort of size (4-span, or 2-cubit) are frequently found in Roman levels on sites in the Iberian peninsula (e.g., Numantia and Osuna).[24]

The counterplate came from a catapult that did not follow Philon's formula exactly, for the frame at least. We know this because the counterplate's external diameter is 286 mm, which is about 40 mm more than twice the internal diameter. The ratio for the width of the hole-carrier in a formulaic Philonian machine is 2D in the middle and 1 1/2 D at the ends, and since D is here less than 122 mm, that means a hole-carrier, to which the counterplate was attached, of not more than 183 mm wide at the ends and 244 mm in the middle. Obviously that was not the case with this particular machine, given the counterplate's 286 mm external girth, which would seem to indicate a hole-carrier at least that wide where the washers were. Therefore, the hole-carrier must have been at least about 290–300 mm wide, while D was, as we have seen, probably in the region of 100 mm. So the width of the hole-carrier on this machine was more like 3D than the Philonian 1 1/2–2D. We might be tempted to suppose that this counterplate is so badly made (pin holes spaced irregularly with regard to the central hole and to each other, as well as the excessively wide external diameter) that it was discarded, but the eight surviving nails that once fixed it to a frame speak loudly against that idea. Perhaps what we have here is one of those catapults that was not even trying to be a formulaic scorpion or a ballista, but was rather one of the many other types of catapult to which the historical sources keep referring but about which we have no details.[25]

More and bigger catapults were also in use by Pompey's forces at Azaila, as demonstrated by a sample of near-spherical limestone ballista balls from the site that were collected and are preserved in the National Museum in Madrid. They vary from 12.4 cm diameter and 1.95 kg to 19.4 cm and 7.3 kg.[26] Specifically, (i) one at 1.9 kg (appropriate for a 5-mina engine), (ii) one at 2.6, (iii) 2 at 3.25, (iv) 2 at 3.5–6, (v) 2 at 4.8–9 (for a 10-mina engine), (vi) 2 at 5.3–4, (vii) one at 6.3, and (viii) one at 7.3 kg (for a 15-mina engine?). We appear to have eight different calibers, or perhaps seven, if group 3 and 4 belong together. As Díez points out, ballistae with significantly bigger spring-hole diameters than those so far found at Azaila would be necessary to launch these stones into the air. A 5-mina ballista, the smallest indicated by the recognized ballista balls, would have a spring diameter of about 170 mm, whereas the largest catapult found had one of about 100 mm. Stones like these, carved out of sandstone, limestone, and even granite, and weighing up to almost 40 kilograms (1 1/2 talent caliber), have been found all over the Peninsula, for example at Arcóbriga, Numantia, Old Cáceres, and Osuna. In the first and last of those sites, as at Azaila, the ballista balls were found around the walls against which, or against the men posted on which, they were shot. Over 300 were also found at Calagurris, further west along the Ebro.

Catapults and scorpions are described as "normal weapons of defence" in the *African War* 31, and certainly appear to be regular elements in Caesar's and his opponents' arsenals, but they were not necessarily permanent fixtures. Catapults were singled out as having been particularly useful in the defense of a camp manned only by a small force, and they played a significant role in the siege of Alesia. Domitius, for example, put catapults on the walls of Corfinium in anticipation of an attack in 49 B.C. Caesar had a ballista big enough to demolish a wooden tower at Ategua during his campaign in Iberia, but a subsequent comment perhaps suggests that a machine of this size was unusual in this time and place.[27]

No extensive training was needed with catapults by this time; novices could be expected to learn quickly how to handle them. Caesar is said to have prepared a plentiful supply of catapults for the African War because his army at the time was small and inexperienced (*African War* 73). This is a sign that the catapult was a mature technology by Caesar's time. It reminds one of what is often said of the crossbow's chief advantage over the longbow, and the modern handgun's advantage over, say, the sword: it required very little skill or training to use. This is a characteristic of developed mechanical technologies; over time and through service they get simpler to use as redundant features are dropped, problematic or fiddly features are designed out, and key features are brought to the fore.

There is also clear evidence for the use of catapults on prepared battlegrounds now, as well as in attack or defense of urban sites. Range was one advantage, if not the chief advantage of the catapult over manually discharged weapons. Repeatedly in the historical sources, catapults are mentioned precisely for this feature. For example, in the last days of the Third Macedonian War an engagement took place between Perseus and Roman forces in a dried-up riverbed. This riverbed is described as being a little over a mile wide (i.e., 1,000 paces), and the slope of each bank from top to bottom was about 300 paces. A Roman pace was significantly more than a modern yard or meter, 1,000 paces being equal to 1,618 yards or 1,480 meters. The camps were so located at the top of opposite banks that both could see down into the riverbed from their rampart walls. On the second day of fighting down there, the Romans came under fire from "all sorts of missiles and stones" shot from the towers of the Macedonian fort, and "as they approached the enemy's bank, the shots from the catapults even reached the rear ranks," leading to far more severe losses that day for the Roman forces.[28] Macedonian *tormenta* are mentioned here only and specifically because they could reach the parts that other range weapons could not reach. Likewise, one of Pompey's camps at Dyrrhachium was said to have been located on a hill that was so far away from Caesar's camp that "a missile shot from a catapult could not reach it." Caesar invited Pompey to fight a pitched battle stopping "only so far from the rampart that a missile shot by a catapult could not reach it." When Caesar besieged Uzitta (which evidently lacked catapults), he built a camp "just out of range of a spear-cast from town," and on his camp walls placed catapults and scorpions at close intervals, so that his troops could harry the defenders, but they could not reciprocate.[29]

In 86 B.C. Sulla appears to have placed stone-throwers behind his main block of infantry to lob fire bombs over their heads at Mithridates' troops at the battle of Khaironeia. In 57, Caesar placed catapults in temporary forts designed and built to protect his flanks in a field engagement against the Gauls. In 51, he drew up his line of battle specifically where "the missiles of the catapults could be discharged into the hordes of the enemy." Catapults were again being used as field weapons in 48 B.C., when Pompey brought them up to an area outside Dyrrhachium (Illyria) that Caesar was trying to fortify, forcing him to withdraw from the field under fire.[30]

I think we can see hints of this sort of use earlier too. The Macedonians had a reputation for deploying and using missile weapons very well. In 198 B.C. Philip's forces included catapults and ballistas ranged on cliff tops, as if on walls, apparently in front of the camp overlooking a plain on which battle was expected.[31] So these machines appear to have been deployed to be used in the field, rather than in a siege situation. If they were little catapults this is not a significant issue, as such catapults could be carried by the men who shot them and mobility is to be expected. But larger caliber weapons need either transport or their own carriage. Field use would surely encourage the development of mobile artillery units. The carroballista is a cart-mounted catapult whose origins are uncertain. We saw in Chapter 6 that a Spartan commander once tried to use catapults on carts in the field, but his opponent saw the threat and acted quickly to smother it. Catapults and carts to carry them were considered separate items at the beginning of the Third Macedonian War. For when, in 171 B.C., Perseus prepared to engage the Romans, the only thing in which the royal stores were said to have been deficient and needed supplementing from elsewhere was vehicles, "to carry the catapults and a huge supply of missile weapons and other war equipment that was ready." Evidently he did secure them, for in 168 B.C. the banks of the river Elpeus were fortified and had catapults "everywhere" (*ubique*).[32]

Catapults continued to be mounted on warships; they were used in support of the invasion of Britain in 55 B.C., where they helped to drive back the Britons from the water's edge, allowing the Roman legionaries the space (and confidence) needed to disembark. Octavian mounted towers and catapults on merchantmen in a naval spat off Brundisium. Catapults sometimes launched irregular loads—more bombs or grenades than stones. For example, during a naval engagement in which Hannibal fought with Prusias of Bithynia against Eumenes of Pergamon and (because of) his Roman allies, stone-throwers were apparently used to launch pots of snakes that smashed onto the deck of Eumenes' ship. This probably made any snake that survived the impact pretty dazed, but frightened the crew very effectively once they realized what the crockery bombs contained. (More gruesome loads were possible: Hasdrubal's head was shot into Hannibal's camp in 207 B.C. Another unusual load was pots of stinging insects, which the Hatreni used effectively to drive besieging Roman forces from the walls.)[33] At Actium, stone-throwers were used to launch pots of charcoals and pitch, which set fire to Antony and Kleopatra's ships, with terrible consequences for their crews. Plutarch reports that catapults on Antony's ships were even mounted in wooden towers. Ballista balls of about 5-mina caliber have also been found on the seabed off Cape Actium, indicating the use of medium caliber ship-borne stone-throwers by the fleets of Antony and Kleopatra or Octavian (or both) during the battle that left the last man standing Emperor of Rome. Cassius Dio describes a naval engagement between smaller, quicker, fully plated ships using traditional ramming methods to try to sink larger, and in particular higher, ships that were well armed with engines and catapults which could destroy the smaller ships if they could hit them with stones at a distance, or catch them with grapnels and bring them close. He likens it to an engagement between cavalry dashing forward and back against heavily armed hoplites guarding against those who approach them and trying to bring them to close quarters. He describes dense showers of stones and sharps thrown at ships charging in to ram. For all this, Dio throughout uses only the terms "bow" and "machine" in his description of this battle, and only "slingers" and "archers," "hoplites" and "peltasts" apparently constitute the fighting forces on deck. Machine, μηχανή, is a perfectly legitimate word for catapults, in particular for powerful ones needing crews to work them according to Vitruvius (10.1.3), but it can mean a host of other heavy devices too. Again we face a situation where catapults are almost hidden by the vague language of the source—Dio describes the battle without mentioning catapults explicitly—but their presence at Actium is not in doubt in view of several points: the firepots in the story; the "machines that can damage at a distance" of Octavian's pre-battle speech (which are clearly not cranes or suchlike); the other sources' explicit mention of catapults; and the ballista balls on the seabed. Catapults on ships, although again not specifically mentioned, explains most easily an engagement on the coast between the forces of Otho and Vitellius, contending for the throne in A.D. 69, in which the fleet on one side, and cavalry on the other, played significant roles. Ships are not ordinarily active players in land battles, but these ships, brought close to shore, seem to have used the catapults on their prows to very good effect against the enemy troops on the beach.[34]

The largest caliber sharps seem to have been used in this period. We find enormous stakes being launched by catapult. The Massiliotes apparently possessed 7 1/2-cubit-long sharps, iron-headed stakes some 12 Roman feet (11 1/2 British feet or 3 1/2 meters) long, which could penetrate any ordinary siege mantlet or penthouse, driving through four layers of hurdles before burying themselves in the earth. Lucan describes their sharps driving through armor and bones and flying on, leaving death behind, piercing more than one body before stopping. And of their stone-throwers, he says that the Massiliotes' shot not merely crushed the life out of their victims, but disintegrated flesh and blood.[35] Massilia, in the south of France, a Roman ally for about a century by the time Caesar attacked it, apparently had long invested in its military hardware, and possessed every kind of war machine and a multitude of engines. This, in combination with the rarity of the very large sharps, and the various other contrivances and stratagems that they employed during this extraordinary siege, suggests that the Massiliotes had their own engineering traditions, which may or may not have converged with those of the Eastern Mediterranean reported and preserved for us by Philon. In response, Caesar's forces adapted their siege machines to deal with the defenders' mega-sharps, such as using foot-square timbers for the penthouses, instead of traditional beams and planking, and two-foot-square beams, plated and bolted, for a gallery that had to go right up to the wall! To deal with the Massiliotes' stone-throwers, as well as sharp-casters, which were directed against a brick tower that Caesar's forces were building in range of the wall, the Roman engineers devised an ingenious roofing system. This involved a sturdy timber frame covered with extended planks, which were in turn covered with a layer of bricks and clay to counter the incendiary missiles, and that layer was covered with mattresses, so that the stones' impact would be cushioned and they would not crash through (*Civil Wars* 2.9). The gallery that Caesar's forces built to go right up to the Massiliotes' wall was even designed to deal with the possibility that the defenders might try to dissolve the mudbricks on its roof by pouring water on it from above. Never let it be said that the Romans lacked imagination.

The Massiliotes' extraordinary sharps were perhaps the inspiration for a novel catapult missile reputedly invented by Agrippa for the great naval clash between Octavian's and Pompey's fleets off Sicily. This was something like a harpoon for catching ships. It is described as an iron-band-reinforced pole 5-cubits long (about 2 1/2 meters or 7 1/2 feet), with metal rings at each end, to one of which an "iron claw" or builder's grasp (see Figure 6.2) was attached, and to the other numerous ropes. Its lightness is offered as an explanation for its good range; one assumes that it was launched by a large sharp-caster mounted on the prow of a warship. Once this claw had grasped an enemy ship, which happened "as soon as the ropes were pulled on it from behind," the device was retracted by machine—a windlass, presumably—and the target vessel was reeled in like a whale. Since the weapon was new and unexpected, the crews attacked were not equipped with "scythes on poles" which, had they been present, could have been used to cut the ropes.[36]

To summarize, we have seen that from around the end of the Republic, if not before, catapults seem to have joined the regular equipment of Mediterranean armies as well as of garrisons and town defenses. Both Caesar and Pompey had them as standard, apparently taking them wherever they went or having them delivered at a specified destination, including for what we might call short-term stop-overs.[37] This development in military structure and catapult function continued into the next period, for at some time under the emperors, every legion came to be assigned a certain number of catapults. The legions apparently had some freedom to choose the caliber of their ballistas, because the Fifteenth's one-talent machine at the siege of Jerusalem (which we will discuss in the next chapter), is specifically said to have been the biggest of all those present, yet there were two other legions there as well.

### Technology Transfer

Military technology is likely to be transferred to the enemy whenever it is used against them. Through battle the enemy at least learn of the existence and capabilities of the weapons and techniques used against them, and may attempt even on that basis to reproduce them. Thus Cato was said (by Pliny, *NH* pref. 30) to have been educated by Hannibal, as well as by Scipio. Or they may capture a specimen and/or people who know how to use it, and copy that with advice from the captive(s). If they do secure a specimen, they will also know more of its shortcomings (every weapon has some). The Nervii learned how to make siege-works by watching the Romans and being instructed by prisoners of war; Caesar elsewhere commented that they were very good at copying, and were inventive too, and some technologies could be transferred simply by intelligent copying.[38] That, no doubt, was one method by which catapult technology was diffused, and would explain why some worked well and some did not.

The Romans were extremely adept at adopting technologies from peoples they conquered. In this process pragmatism was apparently unhindered by prejudice, and the best of ancient technology, wherever it originated, was absorbed into Roman traditions, where it met, modified, and was modified by other technologies, old and new. Their armor and weapons were in constant evolution because of contact and conflict with other peoples. For example, most forms of "Roman" helmet seem to have been based on Celtic designs, the pilum may have originated with the Etruscans, "Moorish" javelins came from Africa, and cataphract cavalry-archers were adopted from the Persian/Parthian tradition. The emperor Antoninos was nicknamed Caracalla after the Celtic or Germanic word for a type of cloak that he adopted and adapted—a full-length hoody. Sometimes the debt was explicit and acknowledged, for example, Polybios tells how the Romans first learned to build a fleet by copying a Karthaginian vessel that ran aground, and later copied Rhodian ships. Centuries later Vegetius added that experience in battle showed the Romans that their Liburnian allies' warships were of a better design than anyone else's, including their own, and as a result the Romans copied both the design and the name.[39]

The Egyptians were reputed by Caesar so good at copying Roman tools and techniques that "no sooner had they seen what was being done by us than they would reproduce it with such cunning" that they seemed to be the originators.[40] Away from the southern Mediterranean, in northern Europe, the Batavians, who were evidently much less skilled than the Egyptians at copying by sight, used deserters and captives to teach them how to make and use Roman siege engines and sheds. The Romans' expertise with catapults meant that such "crude" machines were soon destroyed by Roman shot or firebrand, but the Batavians in due course became the Germans' artillery experts, and other German tribal leaders asked them to build machines and siege works for later campaigns.[41] It is likely that engineers, usually stationed by their machines, were captured if not killed whenever machines were captured; thus we are told that one of Pompey's chief engineers, one L. Vibullius Rufus, twice fell into Caesar's hands.[42] Another method of technology transfer through battle was by accident, so to speak. Hanno moved Utica's artillery to his camp, after (as he thought) chasing away the mercenaries who were besieging the town, but while he and most of his forces were celebrating in said town, the mercenaries came back and seized his camp, and thus obtained both his and the town's artillery. Some Spaniards, after causing a Roman force to abandon its camp under cover of dark, entered the deserted camp and armed themselves with the equipment that had been left behind "in the confusion." The Iapydes of the transalpine region used against Augustus Roman machines that they found lying around some years earlier, after a civil war clash in their territory between Brutus, on the one hand, and Antony and Octavian on the other.[43] In Chapter 10 I will tell the tale of Bousas, who taught the Avars how to build *helepoleis*, siege towers.

Another vector is the arms dealer, moving either between allies, or between suppliers and customers whoever and wherever they may be. A *negotiator gladiarius*, procurer of swords, who pops up in the records of Mainz, was perhaps such an arms trader. This trade is notoriously secretive, as well as dangerous, and it would be naïve in the extreme to think that the paucity of evidence for arms traders in antiquity was an accurate reflection of the state of the business at the time. The largest businesses known from classical Athens were arms manufacturers (shield workshop and blade maker, respectively), and it is inconceivable that the arms trade was not flourishing in a world where warfare was more common and regular than tax collection. The Codex Theodosius laid down capital punishment for anyone caught teaching the barbarians how to build ships, but the Vandals (who had moved down from inland Eurasia) nevertheless had found out how to build a decent navy by A.D. 419. Cassiodoros meanwhile lamented that Italy (being governed by the Goths when he was writing) lacked a navy despite the abundance of timber.[44]

To the victor went the spoils, and on surrender of a town, their catapults, along with their arms, ships, and money, were usually handed over. This could be a very effective method of acquisition of new technologies. Consider a comparative case from Hawaii. In 1790, the lightly armed American merchant ship *Eleanor* fired a broadside and killed about 100 natives. The natives responded by seizing the next American ship to reach the islands. They unloaded its weapons, and captured a white man (*haole*) who was able to teach them how to use the guns. Over the next few years they captured more ships and seized their weapons and gunpowder stores. By 1804, one of the chiefs could deploy 600 muskets, fourteen cannon, forty small swivel guns, and six mortars.[45] Returning to antiquity, paperwork might be an asset to be exploited by the conquerors, or not, depending on the particular conqueror's appreciation of the contents. Thus it appears that the Karthaginian libraries were given away by the Romans in 146 B.C., when they destroyed the city, making an exception only of Mago's twenty-eight volumes on agriculture, of which the senate ordered a Latin translation be made.[46] Polybios records that Philip V, not relying on another bout of such absent-mindedness by the Romans, kept his head in such an emergency and ordered the burning of his papers before the Romans could get their hands on them (18.33). Papers might include blueprints or other scientific or technological information of use to the enemy, as well as diplomatic documents. Certainly, Pompey recognized the value of Mithridates' toxicology results, which arose from a program of research into poisons so successful that Mithridates was reputedly immune to all known venoms and toxins so that when he wanted to commit suicide he had to fall on his sword. Pompey ordered a Latin translation be made of them, apparently for his own use rather than that of the Roman reading public, since there is no hint that it was ever published either in its original Persian or in Latin.

Allies will copy good technology too, of course. Polybios' belief in Greek superiority over the Romans leaks out through his text here and there. At one point he compares in detail the Greek and Roman methods of cutting and setting stakes around a palisade, and having concluded that the Roman way was better, said that if any military contrivance was worth copying from the Romans, then this was it (18.18.18). One could be forgiven for thinking (erroneously) that the phalanx had beaten the legion.

It is not just technology that is reproduced by enemies; fighting techniques are, too. Caesar observed that troops adopt the fighting techniques of the enemy if they fight them continuously over a long period. We are reminded of the Spartan Antalkidas's criticism of his king, when he said that by persistently fighting the Thebans, Agesilaos had thereby provided them with the means necessary to defeat the Spartans. Agesilaos had apparently ignored a decree of the legendary Spartan lawgiver Lykourgos that forbade campaigning frequently against the same people, for that very reason.[47]

Rome's enemies did not always want captured ordnance, of course, and might destroy it instead. Thus the Parthians destroyed the siege engines apparently left behind by Antony in his haste through their country, including an eighty-foot-long battering ram and other equipment whose scale is indicated by the fact that it was being transported on three hundred wagons and protected by more than 10,000 troops.[48] In the third to sixth centuries A.D., the Goths, Vandals, Huns, and sometimes even the Romans themselves destroyed more than they copied, and the western empire descended into the Dark Ages.

NINE

# Strikes Back at the Empire

*What are our tormenta compared to those with which Etna hurls her missiles?*—[Virgil] *Aetna* 555

THE ROMAN EMPIRE TRADITIONALLY REFERS to the period after the battle of Actium in 31 B.C. It has become common to divide imperial history into two parts, the High Roman Empire and the Late Roman Empire, to acknowledge in words the changing nature of that empire. However, there is no very obvious place to draw the line of separation because the change was a long, drawn-out process, not an event. The significant period of change was the third century A.D. I have opted to make the break at A.D. 284, one of the several conventional starting points of the Later Roman Empire, and Chapter 10 will take up the story from there.

After the prolonged civil wars that brought the Republic to a close and which were the focus of much of the last chapter, the winner, Octavian, henceforth called Augustus, employed his veterans on completing the conquest and pacification of Iberia, the Alps, the Balkans up to the Danube, and Germany up to the Elbe. The last proved to be temporary and three legions were destroyed in the Teutoburg forest, in one of the Roman army's greatest ever disasters, leading them to fall back on the Rhine as frontier. Some areas in the east that had been client states, including Judaea, were turned into provinces in fits and starts through the first century. Much of Britain was added to the empire from A.D. 43 when Claudius invaded. Trajan campaigned aggressively at the turn of the century to add Dacia (modern Romania) solidly and Mesopotamia temporarily to Roman dominions. The enemies of Rome responded to the challenge, which included their adopting and adapting her technology to fit their own purposes and styles of fighting, and Hadrian (emperor A.D. 117–138) consequently focused on consolidation. Territorial defense was the main mission of imperial forces through most of these centuries, and the *fabricae* and civilian manufacturers who made and maintained the war material for the troops would have produced ordnance to suit.[1] Meanwhile within the borders civil war erupted again in A.D. 68–69, the year of the Four Emperors, and A.D. 193–197, when, between the death of Commodus and the accession of Septimius Severus, the praetorians and various armies again tried to put their own candidates on the throne. Cassius Dio memorably described the change in Roman society and history after the death, in A.D. 180, of Marcus Aurelius, father of Commodus (in whom he was "hugely disappointed"), as a descent from a kingdom of gold to one of iron and rust. Severus managed eighteen years, but after the killing of his son Caracalla in 217, the empire endured decades of short reigns, quick tempers, escapism, and excess, with few interludes of stability and security under emperors worthy of the name until the end of the century. The Parthians attacked in the east, the Goths in the north, and there were rebellions in Gaul and elsewhere. As Florus observed (2.30.29), it is more difficult to retain than to create provinces, for they are won by force but secured by justice, and justice was in rather short supply in some provinces. Despite this, remarkably little retrenchment resulted, thanks partly to the methods by which emperors were made and replaced in this era, which put a premium on military ability and command.

Josephus (c. A.D. 37–100), Tacitus (c. A.D. 56–after 118), Arrian (c. A.D. 90–165), and Cassius Dio (c. A.D. 164–after 229) retail the histories of Rome's wars with friends and foes, and often provide some context for the archaeological finds, which remain the best testimony to the machines being built at the time. There is a significant amount of archaeological evidence to help reconstruct this part of the story of the catapult, and more is becoming known all the time. Unfortunately, most remains are badly incomplete, but an exceptionally well-preserved torsion frame turned up recently in a gravel-pit dredger's bucket near Xanten in Germany. Another development of this period is the *carroballista*, and yet another represents a striking departure from earlier designs: the all-metal-frame *kheiroballistra*. So let us continue the story of the catapult, picking up the threads during the civil war that erupted after Nero's death.

The large and wealthy town of Cremona in north Italy (Map 6) was arguably the saddest casualty of the civil wars of A.D. 69. This was the year of the four emperors when, following the death of Nero, the armies of the empire put forward their respective generals, to put the reigns of power back into the hands of men with military abilities and interests. Vitellius was third emperor that year, and Vespasian was challenging him for the purple. Troops from eight continental and three British legions, plus assorted picked auxiliaries and detachments of horse, all loyal to Vitellius, were stationed in or near Cremona, and visitors attending a periodic market there at the time augmented the numbers even further. As Tacitus tells it, Cremona was sacked and burned more by accident than by design, by exhausted, resentful, covetous, and battle-hardened troopers and greedy, vicious camp followers. The

FIGURE 9.1 *Cremona battleshield. This decorated metal plate was affixed to the front of one of the catapults that was in action at the battle of Cremona. It protected the springs and woodwork from damage, as a cuirass protected a man's torso, and reinforced the frame. This one belonged to the Fourth Macedonian Legion.*

destruction went on for four days, until nothing was left of a 286-year-old colony, and its surviving inhabitants and occupants were enslaved. Sometime during that process, two catapults were destroyed and became buried in the ruins. The dimensions of the washers correspond to one formulaic three-span sharp-caster and a one-libra stone-thrower. Also found here was a sharp-caster's battleshield, the metal plate that covers the front of the frame almost completely.

From what is written upon it, we know it was made in A.D. 45.[2] Like the Ampurias and Caminreal catapults, but less pronounced than either of them, this too was less high than wide, in technical terms it was *katatonos*, short-sprung, relative to the formulaic 6D x 6D of Vitruvius' "standard" scorpion (Chapter 8).

The Cremona battleshield has the same form as that of the Caminreal catapult, including, significantly, the rectangular cut-out on the bottom edge, so the basic design seems unchanged (Figure 9.1). Wilkins cleverly explained this feature by reference to the use of wedges to assemble the catapult quickly and securely.[3] Although the complete front plating of the little Xanten catapult lacks such a feature (see Figure 9.3), it is obvious from the back that the stock slotted into the frame, through almost the entire depth of the frame, just below the slider channel. This was perhaps another minor variation in design details. The ability to assemble and disassemble catapults had been stressed by Ktesibios and Philon and was no less relevant in the Roman

Empire than it had been in the Hellenistic period.[4] Catapults needed to be moved around as required, and their modular form lent itself to disassembly (see Figure 9.2).

Just as the ability to disassemble and reassemble field guns quickly was necessary in the Boer War (a particularly notable example of which is still celebrated in the Royal Tournament, an annual pageant in Britain), so it may have been in numerous Greek and Roman campaigns. It certainly seems to have been a feature of this campaign, before the city fell; let us rewind history a little, and examine the events leading up to the city's destruction.

After a battle in the field four miles from Cremona, which the Vitellians lost, reinforcements arriving among Vespasian's successful and buoyed troops wanted to go on to the city to receive its surrender or storm it there and then. Their commanding officer dissuaded them with arguments that included: they did not know the lay of the land, nor the height of the walls, and thus did not know whether to attack with catapults and missiles (*tormenta* and *tela*) or with siege works and mantlets (*opera* and *vinae*).[5] He clearly had only anti-personnel catapults in mind with this distinction, the first two being appropriate for low-level defenses, the latter two for serious ones. He went on to add that it would take only one night to bring up "catapults and machines," and so "bring with us the force to secure victory." This leads one to suppose that all catapults and machines are elsewhere, not here on the field with the troops, but as the narrative continues, it becomes apparent that this is a false assumption and the catapults are in the field and thus the time was needed to move them to the city. For shortly after dusk, intelligence was obtained that six Vitellian legions of reinforcements were on the way, and Vespasian's troops were forced to take up position in the field to meet the new attack. Another battle then took place in the darkness, with all its attendant hazards. What is significant for us is that Tacitus explicitly reports that the Vitellians "collected their artillery on the raised road so that they were on free and open ground from which to shoot, after their initial shots had been dispersed and struck trees rather than the enemy." And these were not just small and medium caliber weapons. We are told that the Fifteenth Legion (Vitellian troops) had with them an exceptionally large ballista (*magnitudine eximia ballista*) that threw huge stones into the Vespasians' line, and would have caused extensive destruction had not two brave soldiers, taking up Vitellian shields and passing through their lines, sabotaged the machine by cutting the "bonds and weights" (*vincla ac libramenta*).[6] Tacitus, who Mommsen styled the most unmilitary of historians, presumably meant by this the springs and spanning ropes, which are about the only things on a large ballista that are susceptible to being cut. Protection from physical assault, incidentally, is surely the principal reason why springs were protected with battle-plates or cylindrical covers, protection from the weather being a secondary function.

FIGURE 9.2 *Disassembled ballista. The technical treatises assert that catapults should be easy to assemble and disassemble, for transport and storage. This photo shows the main component parts: spring frame with arms and bowstring (bottom right), case, slider and universal joint (top), and stand (center, with top section, on which sits the universal joint, separate). (Courtesy Legion XXIV)*

The fighting continued throughout the night until dawn, by which time the Vitellians began to waver. And now again Tacitus tells us something astonishing: he says that they could not reform their lines because they were entangled among "the wagons and artillery."[7] The artillery had been moved onto the raised road in order to have open ground from which to shoot (above); there is no suggestion in Tacitus' text that a baggage train was moved to the same place, nor would there be any reason to move one thus. Therefore, the wagons that were, by implication, in the same place as the artillery and that impeded the Vitellian troops were, presumably, for carrying the artillery, or were actually carrying it during the engagement. We appear to have catapults on wagons, in the field, some thirty years before the Dacian wars (and nearer forty before Trajan's column was carved; more on this below).

Finally successful on the field, the exhausted but jubilant Vespasian troops marched on to Cremona, only to find that the city was encircled and thus additionally defended by a military camp built by Otho's troops earlier in the year (Otho had been the second of the four emperors this year). Despite having fought for a day and a night, they now (for sound tactical and psychological reasons) set about attacking the camp. Both sides are said to have deployed, as we might expect, "the familiar artifices of Roman warfare." One would never guess, reading Tacitus, that these included catapults, but it is clear that they did, because in chapter 29 he comments that the Vitellian troops, exasperated by the failure of their missiles (*tela*) to penetrate the Vespasians' *testudo* (tortoise formation), "finally pushed their ballista itself over onto the heads of their assailants below." The size of this particular machine may be gauged from the fact that it apparently took the parapet and the top part of the rampart with it as it fell. Earlier we had been told that both sides were armed alike, and the consequent deduction that the Vespasians had catapults too is confirmed when we are told that a neighboring tower fell under a volley of stones, which unequivocally indicates the use of catapults, this time by the attackers. Cremona then went up in smoke.

Recall that Dionysios, the ruler of Syrakuse under whom the first catapult was invented about 399 B.C., spent time with and took a real interest in people making arms in his city. Interestingly, the same sort of personal interest in arms manufacturers is attested for Vespasian, the victor of Cremona and the first of the Flavian emperors. Immediately after deciding to challenge Vitellius for the throne, he is said to have selected "the powerful towns" (*validae civitates*) to manufacture arms; to have appointed expert agents to oversee the work; and, critically, "Vespasian visited each place in person, encouraged the workmen, spurring on the industrious by praise." His provinces were "filled with din as ships, soldiers, and arms were made ready for their needs"—all of which cost a great deal of money, leading Vespasian's right-hand-man Mucianus to declare repeatedly that "money is the sinews of civil war."[8] The all-metal-frame *kheiroballistra* (see Figure 9.4) was invented sometime around now, and this story of Vespasian's interest in and encouragement of skilled workers and knowledgeable managers among his arms manufacturers inclines me to look for it among his forces. Our principal source for the battle for Cremona and catapults on wagons there is Tacitus, who was a child of these times; he was a teenager when the battle of Cremona was fought, and one of his probable sources for it, Vipstanus Messala, actually fought in it as military tribune in command of the Seventh (Claudian) Legion. This legion had been stationed in Moesia, on the Danube frontier (which happens to be opposite Dacia). So the source is impeccable, the circumstance more than plausible, and the *carroballista* should perhaps be down-dated to at least A.D. 69—but more on this below.

Other new machines are attested for Vespasian, some of which are difficult to visualize. In a battle with the Batavi near the Rhine, his legionaries "who were superior in skill and artifices" are said to have employed new devices against their barbarian foe. One was a "well-balanced machine poised over them," one end of which could be dropped suddenly to catch one or more of the enemy and "with a shift of the counterweight" hurl the unfortunate captive(s) into the camp.[9] This sounds a little like Arkhimedes' claw,

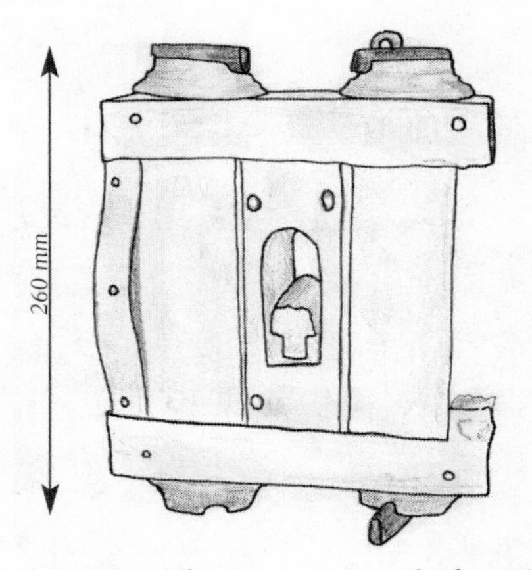

FIGURE 9.3 *The Xanten catapult, front (left) and back (right). This fantastically well-preserved spring frame is a small calibre hand weapon that one man would have held and used. Seven catapults of which traces survive are in this approximate size range; it is the ancient analogue to the modern rifle.* (Author sketch)

scaled down to catch men rather than warships. They also destroyed a two-story siege tower that the Batavians had built and pushed up to a gate, by another apparently novel device that employed strong poles and beams. We met the Batavians in Chapter 8, building copies of Roman artillery for the Germans. About a generation later some of the Batavians had been absorbed into Rome and a unit of auxiliary troops from this area, the *cohors* VIIII *Batavorum*, turns up as a garrison at Vindolanda, on Hadrian's wall (where they occupied levels 2–3), where they appear to be operating Roman *tormenta*.[10]

Just south of the Batavi's home territory is Xanten, ancient Vetera, a Rhine fortress. A mysterious-looking lump was "excavated" from a local gravel pit by mechanical dredger in 2000 and was sent (along with other finds) to the Roman museum at Xanten, as such finds from this gravel pit have been for the last twenty years or so. Investigation of the object began in November 2002, and initial X-rays showed that what lay hidden within a crust of gravel and sand was a small torsion frame, complete but for one side stanchion and a lever (Figure 9.3).

It is the fourth near complete torsion frame to be found, the others being Ampurias and Caminreal, both of the first century B.C. and from Spain, and Hatra, of the third century A.D. and from Iraq. It is extraordinarily fortunate for reconstructing the history of the catapult that Xanten sits geographically and chronologically in between, in first-century A.D. Germany.

Full publication of the Xanten catapult is awaited; it promises to enrich our understanding of Roman imperial torsion catapults significantly. Detailed macroscopic, microscopic and tomographic examination has been undertaken, and casts were taken of cavities in the corrosion "shell" before it was removed, preserving details of the woodwork now massively shrunk and distorted (though enough wood survived to identify it as common ash). Novelties of this specimen include a slightly convex (rather than flat) lateral surface of the side stanchions, and a taper to the cut-out for the arm toward the edge, so that the room for arm travel was increased slightly. The washers—all four survive—are of the pinhole style, four holes in the washer, fifteen in the counterplate, and one pin remains in situ locking its washer in place. The corrosion cast preserved terrific detail on the shape and construction of the slider and case, and showed that the case and frame could be disassembled easily, the stock simply slotting into the frame just below the slider aperture. It was a little catapult, the frame measuring a mere 260 mm high—including the washers—by 220 mm wide, which would correspond to a spring diameter of around 40 mm (the actual figure of this key parameter is not stated in the interim publication). This is another hand weapon, a little catapult, like Ephyra 6, Elginhaugh, Bath, and Mahdia 3.

Vespasian's son Titus, to whom Pliny the Elder dedicated his massive *Natural History*, inherited the throne, and oversaw the completion of the Colosseum, which involved a number of challenging new technologies. He lived in interesting times, having to deal with, among other things, the eruption of Vesuvius and the total disappearance of several populous and wealthy towns, most famously Pompeii and Herculaneum, and numerous villas belonging to the Roman elite, such as the villa Poppea at Oplontis, whose owner may have been Nero's wife. Rescue attempts were entertained until the futility of the task became clear. Before assuming the purple, Titus was largely responsible for putting down the Jewish revolt, which was a long and hard-fought campaign (A.D. 69–70), involving some ingenuity on either side, and some relatively large catapults. This revolt came hot on the heels of the Jewish war, which is described in detail by Josephus and included the siege of Jotapata, which involved the most extraordinary efforts by Romans and Jews alike.

From stories of this war and subsequent revolt, it appears that the legions had a significant measure of freedom in their choice of catapult equipment. During the High Empire, the legions built and maintained their own catapults, along with any siege equipment, buildings, or other wooden structures as required. Josephus, who was present at the time, describes those belonging to the Tenth Legion as "extraordinary," both sharp-casters and stone-throwers being larger and more powerful than the other legions' machines (and this had nothing to do with imperial patronage, for Titus was in charge of the Fifteenth). The Tenth Legion had been famous since Caesar's day, and had a reputation to maintain.[11] Crucially, Josephus notes that the Tenth

Legion's machines were capable of driving the defenders from the walls, and not just repelling "the excursions of the Jews." This implies that most of the Roman ordnance was relatively small caliber. He elaborates that one particular machine operated by the Tenth cast a stone weighing a talent (c. 25 kg) two *stades* or more, which landed a crushing blow on all in its path. A *stade* was about 600 feet (it varied between different states), so this is about 1,200 feet or 400 meters. The range may or may not be exaggerated, but the important point for us is that in Josephus' experience, a one-talent machine was extraordinary, and this is completely consistent with his comment about most of the legionaries' machines being used to repel sallies, rather than to attack defenders on or behind a high and well-built stone wall. It is also consistent with the size of ballista balls excavated from Masada, which fell to the same troops after a six-month siege some three or four years later. In response to the threat of the big one-talent machine, and because the large glistening limestone shot was easily seen, the Jews posted lookouts to watch that particular machine and to shout a warning whenever it was discharged ("the son is coming"), so people could take cover. As a result, the casualty rate dropped to nothing, according to Josephus. The Romans realized what was happening, and responded by painting their big stones to camouflage them, with the result that they were again able to wreak death, destruction, and terror on the inhabitants of Jerusalem, who could no longer see them coming.[12]

Sharp-casters and stone-throwers, presumably the Tenth's, were deployed on the siege ramps as they were being constructed, in order to try to keep the defenders' heads down and prevent them interfering with the construction. In response, the Jews brought their own machines to the walls, machines that they had captured from the Roman garrison in the uprising, but did not know how to use properly—indeed, the rebels were initially so unskilled in the operation of these catapults that they were "relatively useless," said Josephus. After Roman deserters showed them how to use the weapons, they progressed to shooting sharps and stones awkwardly, and began to hinder the advancement of the siege ramps. This story seems to refer to little catapults again. The 300 sharp-casters and 40 stone-throwers being used by the Jews were explicitly said to be lighter than those the Romans in the siege towers were employing, and small caliber, anti-personnel weapons are perhaps what we should expect garrison forces to use. The Romans' technological superiority (their siege towers overtopped the walls) as well as their larger caliber catapults forced the Jews to fall back out of range and abandon the outer wall. It probably did not need many to lose their heads—literally, as did one poor unfortunate hit by a medium caliber stone in the first war—to precipitate such a withdrawal. They took up position on higher ground, and this, together with the experience gained daily in using their catapults so that they

Map 6. *The Roman World*

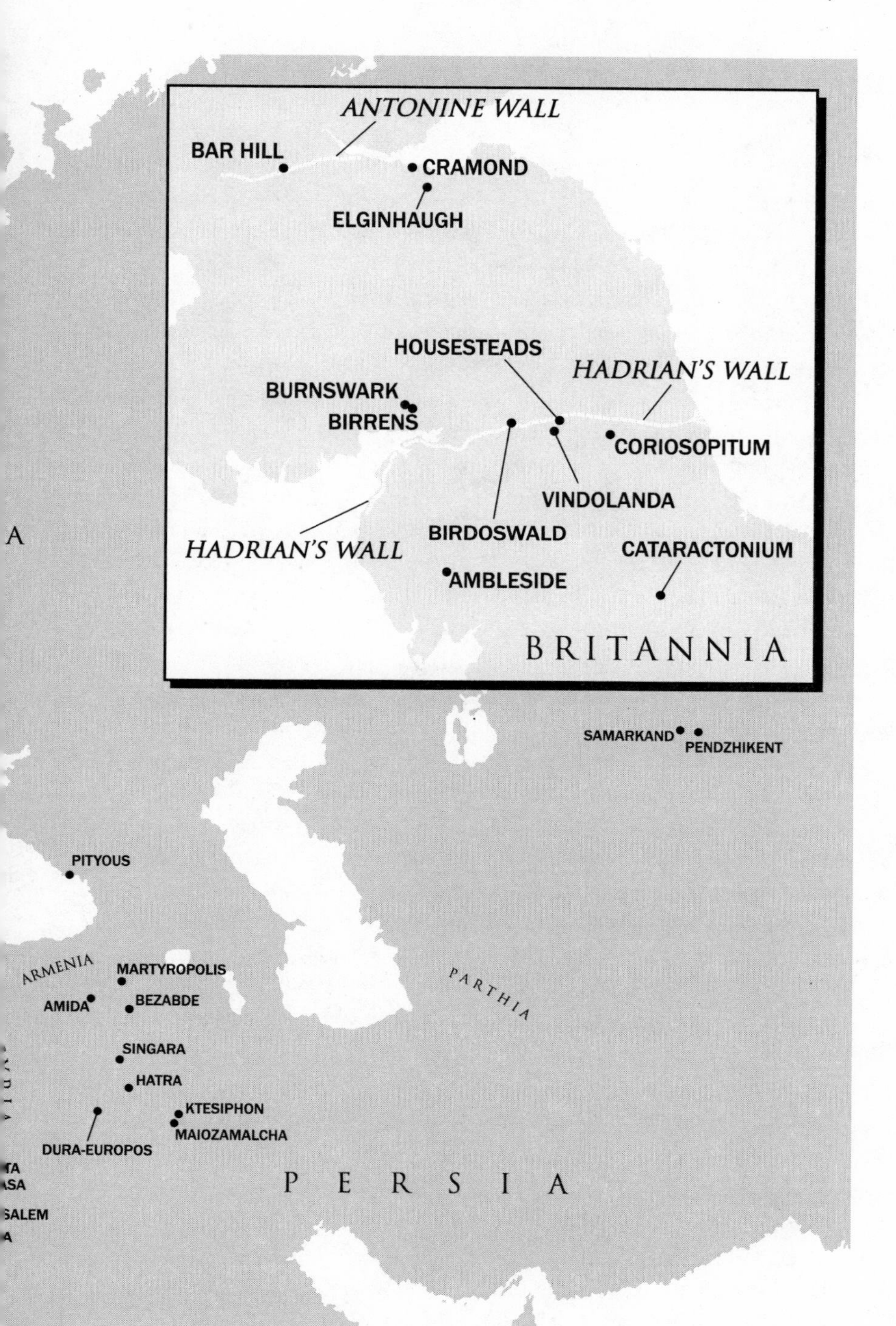
ANTONINE WALL
BAR HILL
CRAMOND
ELGINHAUGH
HOUSESTEADS
HADRIAN'S WALL
BURNSWARK
BIRRENS
CORIOSOPITUM
VINDOLANDA
BIRDOSWALD
HADRIAN'S WALL
CATARACTONIUM
AMBLESIDE
BRITANNIA
SAMARKAND
PENDZHIKENT
PITYOUS
ARMENIA
MARTYROPOLIS
AMIDA
BEZABDE
PARTHIA
SINGARA
HATRA
KTESIPHON
MAIOZAMALCHA
DURA-EUROPOS
PERSIA

became proficient in their use, led to a stalemate. The siege ultimately went on for five months, during which time Titus was hit on the left shoulder by a stone that did permanent damage, leaving that arm weaker.[13]

Titus died, probably of natural causes, in A.D. 81, and Domitian, his younger brother, took the helm. In A.D. 85–92, he launched a war against the Sarmatians, the Dacians, and the Suebic Germans, apparently in (rather belated) response to the Sarmatians' raids across the Danube in A.D. 68–70 when the Romans were preoccupied with civil war. This campaign, despite its length, was inconclusive. After Domitian's assassination in 96, an unrelated senator called Nerva became emperor—the last Italian emperor, as it happens. During his short (two-year) reign, he appointed Frontinus to oversee the Roman water supply, prompting the latter to write on the *Aqueducts of Rome* and thereby produce an ancient theoretical analysis of that greatest of Roman civil engineering feats. Sadly, Frontinus did not see fit to include stories about war machines in his other surviving book, on *Stratagems*, on the grounds that their development had "already reached its limits" and he could see no further development in that area (3 pref.). While catapult technology may well have reached a plateau in his time in near-perfect designs, leading him to think that there was nothing new for him to say about them, "development" is not synonymous with "improvement." The design of functional objects does occasionally reach perfection, but if the function changes, then the fitness declines, and changes are needed to realign the design with the new demands. Thus changes in the availability of materials, or staff training, or enemy tactics, for example, could easily have prompted changes in military hardware. And thus, contrary to Frontinus' expectations, war machines did change during Roman imperial times. They may not have been able to "improve" in terms of mechanical engineering over the palintone design described by Philon, but they met the challenges of the day better in other ways that were then more important.

Childless and elderly, Nerva adopted Trajan, son of a distinguished consul, who had a fine military record in Spain, Syria, and Germany and had himself been consul in A.D. 91. Trajan was one of Rome's best emperors, and we have a unique insight into the administrative aspect of his imperial duties through the letters of the Younger Pliny (nephew of the Elder Pliny), who was governor of Bithynia (south of the Black Sea) during his reign, and who wrote to Trajan frequently, seeking advice, information, or favors. Trajan took a real interest in military matters, marched on foot with his troops on campaign, founded two new legions and, a decade or so after Domitian's campaign, led two expeditions into Dacia (in 101–102 and 105), which resulted in the conquest and annexation (in A.D. 106) of the area, most of which is now known as Romania. There were further hostilities in the region in A.D. 117–119, during the reign of Trajan's successor Hadrian.[14]

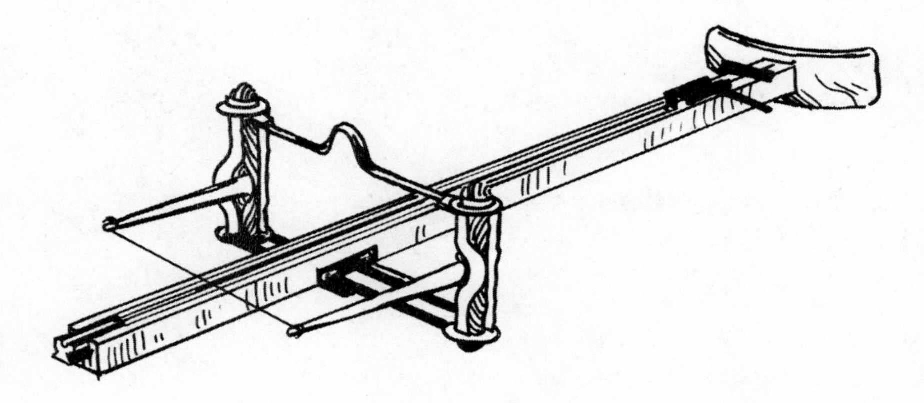

FIGURE 9.4 *The kheiroballistra reconstructed as an inswinger. The framework in this design is made of metal not wood. The case and slider are still wooden. The orientation of the curve at the back is debatable; it may lay horizontal (as shown) and be intended to fit, say, the thigh for loading, or vertical and intended to fit the shoulder for aiming. (T.D. Dungan after Iriarte)*

Trajan's column was created as a monument to his achievement in the Dacian Wars, and on it we see a new catapult design, one distinctly different in a number of respects from earlier models. It has been identified with the sort of catapult described by Heron as a *kheiroballistra*, though the identification is not without its problems (figure 9.4).

Wood continues as the material for the stock and the arms (though the latter have metal inserts), but the frames that hold the springs, and the struts that hold the frames in place, are made of iron. The metal frames are called *kambestria* (καμβέστρια, *campestria*) or field-frames. Note that Trajan's column depicts the Dacians, as well as the Romans, with *kheiroballistrai* during the Dacian Wars (Figure 9.5).

They were perhaps not newly invented when Trajan invaded Dacia, and there is certainly no good reason to assume that the sculptors were carelessly depicting barbarian enemies with Roman weapons. As we saw above, *carroballistae*, which on Trajan's column seem to be *kheiroballistrai* mounted in carts, may have been around for a least a generation—plenty of time for the Dacians to capture men and machines and learn to copy and deploy them, and for variations in the basic design to emerge (again, as depicted on the column). In addition, according to Cassius Dio, machines that had been captured by the Dacians "in the time of Fuscus" were found on fortified mountains and recovered by Trajan. This apparently refers to a campaign during Domitian's reign, about twelve years earlier (A.D. 89), during which standards were lost too. As part of the settlement of that campaign, Domitian had agreed to send "artisans of every craft of peace and of war." One of the conditions of the peace that concluded Trajan's first campaign was that the Dacians surrender "the machines and the machine-makers." So *carroballistae* and perhaps *kheiroballistrai* may have existed as much as a generation before Trajan's reign, perhaps as early as A.D. 69.[15]

FIGURE 9.5 *Dacians with kheiroballistrae, Scene 66, cast 169, Trajan's column (catapult enhanced). These machines might have entered Dacia with Domitian's troops in 89, or their builders may have entered as part of the peace deal thereafter. Note that the spring casing bottoms are flat, not conical, and are level with the lower crosspiece. Although it is lost here, the arch spanning the top would have been whole.*

What we call the *kheiroballistra* is apparently twice depicted mounted on carts—i.e. as a type of *carroballista*—on Trajan's column, in smaller (Figure 9.6) and larger (Figure 9.7) versions, just as ballistae and scorpions and catapults came in two standard sizes.

Although widely supposed to be *kheiroballistrai* of the type described in Heron's treatise, it should be noted that these catapults differ from one another and from the text not only in respect of their size, but they may not be *kheiroballistrai* at all. The flat bottom (in contrast to the conical top) of what appears to be the spring housing, which is very obvious in scene 66 (Figure 9.5), is a feature that has yet to be explained.

Scholars and archaeologists invoke images of Trajan's column when it suits their argument, and they point out the unreliability of the images, or the possible ignorance of the sculptor, when they don't. Vision is a cultural phenomenon, and we all learn to look at (or "read") the world in our own time and culture; so did ancient sculptors. No doubt sculptors were sometimes required to carve unfamiliar sights, but an ignoramus of military matters would not, I submit, be commissioned or allowed to work on a celebration in stone of the emperor's military success and his conquest of a new province. Trajan was no armchair general and he would not, I think, tolerate incompetent work on a massive stone that was to be a memorial to his military might, that was commemorated on coins of his reign, and at whose base, moreover, he intended his ashes to be deposited, as in due course they were.[16] Nor, I

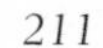

FIGURE 9.6 *(top) Small carroballistae, Scene 40, cast 104-5, Trajan's column (catapults enhanced).* FIGURE 9.7 *(bottom) Large carroballista, Scene 66, cast 163-64, Trajan's column (catapult enhanced).*

think, would Apollodoros, who superintended the building. Appeals to depictions on Trajan's column should start with the presumption that the image is accurate, as we start with the presumption that a text is correct. We should then try to explain the image, rather than explain it away. In other words, let us try to save what phenomena we can, and admit defeat where we cannot. We need to be more sophisticated in our criticisms. So, for example, the hand railings on Apollodoros' famous bridge over the Danube into Dacia are quite out of scale with the rest of the bridge, but if they had been drawn to scale they would be invisible! The sculptor (or Apollodoros himself, qua superintendent of the column) had a choice whether to adapt or to omit this feature, and he chose adaptation. Consequently one can see that the bridge was fitted with railings so that people wouldn't inadvertently fall into the mighty river Danube, but when examined carefully the railings appear to be about the same size as the trusses that support the spans. The flat-bottomed putative "spring housing" of the column catapults remain in need of an explanation.

Three *kheiroballistrai* have been found at Gornea in Romania, and one at Orşova, though both sites were small, square late antique (fourth-century) fortlets on the Danube frontier, with a tower in each corner, so these particular machines were built hundreds of years after Trajan's machines and are at best descendants of them. They are the most important of a number of traces of the Roman occupation that have been discovered in this region. Orşova is on the left (northern) bank of the Danube just above the feature known as the Iron Gate, where the Danube cuts through the bottom of the Carpathian mountains, and which is impassable to ships. Gornea is about 40 km as the crow flies west, further upriver on the left bank, opposite the legionary base at Novae (Map 6).[17]

Trajan had a 3.2 km long, 14 m deep canal cut to bypass the Iron Gate in A.D. 101, and founded a garrison here, which continued in existence throughout the Roman occupation of the area. His other famous monument on the Danube, the bridge over it, was 1,127 meters long, and had twenty piers at 38-meter intervals; the ends survive. It has been suggested that the engineer who designed that bridge, Apollodoros of Damascus, whom we have met before as the building superintendent of Trajan's column, who also wrote a treatise on war engines (but the surviving text of which does not include catapults), may have invented the *kheiroballistra*.[18] This is quite plausible. Recall that architect and catapult builder were often roles performed by one and the same man, as in the case of Vitruvius.

The Gornea and Orşova field-frames (Figure 9.8) correspond closely in design (but not in size) to components described in a treatise that survives only in part and is attributed to Heron of Alexandria, known as the *Kheiroballistra*, which literally means the hand-ballista (Figure 9.4). Debate continues over key issues and elements as well as over details. In particular, most scholars who have published on the text amend it to make the machine larger. This is perhaps because of the tendency to think of all catapults as "artillery," that is, large calibre weapons, or of the desire to identify the machine of the literary text with the machine depicted on Trajan's column, which seems to have a two-man crew—and assuming that all column depictions show one and the same machine.[19] Marsden, for example, who was writing about the text before much of the archaeological material had come to light, assumed that every catapult must have a stand and base, and therefore that if none is given in the text, as for this particular machine, then part of the text must have been lost. This is clearly an unnecessary assumption. No base may be described because no base is required. In this case, even the name of the weapon suggests that it is small, a personal weapon, and textual

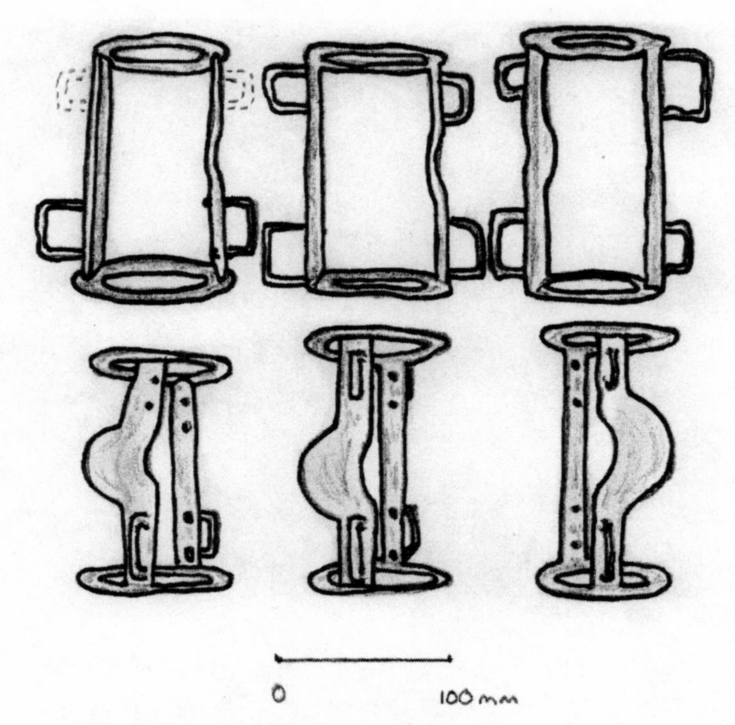


FIGURE 9.8 *The Gornea field frames. These metal components replaced the wooden framework of stanchions and hole-carriers of the earlier designs. The washers sat on the rings top and bottom, the spring was stretched between them, and the arm was inserted into the spring at about the position of the cutaway in the stanchion. The U-shaped components attached to the stanchions probably held the crosspiece and arch that connected a pair of field-frames.(Author sketch after Gudea and Baatz)*

emendation is unjustified, in my view. When dealing with a mechanical system, textual emendation of a key parameter requires emendation of others, and one quickly and unobtrusively moves from a reconstruction of an ancient design to a hybrid of ancient and modern. As Iriarte observed, "the power output should be the conclusion, not the premise, of our reconstruction work."[20] That some of the catapults on the column have stands is only relevant to the argument about the text if and when it is shown that they are the same machine as that described in the text—and not just the same design, but the same caliber. We saw in the previous chapter how in Republican times ballistas, scorpions, and catapults each came in two sizes; there is no reason to suppose that successor weapons would not also.

On the basic design and size of the *kheiroballistra* I have come to agree, after much study, thought, and scale drawing, with Aitor Iriarte and others who have argued in private or in public for the *kheiroballistra* as a radically differ-

FIGURE 9.9 *Vedennius's tombstone. According to the inscription beneath this relief, C. Vedennius Moderatus joined the army in A.D. 60. He served with Vitellius' forces, but was retained by Vespasian. After discharge, he served for 23 years as architect in the imperial arsenal in Rome, where he would have been responsible for building catapults that presumably looked like this one. He was decorated by Vespasian and Domitian.*

ent design from its Philonian and Vitruvian predecessors. I think the *kheiroballistra* was basically a small weapon, the version in the text and the machines found at Gornea resembling a crossbow with torsion springs, set wide for in-swinging arms; the machines found at Lyon, Sala, and Orşova and depicted on Trajan's column being larger caliber versions of the same basic model. The in-swinging arms are the really radical idea, and they are so radical that initially one just assumes they cannot be right.[21] Everybody knows what a catapult looks like, and the arms stick out the side!

That is how they appear on Vedennius' tombstone (Figure 9.9), and the Pergamon relief (Figure 6.3). The trouble is, on Trajan's column, they do not; no arms are visible on any of the catapults depicted (Figures 9.6 and 9.7; see Figures 9.5, 9.10, and 9.11 for the other depictions on the column). The flat bottom of the cylinders is also again very noticeable.

Some would say that this means that the column cannot help in the in-swinging versus out-swinging arms debate. I disagree, because it is not the case that the sides of the machines are obscured or missing. Catapults appear in five scenes, two of which (both cart-mounted, Figures 9.6 and 9.7) show "bulbs" or bulges on the outside of the spring cases where one would expect to see arms, and three of which (on walls and in a "nest," Figures 9.10 and 9.11) show nothing but a straight side where the arms should be. Now I think we can eliminate the idea that these catapults had no arms. Therefore, we should ask, where would the arms have to be in order to be invisible from the vantage point of the viewer? It happens that all machines are shown from essentially the same angle—the left spring (as one looks at it) is foregrounded and the right spring backgrounded. It seems to me that it is only if they were spanned in-swingers that the arms would not be visible from the viewer's angle. In all of the stationary depictions the arms could then be hidden by the left spring or the slider. The *carroballistae* are less easy to explain. The bigger machine

FIGURE 9.10 *(top) Kheiroballistrae on walls, Scene 66, cast 165, Trajan's column. Standing on the walls, these appear to have flat bottoms to the cylinders. No lower cross-piece is visible, but if it was level with the bottom of the cylinders, as in the scene in Figure 9.5 above, then it could be invisible from the viewing point (catapult enhanced).* FIGURE 9.11 *(bottom) Kheiroballistra in a "nest," Scene 66, cast 166, Trajan's column. Two legionaries operate a medium-calibre kheiroballistra in a well-built timber structure akin to a modern "gun emplacement" or "machine-gun nest" (catapults enhanced).*

being transported in a cart (scene 66, Figure 9.7) apparently has no slider and may be depicted disassembled.[22] The smaller *carroballista*, which is shown in action, shooting sharps—perhaps even two at a time—has its legionary operator's right arm apparently extended through the central aperture, which would interfere with the left arm were it anywhere except pointing forward, out of the front of the machine (Figure 9.6). And that may be where it is shown; I can imagine the legionary's hand holding the arm, and his posture reflecting the effort needed to pull it back. Those who believe in in-swinging arms will find this "reading" more plausible than those who do not, of course.

But Trajan's column is only supporting evidence for in-swinging arms; the real case for them is the design of the frame. Whereas Hellenistic (Ktesibios, Philon), Republican (Ampurias, Caminreal), and Imperial (Cremona, Xanten) designs had close set springs, little further apart than the width needed for the case and slider, the *kheiroballistra* of the text and the metal catapult frames from the earth have springs set wide apart, at opposite ends, in fact, of the frame. If the arms still protruded out the sides, I can find no good reason for this change and many reasons against it. The most economical hypothesis for the switch to wide-set springs is that someone spotted that if the relative positions of the arms and their power source were reversed, so that the spring was on the outside and the bowstring when braced at the front, they could increase the draw length significantly, and with it the launch velocity of the missile. With out-swinging arms, the bowstring will foul if the arms come horizontal with the frame, limiting their maximum travel (on the simplest point and line reduction) to something less than 90°. In-swinging arms can pass through more than 90°, however, from pointing directly forward to pointing nearly backward (depending on the length of the stock) without anything fouling the bowstring. (Figure 9.12)[23]

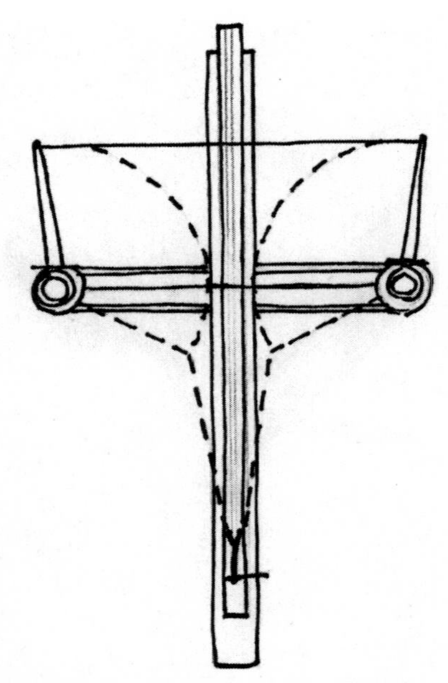

FIGURE 9.12 *Inswinger mechanics. After Iriarte, shown braced. The dotted lines show the position of the arms and bowstring when the catapult is drawn. (Author sketch)*

If the *kheiroballistra* had in-swinging arms, then the springs would have had to be at least two arm-lengths apart, plus some space for two widths of bowstring and the missile to pass without fouling one another. The ratio of the height of the springs to their distance apart is about 1 : 1/2 in the formulaic palintone and euthytone models of the Philonian variety. Vitruvius allowed for flexibility with shorter and longer versions of springs and arms, and there was always a range of sizes for the components, but *this* ratio changes dramatically with the *kheiroballistra*—it becomes 1 : 2 1/2 in the *Kheiroballistra* text, and 1 : more than 3 1/2 on the Orşova catapult (assuming field-frame and crosspieces are from the same machine).

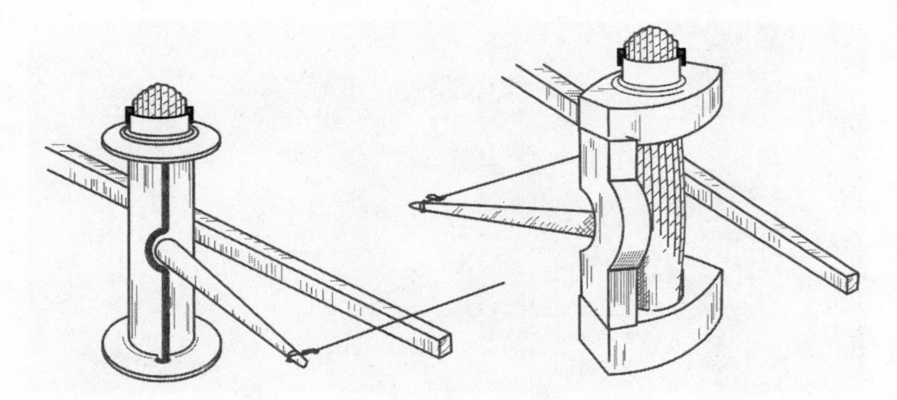

FIGURE 9.13 *Stanchions on wood and stanchions on the kheiroballistra. Cutaway stanchion, arm, part of the bowstring, and front part of the slider, on the metal kheiroballistra (left) and wooden palintone (right). The kheiroballistra is shown as an inswinger, the palintone as an outswinger. The direction of the cutaway and bulge in the stanchion corresponds to the position of the arm: for an inswinger the stanchion must be at the front and bulge outwards; for an outswinger the stanchion must be at the front corner and bulge forwards. In short, to turn an outswinger spring housing into an inswinger spring housing, one needs to rotate the stanchion, viewed from above, 90° clockwise. For stability and rigidity one then needs to move the stanchion from the corner towards the middle of the spring, and to make space for the arms one needs to move the whole spring at least the arm-length away from the slider. (T.D. Dungan after author sketch)*

The position of the stanchions then becomes the limiting factor. Within the field frame the *kheiroballistra*'s stanchions are set opposite one another, as on the old Philonian models. One of the two metal stanchions has the curving bulge midway, allowing the arm to travel that bit further, just as the old plated wooden side stanchions had a cutaway behind and reinforcing bulge in front, which bent toward the front, towards the enemy. With the *kheiroballistra*, the stanchions' positions are naturally to front and back rather than side to side because the simplest interpretation is that they were fixed at right angles to the cross-pieces, which fitted into the handles (see Figure 9.8; precisely how is much debated). Of course the field-frame could have fitted on either end of the cross-pieces, and we do not know which was the left-hand frame and which the right, so we do not know if the bulge was to front or back. If the bulges were at the front, and they pointed outward from the machine, then the natural interpretation is that the machine was an inswinger. If the bulges were at the back and faced outward, then it could have been an out-swinger or an in-swinger (Figure 9.13).

If the bulge was at the front, facing outward, it allows the arm to point directly forward when braced, and to rotate in toward the slider as it is

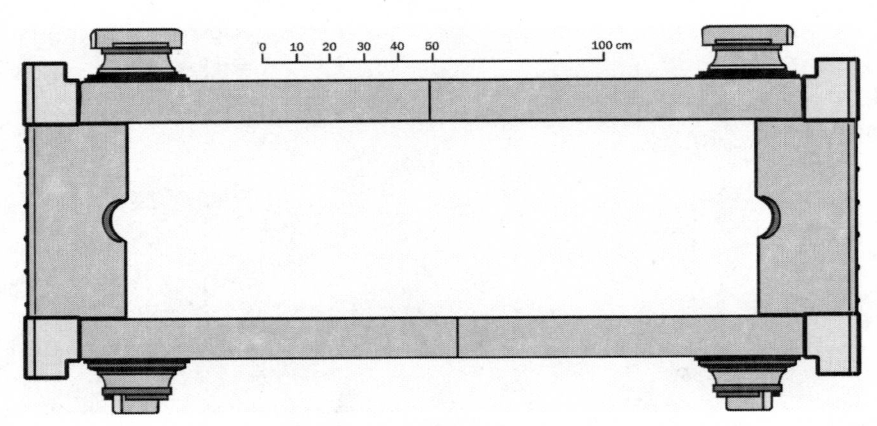


FIGURE 9.14 *The Hatra catapult from the front. The position of the cutaways for the arms implies that this catapult was an inswinger. Further, the shape and position of the cast bronze corner reinforcements strongly suggests that there were wooden stanchions at the back outer corners of the frame as well as at the front outer corners. These would have been incompatible with outswinging arms. (T.D. Dungan after Iriarte 2000: 61)*

spanned. The position of the other (straight) stanchion is opposite on the field-frame, which now means near the middle at the back of the spring. The Hatra ballista, though a wooden plated engine, incorporates these features, with the cutaway in the inner edge of the forward side stanchion, and I think it must have been an in-swinger (Figure 9.14).

The fact that the front of the frame plating survives, and shows clearly that the cutaways are positioned in the inner edge of the front stanchion, to my mind removes any doubt where the arms were when the device was braced. They must have pointed forward, and have been drawn back inward when spanned. We shall look at this catapult in more detail below.

The *kheiroballistra* apparently offers its crew very little protection. The battle-plate that characterizes the Cremona and Xanten catapults has completely disappeared from the design. A wide-open aperture means a wide-open target for enemy snipers, just as a house with magnificent views also has magnificent exposure to the elements. Perhaps the improvement in performance of in-swinging arms was judged to outweigh the risks of being hit. Perhaps misfires were reckoned more dangerous than enemy sharp-shooting, and that it was better to risk enemy sharp-shooters than self-destruction, for, all other things being equal, a misfire ricocheting back off the frame would probably wreak more havoc on the operator than would an enemy sniper's projectile, especially if they were using relatively short sharps, like the Qasr Ibrim shortshaft (see Figure 1.4). Or perhaps the *kheiroballistra* was preferred

FIGURE 9.15 *Catapults aboard a ship. This is probably a Rhine river vessel, built for frontier duties. (T.D. Dungan after Bockius)*

when the enemy had no ordnance capable of sniping at its operators, lacking either technique (no sharp-shooting tradition or training) or technology (nothing sufficiently accurate). A cynic might suggest that those who built and designed the machine were concerned about its performance more than crew safety, since in the imperial Roman army (unlike in Hellenistic armies), the designers would not normally have operated it in the field. However, these machines do not work automatically, so the safety of the crew is essential to their function, and a good designer would consider both aspects.

The only representation of a catapult mounted on a warship (Figure 9.15) seems to show *kheiroballistrai* (there are two) that have some kind of protective structure for the gunners.[24]

Bockius suggested that the catapults are oversized in comparison with the proportions of the ship on whose deck they stand, which appears to have been a Rhine river vessel, perhaps some 30 meters long. Vegetius said that onagers, ballistae, and scorpions (he glosses the last as *manuballistae*) were mounted on ships (4.44), and the term *kheiroballistra* appears not to be in his vocabulary. The two catapults depicted are not identical but appear to be of the same type or lineage, and the cone-like structures descending from the ends of the horizontal crossbeams suggest identification with the *kheiroballistra*. Bockius cites the machine in the "nest" from scene 66 of Trajan's column for the comparison, but a better comparison would be with the larger *carroballista* a little earlier in the same scene (Figure 9.7 rather than Figure 9.11 above), because on the *carroballista* the cylinders project below the horizontal crossbeam, as they do on the ship catapults' depiction, whereas they do not on the machine in the nest (nor the machines on the wall in cast 169, Figure 9.10 above), whose cylinders appear to terminate at approximately the same level as the crossbeam. Similarly, the tripods on which the ship catapults are mounted closely resemble that on the Column's large *carroballista*. The upper sections, however, positively diverge in appearance, since all the Column catapults, including the large *carroballista*, have a clear arch spanning between the top halves of the cylinders, while the ship catapults have instead a kind of crenellated top that Bockius supposed was wooden shielding serving to protect the springs and crew, cut away where the slider lay. His reconstructions do not quite correspond with the depiction on the stone, but that is abstruse and the basic idea of a wooden shield is perfectly plausible.

All the Gornea *kheiroballistrai* were small caliber; the spring diameter of Gornea 1 and 3, for example, was just 55 mm (2 inches), corresponding to a 1-cubit or 2-span sharp-caster on the Philonian formula, shooting something like a Dura bolt or kestros. So White rightly described them as "cross-bows."[25] The Orşova machine was somewhat bigger, but still not a large catapult. It seems to have been the latest version of the popular 3-span caliber. Its overall dimensions, about 2 meters (6´) wide, half a meter (1´6˝) high, and 1.3 meters (4´) long, would have intimidated, for sure, and its 700 mm (2´4˝) thick, heavy bolts could have nailed a couple of men together and terrified any bystanders. But this remains an anti-personnel weapon. It could easily have penetrated wicker shields and barriers, and even solid wood shields, and sent their bearers hurtling back into the ranks behind, but it would not have threatened a plated siege engine. It is, in fact, about as large as one could want for defense of the Danube frontier in the fourth century A.D., where the enemy's best technology was whatever they could seize from the Romans, together with, initially, a member of the Roman forces to show them how to use and later maintain it.

The invention of the *kheiroballistra* did not spell the extinction of its wooden predecessors—indeed, they continue to appear in the archaeology in exuberant variety. The Bath catapult, from the second century A.D., the Hatra catapult, built in the third century, and the Pityous catapult, from the fourth, all prove that the wood-frame torsion catapult continued in production in a range of calibers, from third-smallest (Bath) via medium (Pityous) to largest known (Hatra), for centuries after the invention of the *kheiroballistra*. Confusion over the terms used in the ancient sources for catapults arises when moderns assume that the old models were gone, whereas some at least were still being made, alongside new designs, and the sources may be referring to *them*. For example, Cassius Dio tells us that "there was a huge variety of machines on the walls of Byzantion" in about A.D. 195, most of them designed and built by a local man called Priskos. Some "threw rocks and beams" at short range; others "stones and missiles and spears" at long range; others had "hooks" to catch and reel in ships or siege machines (75.11.1–2).

Some of these may have been built to traditional designs, some novel. Another potential source of confusion is the fact that large sharp-casters were also called ballistas. Another source of confusion over names of machines stems from the assumption that stone-throwers are one type and sharp-casters are another type, and that there is a given name for each type. While this is true in part, it fails to recognize that the best ancient designs could shoot both types of missile. It is a common mistake to equate ballista with stone-thrower, just as, in antiquity, people equated palintone with stone-thrower (recall Chapter 7 and Ktesibios' clarification). All catapults obey the same laws of mechanics, and they are optimized in the formula for the palintone, which could shoot sharps or stones. The basic palintone design was called a "ballista" in Latin. Ballistae could shoot both stones and sharps—the term did not change its meaning over time to mean stone-throwers at an earlier date and sharp-casters later, even if the sources that happen to survive fit such a pattern; that is mere coincidence. Compare the earliest firearms in Europe, which appear to have been designed to shoot both shot and sharps, with special wood plugs fitted to the latter so that "large arrows" could be shot from "Milimete-type guns." Evidently being able to shoot both kinds of projectile from the same weapon was considered an asset not just in antiquity, using torsion and tension catapults, but as late as the fourteenth century A.D., using firearms. But there remains much room for confusion because language inevitably changes in usage, and by the early third century, if not before, for example, the term "scorpion" was also used to refer to the *operator* of a catapult (and that might be true of other design names too), which obviously creates potential for yet more difficulty in the interpretation of the literary and epigraphic sources.[26] With these warnings in mind, let us return to the historians.

Arrian lived in the second century A.D. After studying under the slave-turned-stoic philosopher Epictetus (whose lectures he published, thus preserving them for us), Arrian pursued public office, found favor with Hadrian, and achieved the consulship in about A.D. 130. Thereafter as legate of Kappadokia in Anatolia he campaigned against the Alans (a Skythian tribe) when they invaded in 135. In his *Taktika*, written probably in A.D. 136 or thereabouts, and probably to satisfy Hadrian's interest in military matters, he mentions the Roman cavalry using a "machine" instead of a bow for shooting missiles or light spears.[27] From the context, it is clear that the exercise he describes is meant to be performed on horseback, ideally at speed. This hitherto singular bit of evidence for catapult-carrying cavalry has recently been supplemented through the excavation of the Elginhaugh fort in Britain. Hanson interprets this fort as one built predominantly for cavalry. There are no separate barracks (riders seem to have slept with or above their mounts) that one would expect if legionaries were stationed there. Found among a lot of demolition debris in a well was a small bronze washer, with a spring diameter of just 35 mm, a fraction bigger than Ephyra 6. This indicates a torsion catapult of nominally the same caliber as Ephyra 6, to wit, if a euthytone, shooting bolts about 1 foot, or about 310 mm, long; if a palintone, then shooting bullets weighing about 4-drachmai, or about 18 g, the smallest standard size for lead "slingshot." Some 133 lead shots have been found at Burnswark in Scotland, together with bolt heads and other sharps. This, in stark comparison to the 16 or fewer lead slingshot recovered from any other fort in the north of England,[28] is the chief reason why Burnswark is interpreted as a missile weapon training camp. I would hazard a guess that the site was specifically for catapult training, not training with hand slings, and that some of the people wielding such catapults were riding horses. Also found at Elginhaugh was a linear ratchet, of a size suitable for a 1-foot/ 4-drachmai caliber catapult (which would have a case about 540–640 mm long), which differs significantly from the simple flat pieces of metal described in the literary works. It resembles a solid bar with a T-shaped channel down one face (Figure 9.16).

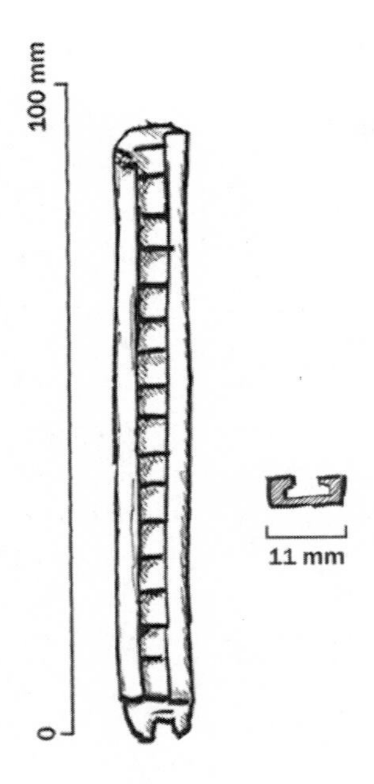


FIGURE 9.16 *The Elginhaugh ratchet (T.D.Dungan after Elizabeth Lazenby's drawing for L. Allason-Jones, with permission.)*

The ratchet, which is broken at both ends, but apparently only where the bar flattens and thus not actually affecting the ratchet part of the bar, is 92 mm long, 11 mm wide, and 5.5 mm deep. I imagine this would have fitted centrally on the underside of a slider or the topside of the groove of the case, but other arrangements are possible. The pawl would have to be placed accordingly, of course.

Ratchets had been a feature of catapult technology since the first *gastraphetes* in the fourth century B.C., but in due course—by the second century A.D. if not before—a nut within the stock replaced the slider with ratchet and pawl in some models. This nut, in a few known cases made of antler, lay in a socket on the top of the stock. It could rotate about a horizontal pin (or taught strong cord threaded through) which secured it in place in the

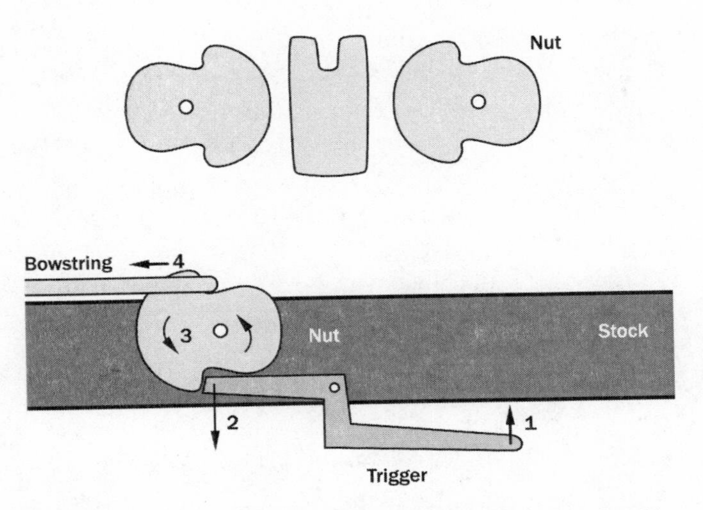


FIGURE 9.17 *Trigger and nut operation. (T.D. Dungan after author sketch)*

stock. The nut was carved so that, in profile, it resembled a circle with two slices removed from one side.

The sear of the trigger was held in the lower cut-away. The bowstring was held in the upper cutaway. The upstanding side was carved away in the middle to make space for the nock of the shaft to engage with the bowstring, leaving the remaining upstanding portion of the nut resembling two fingers. When the machine was spanned, these pointed upward, out of the stock. The simplest design of trigger would have pivoted in the stock, the sear holding the nut and the handle emerging from the bottom of the stock. When the trigger was squeezed or the handle (if present) was depressed (lifted against the stock), the sear moved down and released the nut to rotate under the tension of the bowstring, launching the sharp forward at speed (Figure 9.17). Such nuts are illustrated on Roman crossbows depicted on two bas-reliefs from France that date to the second century A.D., and two, possibly three, nuts survived in British soils. Nuts sit in the stock and the slider becomes redundant, so spanning weapons with nuts requires that the bowstring is grasped and pulled back directly. The simple crossbow is just such a weapon. In weapons too strong for manual spanning, we find the development of bastard strings, which are longer than the bowstring, and which are slipped over the arms to span the weapon, and removed once the bowstring is engaged by the nut. Alternatively, a variety of devices were developed that clipped onto the weapon just to span it and were then removed, such as the claw and belt, the cord and pulley, the screw and handle, the goat's-foot lever, and the cranequin. The nut continued to serve as the bowstring release mechanism for the springald and the last torsion catapults.[29]

In his *Expedition Against the Alans*, Arrian tells how he deployed catapults behind infantry and cavalry, either side of himself and his guard (he was the general), whence they could "cast sharps great distances into the advancing enemy" over the heads of the "phalanx" (legion).[30] Silence was maintained until the enemy were within range, whereupon they were trained to send "missiles from the machines, and discharge stones and missiles from the bows," ideally in such numbers—together with hand-thrown range missiles—that the Alans would break off an attack before coming into contact with the legionaries. This is apparently field artillery again, but Arrian says nothing specific about the *types* of catapult he used in this way. Vegetius, whose fourth-century work was based on earlier war commentaries such as this, proposes a similar arrangement, and is more specific: he put *carroballistae*, carriage-ballistae, and *manuballistae*, hand-ballistae, along with slingers and staff-slingers in the fifth line of battle (3.14).

Arrian's "discharge stones and missiles from the bows" is normally assumed to mean throwing stones on the one hand and shooting missiles from bows on the other. However, he might be referring to stone-thrower tension catapults, that is, crossbows shooting bullets. Recall Chapter 5 and the discussion of little catapults. The same may be true of, for example, Cassius Dio's description of the battle at Issus between Septimius Severus' forces and those of Niger in A.D. 194, where stone-throwers and bowmen shoot over the heads of the infantry (75.7.2).

Part of a sling-string ("bowstring" for a stone-thrower catapult) pouch made of thick leather appears to have survived in the subsoil of Vindolanda, which pickled much of the organic material deposited in it, and is, consequently, the source of so many wonderful letters and memos written by members of the garrison. This sling pouch, apparently now lost, would have been about 75 mm wide (or "high") when complete, and was perhaps over 220 mm long—considerably bigger than the normal range of a hand sling. Another, somewhat smaller, is also much bigger than one would expect for hand-slinging (top specimen in Figure 9.18).

Large cut-out holes in the centre of the big sling pouch were obviously smaller than the shot placed in it, and were perhaps introduced to reduce stiffness, weight, or wind resistance. The leather was made of "very thick" cowhide, according to Carol van Driel, who examined it. She suggested that catapult slings might be made of rawhide normally, since it is an elastic material. However, that sort of elasticity would not be advantageous in a catapult sling pouch, and rawhide would not have the strength necessary for the job on any but the smallest machines.[31] Biton was aware of the issue when he recommended for his approximately 5-pound tension stone-thrower (*gastraphetes*) that the sling pouch be made of hair "strong enough to bear the stone on dispatch" (47.2–4). One wonders if the sling-string in that case might resemble hair spring cord, or even whether the same material might be used for both springs and sling-string. Sling pouches that are woven as one continuous item out of the same material as the sling cord would be relatively homogeneous in performance and robust. Leather sling pouches appear to have been attached to the catapult via *two* bowstrings, strictly speaking, one attached to each arm, and tied to the sling via a hole in each side of the pouch (obviously, both strings must be practically the same length when tied and knotted). The iron ring that the trigger held was sewn, woven, or otherwise attached firmly to the pouch. Another sling pouch from Vindolanda, also made of thick cowhide, is beautifully decorated, but its size is such that it may be a hand sling pouch.[32]

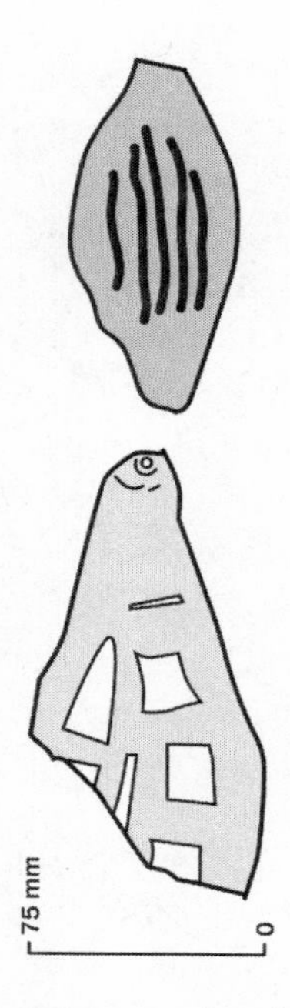


FIGURE 9.18 *Large and medium sling pouches from Vindolanda (T.D. Dungan after van Driel)*

We turn from the northwest edge of empire to the southeast boundary. Hatra, south of Mosul in Iraq, was a well defended and, for most of its history, apparently an independent, relatively multicultural desert city overlooking camel trade routes. Twenty-eight large and 163 small towers punctuated its city wall, which stood atop a ridge, and was encircled by a second, outer wall.[33] Trajan's final illness came on while he besieged Hatra, which revolted in A.D. 116—and he was surely not the only one to fall thus; historically, sieges were notoriously unhealthy operations, with as many if not more liable to succumb to epidemic disease than to military action. Septimius Severus attacked twice during the second Parthian War, in A.D. 198 and 200, both times unsuccessfully. Artaxerxes the Persian attacked it and failed about A.D. 227. By about the 230s a Roman garrison was in the city, and then Sapor (or Shapur) I, king of Parthia, attacked successfully, and destroyed it in A.D. 241. Thereafter the city was deserted. The Hatra ballista is the largest yet found, and a significant portion of the plating survived in the dry conditions.

The frame is 2.4 m wide, 0.84 m high, and 0.45 m deep. The springs would have occupied a cylinder 1080 mm high x 160 mm in diameter, though supports for the lever extend into the spring space either side of the washer from top to near bottom of the inner surface, reducing the space for spring cord (the same feature appears on the Pityous washer). The style of the machine appears to be a euthytone, insofar as the springs were contained in a straight single beam. The frame also has an unusual ratio of spring height to width—here 6.7 D, which seems to fall between formulaic stools. Philon and Vitruvius recommended a spring height of about 5.5D for a euthytone and 9D for a palintone. However, this catapult was made about two hundred years after the latest of our formula-authors was writing (Vitruvius). More conspicuously, the springs are amazingly far apart on this specimen; it looks quite different from a "formulaic" euthytone. I think this machine had inward-swinging arms: the cut-out in the plating here clearly parallels a cut-out in the front inner (stock-side) face of the stanchion, and the cast bronze corner reinforcements suggest that the other stanchion was positioned on the back outer corner. This seems to me to preclude out-swinging arms.[34]

It is closer to a euthytone frame than a palintone, so I would expect it to be a sharp-caster, but it was reconstructed (on paper) by Baatz as a stone-thrower, apparently simply because of its size, and Wilkins followed, though halved the caliber.[35] Size is not a sufficient reason for declaring the machine a stone-thrower. All literary sources say all euthytone catapults were sharp-casters (and three large sharps were found near the frame…). Dio singles out the Hatrans' cavalry and their "archers" for special notice, adding that they were effective at very long range "since they hurled some of their missiles (*bele*) from machines," of which they had many. Severus' own guard were targeted and "many" were hit. The Hatrans are also said to have had machines that could shoot two sharps simultaneously (Cassius Dio 76.11.2–3), like Zopyrus' tension catapult of centuries earlier. The catapult that fell from the tower and was left for us to find would not, by any means, have been the largest caliber sharp-caster known. The Hatra catapult resembles a 3-cubiter (D ideally 150 mm, here 160 mm, so oversprung, as usual), shooting sharps about 1.4 m long. The Hatra frame was roughly 8 feet long, which fits a 3-cubiter too. The length of the springs, while shorter than one would expect for a stone-thrower, is more than one would expect for a sharp-caster, which by formula would be about 5.5 D. Through lengthening the springs by about 20 percent, the builders of this catapult increased their volume significantly. For catapults built to formula, the spring-diameter is an adequate guide to the spring volume, since the length of the spring is known and co-varies with the diameter. For catapults with a non-formulaic frame, the spring diameter is not representative of the spring volume, which needs to be calculated individually from the non-standard dimensions in each case. The spring is essentially a cylinder of sinew or hair whose volume is $\pi r^2 l$, where r is the radius and l the length. In the case of a standard euthytone, r = 1/2 D and l = 5.5 D. In the case of Hatra, r = 1/2 D and l = 6.7 D. Perhaps the extra volume was designed to provide the additional power needed to shoot two sharps simultaneously, and what we have here is one of the Hatran's double-sharp-casters which so impressed Dio.

About twenty years after Hatra was destroyed, Dura-Europos met a similar fate.[36] This place gives its name to a type of missile mentioned a couple of times already, the Dura bolt, because complete and near complete specimens were found here, their wooden shafts and flights preserved in the dry desert conditions. In his discussion of the material Simon James observed that "most of the Dura shafts represent bolts about 0.5 m long, from medium or light arrow-shooters, which were small enough to be easily moved, if not wielded by one man like a crossbow: hence their name, *manuballistae*. . . . The unusual leaf-shaped bolt-heads strongly resemble medieval crossbow projectiles. This may not be coincidental. It is entirely possible that conventional crossbows, drawing their power from a stiff, conventional bow rather than torsion springs, were in use at Dura" (2004: 215). Tower 19 at Dura sank dramatically during a battle for the city about A.D. 257. The Sasanids were undermining sections of wall, including this tower, and the Romans were digging countermines to intercept them. The tower did not *collapse* when the Sasanids fired the props, but one of its walls dropped straight down, some two and a half meters into the mine, bringing down the timbers of the upper floors and the roof, which sealed what they buried and protected it from the flames at the time and the environment thereafter, until excavation in the 1930s.

What emerged from the debris was a picture of a desperate fight for the walls underground, in the mines and countermines. One man died near the site where the fire was started, with sulfur, pitch, and straw in the vicinity; another sixteen or eighteen soldiers died nearby in a huddled mass when the section of the countermine they were in collapsed after being fired, and coins found on the bodies indicate that these were Romans—auxiliaries probably, and certainly not legionaries. Then the tower wall dropped into the mine, and the fire below seems to have been extinguished in the process. What was in the lower part of the tower, now fallen underground, was preserved to the extent that flammable and delicate materials such feather flights, reed arrow shafts, cloth, and even baskets survived, together with a painted legionary's shield, horse scale-armor, and some Dura bolts. This tower and adjoining curtain wall were apparently attacked with a view to creating a breach; meanwhile Tower 14 seems to have been targeted to take out of action the one tower that could shoot down—presumably with more Dura bolts—onto the unprotected right sides of those on the Persian assault ramp, which was being constructed nearby. Ballista balls found in the ruins of the city suggested to the excavators that the Persians were also employing stone-throwers catapults against the defenders, so this battle seems to have involved all the classic ingredients of a siege in the imperial Roman period.[37] The assault was successful, and the city was abandoned after its capture.

### Range

In a vacuum, a projectile will attain the maximum range if it is launched at an angle of 45° from the horizontal. The range of an arrow in air varies from 60 percent to 90 percent of its theoretical range in a vacuum. This is because in air there is resistance and drag. As a result, the maximum range is achieved with slightly lower elevation.[38] Marsden suggests that sharp-casting catapults would reach their maximum range if they shot at an angle of 30°.[39] For shot, the higher the density of the shot, or the mass to surface ratio, the closer it will perform to vacuum conditions. That is why lead shot performs better than stone and stone better than clay, and why big stones perform better than little ones (until the weight starts to interfere with the thrower's action). The range depends mostly on the velocity of the missile. If you double the velocity, the range achieved will quadruple—all other things being equal, of course (which they never are in reality).

Modern estimates are fairly consistent. For example, McNicoll believed that the maximum range for ancient stone-throwers was probably of the order of 350–400 m, shot at high angles of elevation; somewhat further for sharp-casters. Marsden thought that catapults would normally operate at ranges up to 400 yards, that the most powerful machines would make 500 yards, and that Roman arrow-shooting ballistae from around A.D. 100 would be effective at the same distance.[40]

For a point of reference, and to provide a reality check for the skeptics, let us note that the current distance record for a bow, made of modern materials of course, is over a kilometer, at 1,145.09 meters; and the modern crossbow can shoot over 1,800 meters.[41] With his reconstruction of catapults with horsehair springs, Schramm shot a two-cubit bolt 370 m, and a one-pound lead shot over 300 m with a small palintone ballista.[42] At Jerusalem, according to Josephus, 1-talent shot reached more than 2 stades—about 360 meters, or 1,200 feet.

A distance of 120 yards is virtually point-blank range for a heavy bow. Vegetius recommended target practice with the bow and staff-sling at 600 feet (180 m). The distance record for a longbow is 358 m, but a bow only one third as strong shot 311 m. More ordinary ranges for the longbow are supplied by a replica Tudor specimen, 130 lbs draw weight, shooting an arrow of 52 g weight, which exceeded 240 yards on a still day, and a slightly heavier arrow, of 68 g, which traveled in excess of 220 yards.[43] The latter arrow had an initial velocity of 112 mph; the former would have been quicker. A composite bow, however, could do considerably better. In 1794 a Turkish diplomat in London, who by his own admission was a second-class archer in Turkey at the time, shot an arrow 379 m into the wind, and 440 m with the wind, much to the astonishment of the members of the British Toxophilite Society, who were familiar with the performance of the longbow. The longest record for a traditional Turkish composite bow is over 800 m, but an attempted reconstruction of such a bow in 1910 achieved "only" 434 m. An inscription from Olbia, a Greek colony on the Black Sea, which was inscribed in the third century B.C., claims that "Anaxagoras, son of Demagoras, shot 282 *orguias*." An *orguia* is an ancient Greek measure, the distance between the middle fingertips of a man with arms outstretched. It corresponds, therefore, to the Imperial fathom. The inscription thus asserts that Anaxagoras' shot flew approximately 560 yards, or 520 m.[44]

According to Strabo, the range for the sling was a function of the length of the cords. One modern enthusiast has taken his cue from this and tried slinging with superlong slings, rotated from a high platform, but it appears from his website that he has not quite got the hang of it yet. However, I know of no ancient evidence to support the idea of superlong slings, and it seems to me more likely that Strabo was doing what Apollodoros and Livy and others were doing, to wit using "sling" to mean slingstring, and he was simply associating a stone-thrower catapult's range with the length (quantity) of its spring.

The material of which the slingshot is made appears to have much more effect on range than does the length of the cords, with manual slinging at least. Dietwulf Baatz examined the ballistics for the three common types of slingshot: lead, stone, and clay.[45] He assumed they were thrown with a velocity of 75 meters per second (= c. 246 feet/second), which is significantly higher than has been achieved by anyone who has attempted to measure the velocity of slingshot. (It is, apparently, very difficult to measure, since the instruments usually have been designed to measure the speed of a bullet, which is much easier to direct past the electronic eye than is a slingshot). The nearest to Baatz's figure that I have found is Michael Gilleland, who reports that modern sling projectiles have been clocked at about 58 meters per second (190 feet per second or 130 mph), but he gives no details or references.[46] Thom Richardson enjoyed the facilities of Vickers Defence Systems for his ballistic testing, and consistently measured his slingshot velocities in the low 30 meters per second (approaching 70 mph), with relatively little variation irrespective of shot weight or shape except that lead consistently performed a little better than stone. However, he was by his own admission a novice, and in order to shoot his slingshot past the measuring "eye" he thought he had sacrificed velocity for attempted accuracy. Richardson's tests on bows, using three different self-bow replicas with draw weights of 72–90 lbs., three different types of arrows, and modern defense establishment ballistics equipment to measure the speeds, found average velocities of 83–99 mph (37–44 meters per second). He measured two crossbows, one of yew, the other of steel, with a variety of bolts, and found their average velocities to lie in an almost identical range, between 81 and 100 mph (36–44 meters per second).[47] Karger et al. report average velocities of 45 mps for a longbow and 67 mps for a compound bow (1998).

Baatz arrived at his launch velocities by working backward from the attested ranges, and as long as we substitute catapults as launchers, instead of human slingers (who, it seems, are simply not capable of achieving these speeds), then his results are still useful. He emphasized that with a launch velocity of 75 meters per second, the maximum *effective* range for clay slingshot is 65 meters, for stone 85 meters, and for lead over 100 meters—in fact, lead shot could be efficacious up to an impressive 200 meters. Richardson's results are consistent with these insofar as he too noted that lead shot performs much more consistently than stone, and ovoid shapes had greater range than spherical.[48] Projected at a trajectory of 40°, clay will reach 200 meters, and lead over 300 meters, but at that point they are simply falling to the ground, the energy of the throw having been dissipated. With a flat trajectory, the missiles fall shorter, of course, but it is also easier to aim successfully at someone or something if the missile is being launched at a horizontal or near horizontal trajectory, and the missile hits its target with more force too.

The effective range of ancient catapults was significantly below their maxima. I believe that they were normally used with a flat trajectory, so at considerably shorter range than they *could* achieve, because at that shorter range they could hit with tremendous impact. The catapult's raison d'être was its ability to perform better than manual weapons in terms of both distance *and* force delivered to the target; it was not to set flight records. And while the ability to out-range the enemy was advantageous, in such circumstances it would have been in the interest of the party with poorer weapons to close down the distance between troops as quickly as possible, rendering the advantage an ephemeral one.

Since the drawback is ratcheted, as long as the catapult keeps the same elevation, the range achieved by the missile is consistent, extending discontinuously with each tooth of the ratchet. In other words, suppose that at, say, ratchet tooth 10 the range was 100 meters, and at ratchet tooth 11 it was 105 meters, missiles would land at 100 meters or 105 meters, but it would be impossible to make the missile land at 102 meters. This phenomenon is not confined to lab conditions: I have observed it in action with a reconstruction of a Roman *kheiroballistra*, which, for the cameras, was aimed at a target that happened to fall between two steps of its range. After finding the approximate range, the catapult then shot short, and long, and short, and long, for what seemed like forever, as the operators suspected, and then found, and then knew absolutely, that the range of the target fell between two particular ratchet teeth. The director of the TV crew thought that if the catapult just kept shooting it would sooner or later hit the target. In fact, what was required was for the operators to move the catapult (or the target) up or down or back or forward a pace or two; then they would have hit it as consistently as they in fact missed it! No doubt their ancient counterparts discovered this too, probably early in their careers as artillerymen.

TEN

# Late Antique Catapults, East and West

*Very few of the new manifestations in war can be ascribed to new inventions or new departures in ideas.*—Clausewitz *On War* 6.30

THE LATE ROMAN EMPIRE IS DATED, conventionally, from A.D. 284 (the accession of Diocletian) to the fifth century A.D. At the beginning of this period it was still the case that a man who joined the army could become not just a citizen, as in times past, but emperor, if he had the ability, the luck, and the determination to get there, like Diocletian himself. In 285 it occurred to him that he might live longer if he shared power, and the Tetrarchy (literally, four rulers) was born. It worked for his generation but the next could not agree what or where to share. As Herodian observed, supreme power is indivisible, a desire for sole rule is typical, therefore multiple emperors are almost bound to want more than their share, which leads to their downfall, and more civil war waged on and off until Constantine ("the Great") established dominance in A.D. 324. Byzantion was rebuilt (Septimius Severus had sacked it and demolished the fortifications during a civil war in A.D. 194) and renamed Constantinople in A.D. 330, though Prokopios still called it Byzantion two centuries later, and the de facto capital of the Roman Empire was shifted thousands of miles east, from the eternal city to Constantine's city. The basic idea of the tetrarchy reappeared, just as unsuccessfully, in his will, wherein he expected his three sons, unimaginatively or conceitedly called Constantine, Constantius, and Constans, to share the Roman Empire between themselves.[1] They ended up fighting each other and some of their cousins unto death, and by A.D. 363 all members of Constantine's dynasty were dead, less than forty years after he secured sole place on the throne.

The empire was divided in 395, and the Western Roman Empire is the part that "fell" to the barbarian hordes, though it had been losing control of small and large areas at the margins since 260. Alaric invaded Italy in 408, and the Goths captured Rome on 23 August 410. It is perhaps worth repeating here a point made very effectively by Peter Heather that, given the time it took people to get from A to B, running the Roman Empire then "was akin to running, in the modern day, an entity somewhere between five and ten times the size of the European Union."[2] The western empire shrank by fits and creep in the following decades, until in 476 the last, and by this time fairly notional Roman emperor was replaced by the first king of Italy. Two generations later, in the 530s, troops of the Eastern Roman Empire, led by the very talented general Belisarios, came to liberate Rome and much of Italy from its Ostrogothic rulers, but the light was soon extinguished again. Protracted fighting led to the collapse of institutions and technologies, as people began to move to more easily defended sites on hilltops and built for security rather than comfort. Pandemic plague akin to the Black Death arrived in Constantinople in spring 542, spread along trade routes in one or another outbreak (finally reaching Britain sometime between 634 and 642), and devastated society and economy wherever it went. The first effects were inflation and famine, and each of those brought more trouble in their wake. Dense concentrations of people, such as found in cities and armies, were worst hit, and the scale of the catastrophe is indicated by Justinian's order to dig mass graves, outside Constantinople on the other side of the Golden Horn, big enough to hold 70,000 bodies each. Some settlements were completely wiped out.[3] The lack of manpower that is seen to bedevil the Byzantine army also bedeviled society and economy as a whole; pandemic plague forced rapid downsizing and dislocated many institutions, striking people (and sometimes animals too) with no apparent discrimination.

The Eastern Roman Empire is better known as the Byzantine Empire and it lasted until 1453. The Byzantine Empire is a fuzzy historical concept. Some would emphasize the Roman elements, others the medieval elements, some the Eastern elements, some the Greek elements. Some historians would start it in 284, others in 324, others in 395, 476, 565, 610, or 717. The ending, by contrast, is clear and precise: 29 May 1453, when Byzantion, aka Constantinople, fell to Mehmet (II) the Conqueror. Within this period there is, of course, a huge amount of continuity, wherever one wishes to draw lines for convenience of historical analysis.

I believe that what we see in the late empire is a resurgence of variety and adaptations in catapult design. The Roman army's loss of dominance inevitably entailed the erosion of their catapults' dominance. Many areas of the empire were left to their own devices for years, sometimes decades on end. In the absence of Roman forces, or in their presence but non-deployment (there is a noticeable failure of nerve on the part of the Roman military as time slides on), space opened up for the development of other weapons. Isolated Roman forces were in a similar position to most of their enemies.

Weapons that were technically inferior to the best Roman models could, in their absence, be developed and deployed successfully. Adaptations to meet contingencies—especially to use available, even if not necessarily ideal, materials and skills for construction and use—would have proliferated during unstable periods (which were distressingly numerous through these centuries). In many ways, the pattern would have been similar to the Hellenistic and Republican period, for the break-up of empire was structurally akin to the creation of empire, but in reverse. Rarely now did an emperor take forces from one side of the empire to another; indeed, the attempt to move troops long distances could spark rebellion, as Constantius found when he ordered Julian to bring Gauls east. The homogeneity of culture that is characteristic of the High Roman Empire ("Romanization") began to dissolve into oriental and occidental, cool-climate and hot-climate constituencies that grew increasingly dissimilar as travel between them became less regular for troops and merchants, less popular for civilians, and more infrequent for everybody.

Inevitably in this process, knowledge and skills were lost—and the burning of books in A.D. 371–372 (and at other times too during these years of ignorance, arrogance, suspicion, and rampant superstition) would have fueled the decline.[4] In such a climate, the natural tendency to diversity in technology (as in biology) is not stifled, and less-than-best designs are not inevitably out-competed while still young and tender. Consequently, novel devices could be developed into weapons that could challenge for dominance in their region, and in due course meet the standard set by their predecessors, or even supersede them. As the monopoly of Roman power gave way before the encroachments of a thousand different self-interests, competition opened up again between various weapons, leading to new designs. Ultimately, Rome ebbed away and domestic and foreign forces transformed the West into the medieval world, while a Hellenized, Christianized, bureaucratized "Rome" endured in Turkey for another thousand years.[5] The successors to ancient catapults can be found in both.

Three authors from the fourth century provide descriptions of individual machines. These authors say little explicitly and intentionally about the development of weapons; they describe, in more detail than do most ancient historians but in much less detail than the ancient technical writers, weapons that they knew directly or through books in their own time, and occasionally they contradict one another. The first, chronologically, is the anonymous author of *De rebus bellicis*. This is full of all sorts of apparently novel ideas for war material, and was addressed to the reigning emperors, who were probably Valentinian and Valens, probably in A.D. 368.[6] The second author is Ammianus Marcellinus, who lived A.D. 330–395 or thereabouts. He was a historian—a good one—and his *Res Gestae* is, among other things, a salutary lesson in the horrible consequences of corrupt government. He fought on campaign and was himself besieged at Amida in A.D. 359. The third is Vegetius, who wrote between A.D. 383 and 450, and therefore was a generation or two later than Ammianus. He was a military theorist, and it is largely through his advice on training regimes and techniques that we hear about weapons thought old in his time. All of these sources' descriptions suffer from brevity and incompleteness, and modern scholars have accused their authors of confusion here and there.

The fifth century is a dark age, when most of the Western Empire and North Africa was lost to barbarian invaders—the Vandals took Africa, the Ostrogoths Italy, the Visigoths Gaul and then Iberia, the Franks northern Gaul. Very little written or archaeological material of use to us survives from this century; Cassiodoros, who lived about 490 to 585 and was advisor to the Gothic King Theodoric, is one of the few lights to shine, but he has nothing to say on our subject.

From the sixth century we have an excellent historian, Prokopios. He was probably born around A.D. 500, and lived almost his entire adult life under Justinian, who reigned A.D. 527–565. Prokopios served with the great general Belisarios and wrote about contemporary events. His *Buildings of Justinian* record the emperor's efforts to rebuild the Roman empire (literally; aqueducts and fortifications were major priorities); his *Histories* describe Justinian's wars against the Persians, Goths, and Vandals; and his *Secret Histories* retail some of the more lurid and slanderous stories told about Justinian and other members of the "Roman" (Byzantine) court. The last apparently were not published until after Justinian's death. In addition to Prokopios we also have a couple of manuals on strategy, one anonymous, and the other apparently written about A.D. 600 by the emperor Maurikios, aka Maurice, whose name derived from the Mauri (= Moors), long since incorporated into the Roman Empire. He was not the only emperor to write on such topics, as he was not the only emperor to fight campaigns and lead troops in the field, and this work and others like it were not royal conceits but handbooks considered by their peers to have "real practical value."[7] There are assorted further sources that can shed a little light here and there, such as Malalas, who wrote a chronicle of events up to his own time in around A.D. 570, and the *Miracles of Saint Demetrius*, which includes a description of the war machines built by the Avars during the first siege of Thessalonike, in A.D. 597.

THE ANONYMOUS AUTHOR OF the *De rebus bellicis*, a treatise on matters of war, is our first witness to developments in late antiquity.[8] Written about A.D. 368, it begins with the effusive obsequiousness that is alien to our own time but was becoming increasingly common in his age, and then promises to advise the emperors on how they could reduce taxes by half, pacify the frontiers and populate empty land, double imperial income without raising taxes, and pay the troops more. Thus it sounds thoroughly modern, at least in its economic aspects, promising cake today and cake tomorrow and gain without pain. The author, whom I shall henceforth call DRB (on the model of AP for the author of the *Athenaion Politeia*), then says that he "felt it necessary to add the prerequisites for victory on land and sea amid the exigencies of war," and he has selected "for description a few mechanical inventions so as to sustain your [imperial] interest" (Preface 7, Thompson). It becomes apparent that he actually has much more to say about the mechanical inventions than he does about the economy, taxes, recruitment, and governance of the empire.[9] Clearly, given their quantity (two-thirds of the text), the descriptions of war machines are not mere literary devices to keep the emperors' minds from wandering off while being advised on political and economic matters of state. Indeed, close analysis of his language and style led Ireland to suggest that our author was a man "of strong mind" and "of large intellectual curiosity with an elementary grounding in rhetoric but little experience of formal written expression [who] had acquired, probably in the provinces, some experience of financial affairs and the fiscal system," and "an army man, perhaps an officer, perhaps connected with the artillery, but now in retirement" (p. 150).

DRB recognized that a written description is not enough to convey information about machines that are unfamiliar to the audience, and so he provided carefully drawn colored illustrations of the *tormenta* and other equipment (6.4). The first such machine he discusses is the *ballista quadrirotis*, which was mounted on a cart with four wheels (hence the name) that was pulled by two armored horses.

> [The four-wheeled ballista's] utility, resulting from its ingenious design, is such that it can shoot sharps at the enemy on every side with the flexibility and dexterity of the archer. It has holes in its four sides by which it is turned and aimed as required, and it is very easily prepared for any attack. By a screw at the front it is very quickly depressed or elevated. Its stock can be turned by rapid and smooth movement to whatever direction is required by the situation and then elevated. It must be noted that this kind of ballista requires two men and shoots its sharps after being spanned not by ropes as others are, but by rods.[10] (DRB 7)

This is sufficiently vague to require the reader's active mental participation to join the dots to create a picture, which simultaneously leaves lots of room for the imagination to maneuver. Since the original *had* pictures, and it would have been easier for the copyist to copy the originals than deliberately ignore them and invent new pictures, I start from the presumption that the illustra-

*Figure 10.1 Ballista quadrirotis. (MS. Canon. Misc. 378, fol. 71v, The Bodleian Library, University of Oxford)*

tions are meaningful.[11] Book illustration began with illustrations for technical treatises like this.[12] The illustration of the ballista quadrirotis on the manuscript in the Bodleian, reproduced here in Figure 10.1, thus probably shows what DRB had in mind, with the qualifications that it was modified by what the illustrator (i) recognized as significant and (ii) could depict within the constraints of a miniature (the image is about 145 x 200 mm).

That the illustration is deficient is immediately obvious; the man at the left does not seem to have his feet on the ground, for example,[13] but exactly what the deficiencies are with the ballista is a much less tractable problem in which the observer is the chief element. I do not believe, as some do, that if we cannot figure out how a machine might work from the evidence that we have, then the machine cannot have existed in reality and is just an illustrator's pipe dream of a writer's pipe dream. That attitude seems to me arrogant rather than critical. The evidence may be faulty, but so may be our understanding of it, and if we knew more, or looked at it from another perspective, it might make sense. For example, it has become clear in recent years that medieval "technical drawings" frequently portray composite diagrams, showing in one picture some elements in plan, some in elevation, and some from a bird's eye view, choosing whatever was best suited to depict the element in question. From that perspective, medieval illustrations were more efficient than ours are, achieving in one picture what we do in three. However, since we are not used to looking at such graphics, we find them baffling. This, however, is not that sort of drawing, but appears to be an honest attempt to render a scene, in perspective, albeit faltering.[14] I certainly find this illustration of the *ballista quadrirotis* baffling, and look forward to the day when someone, somewhere comes up with a convincing explanation of it. What *is* that crank-like rod sticking out of the front of the box, just beside the case and slider? Meanwhile, I perforce concentrate on the text.

Much of the description focuses on the fact that this ballista revolves on its wagon, so that it could shoot in any direction without the whole wagon needing to be brought about. This represented the same sort of advance over the carroballista that a turreted tank represented over the first unturreted ones. I imagine that it was never discharged in the position illustrated in the manuscripts, to wit, with the large sharp exiting the device between the horses' heads, for few equines would tolerate such actions. An improved version appears two centuries later in Maurikios, where a *pair* of revolving ballistae is fitted to a single (presumably larger) wagon, one at either end.[15] The necessary mechanics (a swivel post, for example) would be hidden by the structure of the wagon and therefore should not be expected to appear in the illustration. The four holes by which the ballista was "turned and aimed" refer perhaps to the windlasses by which the catapult was spanned and/or to the unknown mechanism by which it was turned to face different directions.

A particularly interesting part of the description is the last line. Marsden suggested that DRB was referring to toothed iron bars engaging with a cog-wheel on the winch, likening the action to the cranequin (one of a variety of winding mechanisms) on the medieval crossbow. Scholars and re-enactors have neglected this idea, but I think Marsden may have been on the right track. For the recently discovered Elginhaugh ratchet (Figure 9.16) must have fitted centrally to either the top of the case or the bottom of the slider, and thus was in just the position one would expect to find such a rack and pinion device. Moreover, medieval cranequins typically were very short, since the distance the bowstring was pulled back from brace to span positions was as little as six inches.[16] I am not for a moment suggesting that the detachable, double-handled cranequin existed at Elginhaugh in Roman Britain; what I am suggesting is that a ratchet-bar attached to the slider engaging with a toothed-drum pull-back mechanism would explain both the DRB's description and the Elginhaugh artifact. Nor do we have to look *forward* in time for analogous devices, as Marsden did. The rack and pinion concept goes back at least to Ktesibios, in third-century B.C. Alexandria, for his water clock used such a device to make the dial (on a drum) rotate as the figure (on a rack) rose.[17]

I think Marsden was wrong, however, to take *cochlea* as "roller" instead of screw, and the device at the front to raise and lower the stock probably

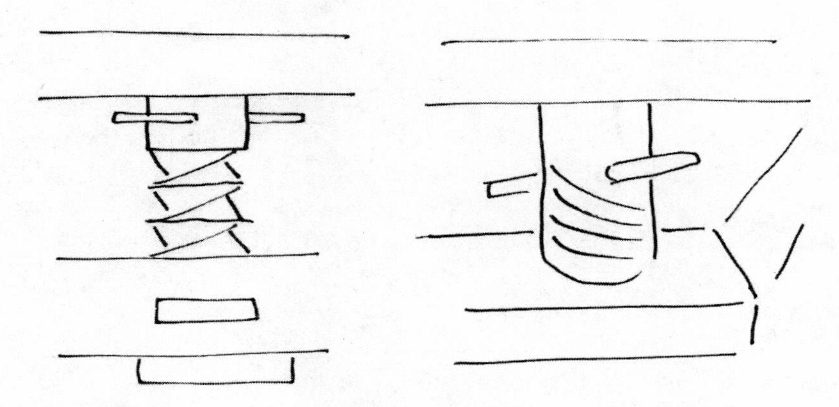

FIGURE 10.2 *Screw elevator on de rebus bellicis MSS, Munich, left, and Paris, right. (Author sketches)*

involved a screw of sorts. Such existed for the *dioptra*—a sighting instrument existing from at least the first century A.D. to the fifth—and it has been suggested that a stronger version could have been made for catapults.[18] I imagine that an elevating screw for a catapult would engage a frame, like an inverted screw press.[19] Indeed, I think that is exactly what the "column" with a stick through it, near the front corner of the device in the Bodleian illustration is. The same sort of feature on the ballista fulminalis is clearly a screw in the illustration in M (Figure 10.2), which, like the Bodleian MS, was copied from the Speyer MS, and it is less obviously a screw in the Paris MS as well.[20]

DRB then gives descriptions and illustrations of a *tichodifrus*, for protecting ballistae while they are advancing and working (Vegetius' *pluteus*, 4.15, is very similar), and a *clipeocentrus*, which is a sort of hobnail shield. This is followed by descriptions of two types of *plumbata* (lead weighted dart), in the first case modified to incorporate a *tribolos* or caltrop along the shaft and called the *plumbata tribulata*, in the second given a sharp conical head and called a *plumbata mamillata*. Heads of the latter shape have been found in London, and identified as ballista bolt heads, and in fact plumbata heads and ballista bolt heads seem to be indistinguishable.[21] We noted in Chapter 6 cross-fertilization between catapult and manual technologies with respect to the *kestros*. *Plumbatae* were similar missiles, and probably transferred between delivery systems too. Next we are told about a *currodrepanus*, which was a sort of horse-drawn scythed chariot without the chariot—just the axle, two wheels, two scythes, and cordage, by which the riders on the horses that pulled this machine could raise or lower the scythes at will as they rode around the battlefield. This machine is given in three versions: a simple one-horse one-rider type, a simple two-horse two-rider type, and finally an armored, two-horse one-rider type that also incorporated whips, presumably attached to the revolving axle, which lashed the horses on automatically—it is the last element perhaps that entices modern readers to write the whole off as fanciful, but I think more caution is needed. The whip may be over-the-top, but it is not an irrational addition, and it would not be the only military technology, ancient and modern, to which this label might be applied. Moreover, the number of versions given for this device speaks strongly in favor of the existence of at least one of them. This device (in its simpler manifestations at least) is said to be particularly effective against broken ranks, and that would be a thoroughly appropriate environment for it. Arrian, writing back in the second century, reckoned scythed chariots obsolete (*Taktika* 2.5), but so too he reckoned elephants, yet we know that they were still (or again) being fielded, and still found formidable, in the fourth century when Ammianus and DRB were writing. Leonardo da Vinci—apparently inspired by the DRB—came up with his own versions of the *currodrepanus*, for use on the battlefields of fifteenth-century Italy.[22] The Roman empire was large, and different enemies were familiar with different tools and techniques at different times. The shock value of any weapon could be restored by resting it for a generation or more.

The chariots are followed by a description of the *thoracomachus*, which DRB specifically says is an ancient invention and was probably right to do so,[23] a felt garment originally worn under armor to protect the skin from the discomfort and abrasion of the metal, and to keep the wearer warm. The anonymous sixth-century treatise on *Strategy* describes the same thing and adds information when he says that defensive armament was, ideally, iron armor over a thick (at least a finger's thickness) undergarment, to prevent chaffing and to make it "more difficult for missiles to penetrate, or at least to penetrate deeply."[24] The last sentence is important. The Conquistadors discovered empirically that a stone-tipped arrow shot from a relatively weak bow (draw weight about 200 Newtons) could penetrate two coats of chain mail, but that padded cloth "armor" offered them better protection against such projectiles than chain mail did.[25] DRB seems to have been aware both of the *thoracomachos*' origin as an undergarment, and of cloth's superior performance over mail to protect against arrowheads. He recommends that a leather coat, of the same cut, be worn atop the *thoracomachos* to keep it dry and light.

DRB then describes the *ascogefyrus*, an infantry pontoon bridge.[26] His last word on this device is that *manuballistae* will need to be deployed on both banks to defend those making the bridge. Next he describes the oft-discussed ox-powered warship, a *liburna*, wherein three pairs of oxen each power a pair of paddle wheels affixed to the sides of the boat. Then he describes his second major catapult, the *fulminalis* or lightning ballista.

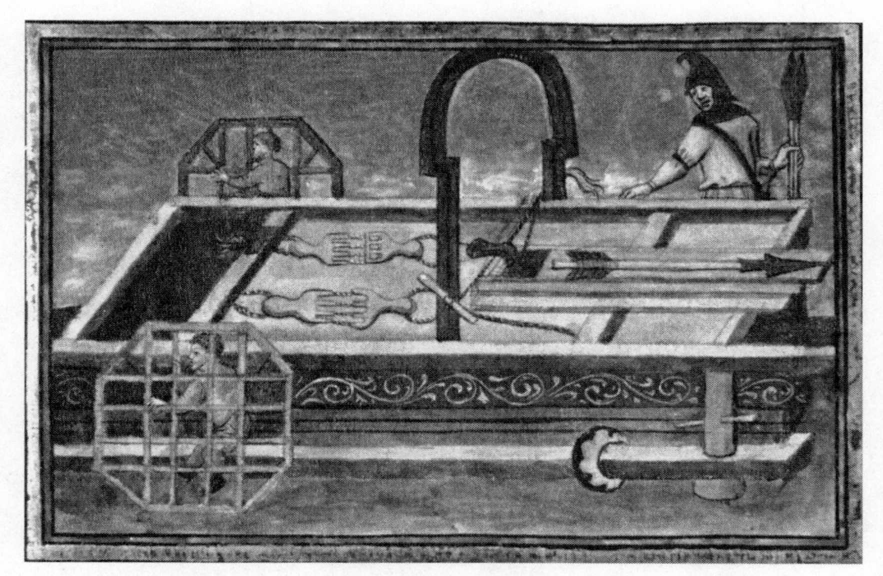

*Figure 10.3 Ballista fulminalis. (MS. Canon. Misc. 378, fol. 76r, The Bodleian Library, University of Oxford)*

> [the lightning ballista], essential for the defence of walls, has been found superior in practice to all others in speed and power. With a steel bow [or With an iron arch] erected above the groove along which the sharp is discharged, a powerful sinew-rope is drawn back by an iron claw and, when released, propels the sharp with great force at the enemy. The size of the machine does not allow this bowstring to be pulled back by the soldiers by hand; rather, two men draw it back by pressing against the spakes of two windlasses, the mechanical force acquired being equal to the difficulty involved. A screw elevates and depresses the ballista, as required, to direct its missiles higher or lower. Here is evidence of its shock value: although it is a combination of so many different components, it is controlled, with respect to shooting the missile, by one man only at his leisure, as it were. The originality of the device would be less if a crowd of people was engaged in operating it. A missile shot from this ballista, comprising so many ingenious devices, travels so far that it can even traverse the Danube, a river famous for its size. Its name, the lightning ballista, testifies to the effect of its power. (DRB 18)

The power source for this (and the earlier catapult) has been the subject of much debate. Like that for the ballista quadrirotis, this illustration, shown in Figure 10.3 and in color on the jacket of this book, does not obviously show any springs, and does appear, on casual inspection, to show a gray bow, but it is shown vertically instead of horizontally, and thus not as we would expect it.

One could argue that artistic license had hidden the torsion springs in the rather slim uprights, in which case the big arch over the top is not a bow but a development of the *kheiroballistra's* "little arch," as Marsden supposed. Despite his attempts to interpret it thus, most scholars have, I think, returned to reading the text as a reference to a steel bow, even if they cannot imagine precisely how the machine illustrated in the accompanying miniature would work.[27] The lightning ballista was probably, they argue, a large caliber crossbow, the bow being made of resilient steel, and sufficiently strong that to span it required two men, operating a windlass, which was in turn connected to a pair of compound pulleys, if the illustration is accurate in this regard. The claw spans the weapon apparently by engaging a recess in the back of the slider itself. The bowstring appears to run behind the back of the slider, and while I can imagine various methods to discharge this weapon, I see nothing resembling a trigger.

The *De rebus bellicis* illustrations, which apparently show the "bow" at right angles to the stock, may represent these machines as other medieval manuscripts have been shown to represent machines, that is, in plan, elevation, and perspective combined, each part of a machine being depicted in the form which best shows its particular features. Consider, for example, a medieval depiction of a springald, whose arms appear to be vertically arranged (Figure 10.4), though we know that they were horizontal. Thus the "arch" of the DRB ballistae may be intended to show a bow actually lying horizontally, not vertically. Remember too that these illustrations are miniatures, fitting onto a codex page. If a horizontal bow were superimposed on the rest of the picture, it would be well nigh impossible to distinguish the different elements. None of the other texts that have been thought—perhaps wrongly—to describe the same sort of catapult—texts that we will discuss later in the chapter—*locate* the posts that they say hold the bow on the catapults they were describing. The obvious place for such posts is central to the bow, where the bow does not bend, where the hand grip is on a manual bow, where the stock is attached on a crossbow, and *not* at the ends of the bow, where they would obviously interfere with its operation. If a bow between posts is indeed what this artist drew, then he clearly took the "bow" to be fixed length-wise, rather than width-wise, between posts. But the depiction of the "posts" at the edges of the carriage, and therefore at the end of "bow," seems bizarre. They appear to be offset uprights, and they seem to be made of the same material as the "bow," and they seem to be distinctly oblong. This perhaps has led more than anything else to the idea that this device or this illustration is unreliable. But if the artist was not depicting a bow catapult, then the other texts do not refer to this device and we are not necessarily dealing with a bow held between posts.

FIGURE 10.4 *Springald from an early fourteenth-century MS of the Alexander Romance. Here the arms appear to be bound directly to a horizontal frame. If the bindings were made of resilient material, this could work. The arms and bowstring appear to be vertical, but one can see from the angles of the frame that the illustrator was trying to render a perspective image and that in reality they would have lay horizontally. (T.D. Dungan after Bodleian 264 fo. 201r)*

Independently I came to the same conclusion as Mark Hassall that the arch is just that, a strong iron structure to hold the ends of *the bowstring* absolutely firm.[28] I think that the "powerful sinew-rope" that is tied between the posts of the arch is both the bowstring and the power source simultaneously. I would see it as pre-stretched sinew just like that used in torsion springs, which would move very little (but very quickly) rather than, as Hassall, *ligamentum nuchae*, which is much more elastic, and which, if used, implies that the illustrator has "drastically reduced" the length of the vehicle (p. 80). In any case, the bowstring is stretched taut across the space between the legs of the arch, a wooden batten apparently securing the end nearest the viewer, and of a design that could obviously be tightened easily. The leverage offered by arms has been dispensed with, and the resilience of the sinew alone is used to move the slider, which carries the missile. The "bowstring" is now a sort of very powerful elastic band that propels the slider; when the slider stops, the missile carries on. Technically, this would be a tension catapult, not torsion, since the sinew is not being twisted by the retraction of arms to store the (evidently tremendous) energy put into the machine in spanning it. The claw appears to hold the back of the slider, and the "bowstring" appears to run behind to slider, so to discharge the catapult would have required the claw to release the slider. This perhaps explains the short rope that runs from the far post to a dark "square" at the back of the slider. If that dark square was, say, a metal plate that could be raised by pulling on the rope, then the claw could be lifted by it to release the slider. Alternatively one may suppose a standard sort of trigger arrangement, as Hassall does, though there is nothing in text or illustration to suggest it.

The *De rebus bellicis* then turns in the pre-penultimate chapter to offering suggestions for deployment of the aforesaid machines, such as using the *tichodifrus* to protect the ballista or to render assistance to the reserves, and for "the long array of machines or tormenta" to protect the flanks in confined situations. His suggestion to have spare teams of horses for the ballistae, to replace those injured or fatigued, shows either great foresight and attention to detail if these machines are genuinely novel, or experience of their potential shortcomings when they had been deployed in the field hitherto. I suspect the latter. The penultimate chapter suggests a system of frontier forts and defences, funded and manned by the locals. The last chapter contains the author's final idea for solving the woes of the time, which is that the emperor put a stop to dishonest litigation.

Let's turn to our next source: Ammianus Marcellinus, who was approximately contemporary to DRB in time and whose history includes a description of artillery in his time which has been used by some to try to explain the DRB machines. Ammianus offers a concise description of siege-engines for those who are unacquainted with them; this digression is one of many which pop up throughout his history on topics as diverse as the cause of rainbows and the transportation and raising of a huge obelisk, complete with translation into Greek of the hieroglyphs upon it. His description of artillery is of a piece with those, that is to say, it is rather woolly and unfocused, and in particular, suggests no understanding of why and how the machines work (I assume he didn't understand hieroglyphs either; neither hole in his competences stopped him writing authoritatively on these topics). It was perhaps inserted as a result of then-current trends in historiography, following one of Lucian's (surviving) recommendations on how to write history.[29] In fact, Ammianus' description of these machines leaves so much latitude that his *ballista* is now variously interpreted as a traditional sharp-caster, a Heronian *kheiroballistra*, and a bow (tension) catapult—which may or may not have been the same or similar to DRB's lightening sharp-caster—so to at least three very different machines. Ammianus' Loeb translator seems to understand the machines no better than his subject did, which actually is useful in this contested area: it renders him a relatively independent judge of Ammianus' rococo Latin.

> [Ballista]: between two posts a long, strong iron bar is fastened, and projects like a great ruler; from its smooth, rounded surface, which in the middle is highly polished, a squared staff extends to a considerable distance, hollowed out along its length with a narrow groove, and bound there with a great num-

> ber of twisted cords. To this two wooden rollers are very firmly attached, and near one of them stands the gunner who aims the shot. He carefully places in the groove of the projecting iron bar a wooden arrow, tipped with a great iron point. When this is done, strong young men on both sides quickly turn the roller and the cords. When its point has reached the outmost ropes, the arrow, driven by the power within, flies from the ballista out of sight, sometimes emitting sparks because of the excessive heat. And it often happens that before the weapon is seen, the pain of a mortal wound makes itself felt. (Ammianus 23.4.2–3 Loeb)

Ammianus seems to have little idea of what actually drives the arrow—what does "the power within" mean? His description allows a variety of possible intepretations. For example, one could take the long, strong, projecting iron bar with smooth rounded surface to be a *barrel*; the highly polished "middle" is then the inner surface of that barrel. If it isn't a barrel but a bow, why is the middle of it highly polished? Elsewhere different problems arise; so he appears to assign "the groove" both to the squared staff and to the projecting iron bar which apparently holds the staff. Were they both grooved, or was he confused? Before spending too much time and effort on this, let's consider his description of the onager.

> The scorpion, which is called the onager these days, has the following form. Two beams of oak or holm-oak are hewn out with a slight bend, so that they seem to stand forth like humps. These are fastened together like a sawing machine[30] and bored through on both sides with fairly large holes. Between them, through the holes, strong ropes are bound, holding the machine together, so that it may not fly apart. From the middle of these ropes a wooden arm rises obliquely, pointed upward like the pole of a chariot, and is twined around with cords in such a way that it can be raised higher or depressed. To the top of this arm, iron hooks are fastened, from which hangs a sling of hemp or iron. In front of the arm is placed a great cushion of haircloth stuffed with fine chaff, bound on with strong cords, and placed on a heap of turf or a pile of sun-dried bricks, for a heavy machine of this kind, if placed upon a stone wall, shatters everything beneath it by its violent concussion rather than by its weight. Then, when there is a battle, a round stone is placed in the sling and four young men on each side turn back the bar with which the ropes are connected and bend the pole almost flat. Then finally the gunner, standing above, strikes out the pole-bolt, which holds the fastenings of the whole work, with a strong hammer, thereupon the pole is set free, and flying forward with a swift stroke, and meeting the soft haircloth, hurls the stone, which will crush whatever it hits. And the machine is called *tormentum* because all the released tension is caused by twisting (*tor-*

> *quetur*), and scorpion because it has an upraised sting; modern times have given it the new name onager because when wild asses are pursued by hunters, by kicking they hurl back stones to a distance, either crushing the breasts of their pursuers or breaking the bones of their skulls and shattering them. (Ammianus 23.4.4–7 Loeb)

The idea that the ropes which are, in fact, the power source, are there simply to hold the machine together and stop it flying apart indicate immediately and categorically that this description was written by someone who had absolutely no understanding of the machine he was describing. He was probably ignorant of the technical literature on catapults, as he seems to have had little, if any, understanding of Greek.[31] Consequently, this nonsense should not distract us. Recall, for example, that the name "scorpion" is documented from about 330 B.C., before even the first certain reference to torsion catapults, and in no source earlier than Ammianus is the name associated with the onager. Likewise the last section has been a bit of a red herring: Ammianus' attempt at etymology, like so many of the ancients', is misguided (compare Prokopios' on "ballista" as a combination of *ballo*, throw, and *malista*, special); onager is more likely derived from *monagkon* (drop the m and the k), which also spawns mangonel and derivatives. This is a one-arm catapult, and evidently a large caliber version in this particular case.

Another of Ammianus' comments concerning catapults has led moderns astray. Ammianus refers to Julian taking only *ballistarii* and cataphracts as his guard "to avoid any delay" when rushing across Gaul to attack the Alamanni (16.2.5), and adds his personal opinion that he thought this choice poor—these troops were "far from suitable to defend a general." However, although Ammianus was a serving officer in the military, and therefore his opinion counts for something, Julian was more than that, and therefore Julian's decision should warrant at least as much respect as scholars are wont to give Ammianus' opinion. Julian was not just a serving officer, he was the *commander* of military forces before and during his reign as emperor, and further, he was especially interested in mechanics. For example, he installed mural artillery (*tormentis muralibus*) and powerful war engines in a repaired Trajanic fortress. He stationed *tormenta* on riverbanks to deter Persians from crossing. He even wrote a treatise on mechanics. Ammianus, by contrast, can't tell his arm from his elbow when it comes to catapults. Therefore Julian probably had a much more informed opinion than Ammianus on the suitability of *ballistarii* as bodyguards. And it was his body they were guarding, so he had a vested interest in the choice. Moreover, moderns usually assume these *ballistarii* to be "artillerymen," imagining large weapons as the English name and Ammianus' description of "ballistae" clearly imply, but they may not have been; these *ballistarii* could have been *manuballistarii* wielding little catapults, personal weapons. The context, emphasising Julian's haste, suggests that he selected only those troops who could move quickly. His progress was compared to that of a meteor and *malleolus*, fire arrow (which are thus compared to one another). It makes sense to see *ballistarii* in this context as shorthand for *manuballistarii*. The same sort of arguments could explain at least some of the units of *ballistarii* in the *Notitia Dignitatum*, a list of officials and units at military establishments across the Empire in the late fourth century A.D.[32] Other clear indications of little catapults in this period are, for example, Ammianus' description of a crocodile's hide as "so strong that its mail-clad back can hardly be pierced by the bolts of *tormenta*." Likewise, he comments that when some Persian infantry's approach brought them within a bowshot, "though holding their shields before them they found it hard to avoid the sharps shot from the walls by the *tormenta*." An infantryman's shield would not be thought even potential defense from anything bigger than a little catapult. So too Maurikios recommended the creation of strong entrenchments occupied by infantry armed with ballistas for defense work during the construction of towers to guard a bridge. These sound like personal weapons, held in the hands of men dug in, in trenches, around a construction site.[33]

Ammianus' descriptions of catapults are not a patch on those written by his near-contemporary Vegetius, whose treatise on war, written probably in the 380s A.D., aimed to preserve, and in some cases resurrect, the best military practice and theory known to him. We will discuss him properly below, but need to consider his description of the ballista and onager here.

> The besieged normally defend themselves against [assault machines] using ballistae, onagri, scorpions, crossbows, staff-slings and slings. The ballista is strung with ropes of sinew, and the longer arms it has, that is, the bigger the machine is, the farther it shoots darts (*spicula*). If it is tuned in accordance with mechanical art and aimed by trained men who have worked out its range in advance, it penetrates whatever it hits. But the onager shoots stones, and throws various weights in proportion to the thickness of the sinews and size of stones. The larger the machine, the bigger the stones it hurls like a thunderbolt. No type of torsion engine more powerful than these two types is found. They used to call scorpions what are now called *manuballistae* (lit., hand-throwers), they were so named because they inflict death with tiny, thin darts. I think it superfluous to describe staff-slings (*fustibali*), crossbows (*arcuballistae*), and slings (*fundae*) which are familiar from present-day use. Heavy stones thrown by the onager destroy not only horses and men, but also the machines of the enemy. (Vegetius *Epitome of Military Science* 4.22 Milner)

There are many noteworthy comments in this short passage.

Note first that onagers and scorpions are distinguished, *contra* Ammianus; scorpions are here identified with manuballistae, which is consistent with pre-Ammianus sources and we have yet another putative explanation for the name. Given Ammianus' patent ignorance of things mechanical, he was probably confused and Vegetius was probably right. However, once Ammianus' views were disseminated, so was the error, and through its reproduction what he said would have become true! *After* Ammianus some people would have called the onager a scorpion, as some still repeat him today. Because of one man's confusion, one device acquires two very different names and one name is applied to two very different devices. The same thing no doubt happened from time to time with those other catapult names that seem to mean different machines at different times. The deep patterns survive, however, allowing us even at this distance to sort the wheat from the chaff in our sources.

Second, we see the length of the catapult's arms serving as marker of caliber: "the longer arms it has, the bigger the machine is." Vegetius' treatise appears to be written for a technically weak audience if not by a technically weak author, but it does acknowledge some correlation of size of parts.

Third, the ballista here is clearly a torsion catapult, not a bow catapult.

More significantly, fourth, either the formulae and scaling rules had been lost and machines of different calibers no longer shared the same proportions, or Vegetius did not understand the mechanics and had not noticed that the range of different caliber machines in the same lineage was essentially the same. I suspect but cannot prove that the former was the case.

Fifth, there is more than a hint of unreliability in performance here; "*if* it is tuned in accordance with the mechanical art and aimed by trained men" *then* it works well. This is repeated in slightly different and even stronger form a little later: "ballistae and onagri, *provided* they are tuned *very carefully* by *experts*, surpass everything else—no amount of courage or armour can defend soldiers from them, for like a thunderbolt, they generally smash or pierce whatever they hit."[34] This suggests to me that catapult construction was returning to the same sort of pattern we saw in the early days of torsion; some catapults worked well and some didn't, and it was not entirely clear to those outside the business (and perhaps some of those in it, too) why that was so. Expertise and experience could be assumed to have a lot to do with it, as Vegetius seems so to presume, but he at least seems oblivious to the fact that the knowledge required for reliable good performance could be (and used to be) expressed mathematically and written down. Similarly, his comment that the onager "throws various weights in proportion to the thickness of the sinews" is particularly interesting—is this an echo of a formula for one-arms? No such formula survives to us for the onager, but it does not follow that none was discovered. Onagers have washers and hole diameters too, but only one spring.

Sixth, "no type of torsion engine more powerful than these two types is found" implies that there still (or again) *were* other types; these are simply the best types of torsion engine in Vegetius' estimation.

Seventh, the implication of the penultimate line is that ballistae, onagri, and scorpions were not "familiar from everyday use" but were rather unusual bits of gear at this time.

Evidently, Vegetius tells us a lot more about "artillery" of the time than does Ammianus, who repeated his (erroneous) belief that "the scorpion type of *tormenta* . . . is called an onager in common parlance" in his description of the siege of Hadrianopolis. One can imagine the lay public being neither able nor interested to distinguish between types of catapult very well, much as people uninterested in vacuum cleaners will describe one as, for example, "a hoover," when the machine they mean may in fact be a Dyson. Ammianus seems not to have been much more interested or better informed. Although he tells us that one of the few imperial war machine factories (*fabricae*) was located in the besieged town, so there was probably a large and comprehensive range of ordnance available, he actually says little specific about their role. In such circumstances, one imagines that the walls were bristling with catapults and the troops were spoiled for choice of mechanized weapons, and were experts in their use, yet they are barely mentioned in his account of the fighting. Instead we read that the Romans held the upper hand "since almost no missile hurled, or bullet from the strap of a slinger, missed its mark." Of course, we may have here another example of what we have seen elsewhere, to wit "missile" doing duty for catapult munitions as well as arrows and javelins, and little stone-thrower catapults passing under the name of slings. Ammianus does describe Goths writhing in pain, mortally wounded, "crushed by great masses of stone or with their chests pierced by javelins," and we may guess that at least some of these missiles were launched by catapult, but he does not actually say so. He says only that sharp-caster and stone-thrower catapults were installed at suitable locations for throwing ordnance "in all directions." He also tells us that one such machine, located directly facing a dense mass of the enemy, threw a huge stone, which although it landed on the ground without hitting anything, nevertheless, so terrified the enemy that they ran away at the mere appearance of this "spectacular novelty." Ammianus introduces this little vignette with the comment that this was "an entirely unexpected chance [that] had great influence in the middle of this battle," but this simply demonstrates his inexperience of large stone-throwers and consequent ignorance of the impact they could have as a "shock" weapon. I expect that the men who worked in the city's artillery factory were much less surprised by the effects their weapons had on enemy forces.[35]

Another city with a factory making *tormenta* was Amida, and it too was besieged. In this case, our principal source Ammianus was directly involved in the engagement. As we would expect of a city making catapults, "all kinds of *tormenta*" were hurling missiles at the forces flying toward the city (which included some friendly forces who were trying to escape to it, such as Ammianus himself), and he found safety from them by getting close enough to the walls to fall inside their operative range.[36] The sort of munitions that were available are indicated by the more detailed description of one of the battles for the city that ensued in the following months. "Then heads were shattered, as masses of stone hurled from scorpions [*sic*] struck many of the enemy, others were shot with arrows, some were hindered by their bodies being fixed to the earth by *tragulae*, others who were only wounded fled headlong back to their friends." His near contemporary Vegetius talks of troops called the *tragularii*, who apparently shot this specific type of sharp from "*manuballistae* and *arcuballistae*," hand ballistas and bow ballistas, respectively.[37]

Some of the catapults on the walls of Amida ("mural artillery") were relatively light and mobile, for when some Persians penetrated the city via subterranean tunnels and occupied the southern tower, and were shooting down on the defenders from within, "five of the lighter ballistae were moved and placed over against the tower." They began rapid shooting, and their missiles were sufficiently powerful sometimes to pierce two men at once. In fact, they frightened some of the Persians so much that they jumped off the tower to fall onto the rocks below. Job accomplished, quickly, "the engines were restored to their usual places." The same sort of mobility is implied for a couple of ballistae placed on the top of each of the Romans' siege mounds at Bezabde (Map 6), when Constantius tried to take the city back from the Persians, for they were moved, not abandoned, when the mound caught fire. These ballistae were placed specifically to attack those who would defend the ramparts, and the Persians were apparently more troubled by them than by the ram, scaling ladders and hand tools being used against the walls. So they sallied out to burn the mounds on which the ballistae sat—successfully. But when the mounds burst into flame the soldiers got away with their engines uninjured, albeit only just.[38]

Evidently some catapults at Amida were sniper's weapons, for when King Sapor, demanding surrender of the town, rode close enough to the walls of Amida to be recognized and in range, he was targeted. He escaped with torn clothing, in a dust cloud. The next day one of his allies, King Grumbates, rode up to demand the surrender of the town; his son, riding next to him, was hit with a ballista bolt that penetrated his breastplate and chest and killed him. Something similar happened at Bezabde: Sapor advanced to the edge of the trenches and became the target of repeated missiles from the ballistae of the defenders.[39] Leaders seem always to have had bodyguards, but we rarely hear of them deliberately "taking a bullet" for their boss. One such instance is told by Prokopios: Unigastos put out his hand to stop an incoming missile before it hit Belisarios in the belly; the missile severed the tendons of his hand and he was never able to use it again. By this time (sixth century A.D.) it was traditional to compensate soldiers for wounds sustained in battle, and though no figure was set and Belisarios' employer Justinian had a reputation for stinginess, Belisarios himself was wont to pay out large sums, so Unigastos was probably well compensated for guarding with his body that of his general.[40]

The Persians besieging Amida also had catapults—all sorts of *Roman* catapults, war engines and missile weapons that they acquired when they took the fortified outpost of Singara (Map 6), which had been well stocked with them in order to try to repulse Sapor's invasion. Twenty-one U-shaped towers, about six meters wide and more than eight deep (on the inside) still survive in part or in trace (the stone having been robbed out), projecting from the walls at roughly eighty meter intervals.[41] Roman catapults were presumably superior to Persian munitions at the time.[42] Thus both sides in this siege were equipped with the same weapons, and we are told, for example, that a ballista was placed at the top of each Persian siege tower to drive the defenders of Amida from the ramparts.[43] As a result of the concurrence of that circumstance and a surviving historian's presence in the town, we have a rare indication of what it was like to be on the receiving end of a barrage from Roman mechanized weapons. The image conjured up in Ammianus' description of this siege mirrors the images that Celsus and Paul offer of missile extraction, discussed earlier (Chapters 1 and 5). When the 120,000 men, women and children in Amida were overwhelmed by a dense missile barrage, "each cured his wounds according to his ability or the supply of helpers. Some, who were severely hurt, gave up the ghost slowly from loss of blood; others, pierced through by arrows, after vain attempts to relieve them, breathed out their lives, and were cast out when death came. Others, whose limbs were gashed everywhere, the doctors forbade to be treated, lest their sufferings should be increased by useless infliction of pain; still others plucked out the arrows and through this doubtful remedy endured torments worse than death."[44]

Clearly, the production and development of torsion weapons was ongoing in the fourth century A.D.; they were not relying on old machines. Catapults were being manufactured in preparation for war by Constantius (337–361). One engine-maker of this period is named: Sericus. Valentinian I (emperor 364–375) apparently took a personal interest in arms manufacturing, like Dionysios and Vespasian. It is natural to suppose that when Ammianus says that Valentinian invented new kinds of arms he was simply following the habits of this period, whereby the emperor was credited with the achievements of his subjects, as victory in battle is attributed to him even when another was the general who actually fought it. However, it is not usual for an emperor to be said to be interested in arms production, still less involved in it. Moreover, Valentinian is specifically said to have been a talented painter and modeler, so he had some aptitude and inclination for design. Like many emperors of this period he was also of humble background, specifically something to do with rope-making, and ropes are of course an essential element of many mechanical devices, including catapults.[45] Therefore, when Ammianus says that Valentinian invented new arms, I am inclined to take him at his word.

Interestingly, Ammianus supplies one of the few notices that catapults influenced ordinary life—they created a standard measure of length that continues on into the sixth century and beyond: the bowshot. He tells us that the Persian siege camp around Amida had outposts "not much further than a bowshot" from the city, and since Amida was well equipped with mural artillery, this must mean a catapult shot, else the besiegers would have been in range. Therefore, when troops came "within a bowshot" they came within range of "the sharps shot from the walls by the artillery." The bowshot as a catapult shot is again unequivocal in the sixth century Maurikios' recommendation that once the construction of bridges had reached "a bowshot" from the opposite bank, the small ballistas mounted on the prows of most if not all warships should be used to clear it of enemy troops.[46] Alexander had used catapults from rafts for the same purpose, back in the fourth century B.C., and the same tactic had probably been in use as required ever since. Of course "a bowshot" is thus an approximate measure, a variable measure, an informal measure of everyday speech and writing, like the modern English measure of "a stone's throw."[47] Thus it is not so very different from our experience. Further, although we have precise measures, in ordinary life we regularly *guess* distances in terms of numbers of them; a "mile" is probably at least as variable in modern spoken English usage as was the range of an ancient catapult bolt. For what it is worth, a bowshot was 400 yards in medieval times.[48]

Our next witness is Vegetius, who was writing probably in the 380s A.D. He sought to preserve the best from earlier military manuals by "abridging authors of the highest repute" (3.9), setting out "the principles that were handed down to posterity after winning the approval of different ages in the test of experience" (3. fin). His principal sources dated from the second century B.C. onward, and include Cato the Censor, Celsus, and Frontinus. His *Epitome of Military Science* deals with recruitment and training (book 1), the traditional composition and administration of the legion (book 2), logistics, strategy, and tactics (book 3), and fortifications, siege warfare, and naval matters (book 4).

Vegetius lists a variety of catapults: the ballista, onager, scorpion, manuballista, and arcuballista.[49] All soldiers were expected to be able to operate the onagers and ballistae at least, for Vegetius advised that "those soldiers who prove less useful in the field" should be equipped with onagers and ballistas to defend cities and forts. The assumption should be that they were all trained to use the onager and ballista along with the other weapons that every soldier was expected to use, as detailed in Book 1. He opined that there should be a plentiful supply of sinew, because the "onagers, catapults and other torsion engines are useless unless strung with sinew." He added two qualifications to the last: 1. that "horsehair taken from the mane and tail is said to be useful for catapults"—"catapults" here apparently meaning a specific category of torsion engine. 2. "Women's hair is as good in such types of torsion engine," he says, citing as an example the siege of Rome (presumably one of those that occurred in the late third century) when the torsion engines' springs broke after continuous and long use, and there was no more sinew for repairs. The Roman women presented their hair to their husbands, the machines were repaired, and they repelled the hostile attack.[50] The word *arcuballista* first appears in Vegetius, and the "straight" reading of it is "bow-ballista," as "manuballista" literally means "hand ballista." The difference between them would appear to be that the *arcuballista* is a small tension catapult, powered by bow like a crossbow, while the *manuballista* is a small torsion catapult, like Xanten. Both *arcuballista* and *manuballista* could probably shoot stones or *glandes* as well as sharps (depending on whether fitted with a bowstring or a slingstring and appropriate stock); medievalists sometimes call a stone-shooting crossbow a stonebow. Interestingly, the shape of the stock on some medieval examples is arched, and resembles very closely the well-known Roman "crossbow" type brooch.

Vegetius is the source of the oft-repeated statement on the numbers of catapults with which each legion was equipped.

> The legion is accustomed to be victorious not only on account of the number of soldiers but the type of its tools also. Above all, it is equipped with missiles which no cuirass or shield can withstand. For each century customarily has its own carroballista, with mules assigned to draw it and a section, that is, eleven men, to arm and aim it. The larger these machines are, the farther and more violently they shoot projectiles. Not only do they defend a camp, but they are also placed on the battlefield behind the line of the heavy infantry. Faced with their attack, neither the cavalry cuirassiers nor the shield-bearing infantry of the enemy can stand their ground. In one legion there are traditionally 55 carroballistas. There are also ten onagers, one for each cohort. They are carried around ready-armed on ox-carts, so that should the enemy come to attack the rampart to storm it, the camp can be defended with sharps and stones. (Vegetius *Epitome Military Science* 2.25 Milner modified)

Note that the carroballistae are highlighted as anti-personnel weapons: "which *no cuirass or shield* can withstand." Note also the assumption that carroballistae vary in size: bigger ones are more powerful. We saw in Chapter 9 that imperial Roman legions seem to have had a measure of control over the caliber of their catapults, and so long as weapons were being made (or commissioned) by troops or legions for their own use, one can imagine that situation prevailing. Once imperial factories were established, however, variation may have been reduced as "standard issue" equipment would be homogenized over a wider area, if not the entire empire. Note also the final comment that onagers were carried around in ox-carts. Evidently Vegetius was imagining or saw machines made to a much better standard than those witnessed by Ammianus, which apparently were so badly made and operated that they recoiled dramatically. (It is perhaps for the same or similar reasons that it is Ammianus who records the catastrophic "misfire" wherein an "artifex"—catapult maker or master—was dismembered so badly that they couldn't find all his body parts. And it is Ammianus who reports men discharging their machines empty, so that by the noise of their operation, a sally force that was unknowingly running into trouble would be persuaded to run back to the city and the pursuers discouraged from chasing them, without actually risking "friendly fire" casualties. This might look like a good stratagem to the uninitiated, but it would be very bad for the machines.)[51]

Vegetius was writing at one of the most difficult times in the history of the Roman empire, and for the next hundred years plus we have a very poor idea of what was happening, largely because few people had the ability or inclination to record it. The fog clears again in the sixth century A.D. when we find ourselves in the hands of the very competent, if sometimes patronizing and tedious historian Prokopios. He recorded the remarkable military comeback of the empire in the reign of Justinian, much of it owing to the outstanding abilities of his best general, Belisarios.

I think the combination of a great general and dire circumstances inspired unusual levels of creativity and inventiveness in the military at the time. Belisarios was an extraordinarily open-minded man, spotting possibilities and creating opportunities freely. Rome had fallen to the Goths in A.D. 410, the same year that Britain was told to expect no more help from the empire and must henceforth defend itself. Constantinople got its fabulous triple wall in A.D. 413, which kept out all and sundry till 1453. The Vandals sacked Rome in 455 and thereafter Rome (the real one, not the "Rome" in the east based on Constantinople) was more or less under the rule of one or another of the barbarians. About eighty years later, on instructions from Justinian, in A.D. 535 Belisarios led an expedition from Constantinople, or Byzantion as Prokopios insists on calling it, to wrest Italy back from them. He had occupied Rome by the end of 536, and the siege of Rome was part of the Gothic fight-back. Prokopios's description of sharp-caster catapults at the siege of Rome in A.D. 537/538 is taken by some to refer to bow catapults.[52]

> Belisarios placed in the towers machines that they call *ballistrai*. These machines have the form of a bow, beneath which a hollow wooden horn projects, which is loosely attached and placed directly on the iron. When men want to shoot the enemy with this, by fastening a short noose they make the woods, which are at the ends of the bow, bend towards one another, and they put the bolt in the hollow horn. The bolt is about half the length of other sharps shot from bows, but about four times as wide. However, it does not have flights of the usual type attached to it, but they give it the shape of an arrow by inserting thin slivers of wood in place of feathers, and fitting a large and appropriately thick head to it. The men who stand each side draw it back tight with certain machines, and [when they release it] the hollow horn, shooting forward, is driven out, and the bolt is driven out with such force that it travels not less than twice the distance of an ordinary bowshot, and should it strike a tree or stone, splits it easily. Such is the machine that bears this name, because it shoots (*ball[o]-*) with special (*-[mal]ista*) force.
>
> And they set up on the parapet of the circuit walls other machines designed to throw stones. These are like slings and are called *onagroi*. And outside the gates they placed *lukoi*, which they make in the following manner . . . . (Prokopios *History of the Wars* 5.21.14–19)

Despite Marsden's apparent confidence that what Prokopios is trying (unsuccessfully) to describe in this passage is a Heronian *kheiroballista*, I think most now take the view that the machine in question is a bow catapult.

Concurring with that view, there are still many other things to note here. I believe that the "hollow wooden horn" (κοίλη ξυλίνη κεραία) is a barrel; that is precisely what "horn" here signifies, I suggest. I think that a wooden barrel has replaced the grooved slider, and that the barrel slides back and fore on an iron stock. The bowstring engaged with the barrel, not the missile (as Dewing, the Loeb translator, thought). When braced, the barrel would project out of the front of the machine, and when spanned, the barrel would project just beyond the bow. The bow was explicitly said to be laid *above* the barrel and stock (and not below it as on the old *gastraphetes*), which some have seen as a problem. If the barrel was almost level with the bowstring and if the handle of the bow was narrow, as some are in relief depictions, it need not have been. The relationship between bow and barrel could have been essentially the same as that between bow and arrow as held by an archer.

Assuming this *was* a mechanical bow catapult, some method of attaching the bow to the stock is required. This is an easily overlooked issue even with Zopyros' *gastraphetes*—there is apparently just a bracket to hold the bow at the front of the stock, yet as Marsden pointed out,[53] the attachment of the bow would be one of the most critical tasks in the entire construction of that machine. The ability to disassemble machines for storage, transportation, and maintenance was still an issue.

The "short noose" used to span the machine perhaps belonged to a bastard string, slightly longer than the bowstring, with a noose at each end, which was slipped onto the notches of the arms and the retraction claw to span the device, and removed once the bowstring was engaged by the trigger. If so, this was a key development. Recall that the nut was in existence from at least the second century A.D., used on manual crossbows (Chapter 9). The bastard string allowed the same sort of design to be used to make weapons that were bigger and even more powerful than hand-held crossbows. It is the sort of arrangement depicted on the springald (see Figure 10.4 above).

The sharp, with its wooden wings (uncrushable and presumably now equidistant, if designed to be launched through a barrel) and similarly sized head, was placed in the barrel. (Hitherto the sharp was merely laid in the groove and the wings were at 90° intervals on three sides only because the fourth side, being laid in the groove, could not take a flight.) Prokopios' *ballistra* could not be depressed, that is, shot downward (5.22.21), for the missile would fall out. When the machine was discharged, the barrel shot forward and then stopped (held presumably by the iron stock), while the sharp within continued with that "muzzle velocity" on the same trajectory (again, as Dewing supposed). Thus this catapult seems to have incorporated an element of the DRB's lightning ballista, in which the missile rested in the groove and the groove was projected. Closing the groove to create a barrel would reduce the likelihood of misfires. The text says that the *stock* was made of iron, which would make sense if it had to restrain the force of the discharged barrel, and the ends of the bow at least were wood. Either the whole bow was wood and we are *not* dealing here with a steel bow, or, and I think more likely, wood was the material of the ears, as horn sometimes served on manual bows, which were presumably firmly strapped to the steel bow. This would make sense for nocking if nothing else; deliberately introducing a crack into the ends of a metal bow to hold the bowstring would invite stress fractures and crack propagation. Making the nock for the bowstring in wood, horn, or antler and then fixing that to the bow would both reproduce traditional bowyers' methods (likely in itself) and avoid the problem of trying to tie the string securely to the metal bow ends.

Bringing all the threads together, I think that Prokopios is describing a new and otherwise unknown type of bow catapult, a barreled bow *ballistra*, a direct or indirect single-barreled descendant of Zopyros' double-barreled sharp-caster. Additional arguments could be advanced, for example the fact that Belisarios' sharp-casters had specialist technicians to operate them, unlike the onagers, which were stationed along the walls, whose operators are not distinguished as such, and note that these were small—even tiny—onagers, since they fitted on the wallwalk. Belisarios' *ballistrai* terrified the Goths, who seem to have been completely unfamiliar with them or anything like them. Since they first started raiding Italy, the Goths had been swapping their own arms for captured Roman ones, but now they had been "governing" Rome and Italy for generations—so had had to depend on their own resources.[54]

Bow catapults were probably never displaced, ever since their invention in Syrakuse at the end of the fourth century B.C. *Gastraphetai* threw sharps and stones and came in a variety of different sizes before the invention of torsion engines (Chapter 3). Although the evidence for their existence thereafter is sporadic and thin, that is explained more plausibly by the idea that their distribution was sporadic and thin than it is by the idea that bow catapults were reinvented every few hundred years. Further, on the subject of steel bows in particular, springy metal had been tried as a motive force as far back as the third century B.C. in Ktesibios' bronze-spring catapult. It may have taken a long time to get the metallurgy and the mechanics to synchronize, but that is precisely what there was between Ktesibios in the third century B.C. and Prokopios in the sixth century A.D.—more than seven hundred years, plenty of time to iron out problems and solve difficulties.

New devices and techniques seem to abound around Belisarios. Another device unknown from earlier writers is the "wolf," which Prokopios goes on to describe in some detail; its heir is sufficiently familiar in medieval siege technology to be reproduced in Playmobil toys.[55] Making ladders the right length for the wall to be scaled was still a challenge. Prokopios' attempt to excuse his hero Belisarios (unlike Polybios' castigation of Philip II for the same mistake) does not quite come off, for even if the ladders were made by men out of sight of the wall, Belisarios should have had them checked before ordering the troops to try to use them to scale the wall of Naples. Vittigis, leader of the Goths, apparently checked and double-checked the height of the wall of Rome before starting construction of his siege towers.[56] When Belisarios approached Panormus in Sicily and noticed that the masts of his ships were higher than the parapet of the city wall by the harbour, he put archers on board small ships (*lemboi*) and hoisted them to the tops of the masts, whence the archers began shooting down on the inhabitants. This ad hoc siege tower caused the besieged to surrender immediately. When the Goths cut the aqueducts of Rome, one of which powered the flour mills on the Janiculum Hill, Belisarios had the idea of putting the mills on boats and tethering the boats in fast-flowing parts of the Tiber, so that their water wheels were turned by the river instead of aqueduct water. And when the Goths, learning from deserters that he had done this, threw trees and Roman corpses into the river upstream and they, carried down by the current, broke the wheels, Belisarios suspended iron chains across the river to catch any more such debris before it hit the mill-boats, and assigned men to remove it constantly.[57] Later, when the Romans had lost Rome again and roles were reversed, with Goths defending and Romans attacking, the same or similar technologies were tried by either side, so now the Goth Totila copied Belisarios and hung an iron chain across the Tiber, to stop Belisarios copying him and sending debris down the river. Now Belisarios employed a *lembos* in another capacity. He set fire to one and "threw" it on top of a Gothic tower that was big enough to hold some 200 men. Prokopios does not explain by what mechanism Belisarios could throw a small boat; an onager is the most tolerant of load shape, but we can only guess. It would appear that this flaming *lembos* was launched from a tower, built on a pair of unusually wide skiffs lashed together, that was apparently much higher than the Gothic tower on the bridge that they were attacking, so it could have been more "dropped" than "thrown."[58]

It is quite possible that technologies preserved and developed in the Roman east were in the process of being lost in the barbarian west. Three particular stories suggest this. First, when one of Belisarios' catapults threw a sharp that, passing through the breastplate and the body of a Goth, sank more than half its length into the tree by which he stood suspending him there a corpse, his comrades fell back out of range terrified. Evidently, Belisarios either brought these machines with him, or made them in situ. The former is vastly more likely, but in any case, they were not part of the city's armament or experience before his arrival. Second, mobile siege towers were no longer iron-plated even in Vegetius' time; fire-protection was provided by rawhide and fire-blankets, but either there was no need to defend against powerful stone-throwing catapults any more, or the additional weight or cost was perceived to outweigh the potential benefit of metal plating. Literary sources of the period suggest that big stones might be rolled down a hillside or from a battlement to attack enemy machinery as they had done ever since sieges began, but they did not threaten the faces of the tower. What are described as "larger" ballistae in Vegetius are sharp-casters, not stone-throwers, and onagers, one-armed stone-throwers, are notoriously imprecise weapons whose payloads would, in any case, descend in an arc from above.

Third, when the Goths made a siege tower and advanced on the city, Belisarios burst out laughing and ordered everyone to get down and keep quiet. The Romans, that is, the inhabitants of Rome, did not understand why, thought it was a nervous laugh, and feared that he was shirking his duty to defend them. Both the Goths and the inhabitants of Rome lacked sufficient knowledge and experience of siege towers to know (and were too naive to realize themselves) that one did not move them into place like wagons, hauled by oxen in front. Belisarios waited until the tower was demonstrably in range of the walls, at which point he told the men round him to shoot the oxen. The tower was stranded in range of every missile from the walls and the result was a turkey hunt. Shortly after, it went up in flames. The Goths didn't make the same mistake twice, pushing it from within when they built a second siege tower at Ariminum, but now they made a different mistake. Again the Roman inhabitants were terrified, but the commander from the eastern empire, completely unfazed, simply went out in the night and dug a big trench in front of it. The next morning Vittigis, the Gothic leader, executed the guards for not noticing or reacting to these night-time activities, and ordered the trench be filled with faggots of wood—making a third mistake, for these were quite incapable of taking the weight of the tower, so that when it was moved forward in due course, it sank and was stuck. By the time we get to the end of Prokopios' narrative, we find Goths operating *ballistrai* from wooden towers on a bridge over the river Drakon near Nuceria (modern-day Nocera); presumably they learned both the technology and techniques from deserting Roman soldiers, of which there were many in this period.[59]

Artisanship does not thrive in mobile, insecure, homogenous communities; it needs stable, populous, diverse, and prosperous cities. My focus here is on the sixth century, but many of the disruptive phenomena I cite occurred sporadically earlier. For example, in about A.D. 206 or 207 a gang of about six hundred men practiced highway robbery in Italy for two years, apparently driven to it by poverty and starvation: the leader is reported to have warned the Romans, "feed your slaves so that they may not turn to brigandage." From most victims they took a part of what they had, and then let them go, but artisans were "detained for a time and [the leader Bulla] made use of their skill, then dismissed them with a present."[60] The large gang apparently lacked artisans, but they could appropriate technologically skilled labor by force. However, by the sixth century, there were few places left in Italy for anyone to parasitize in this way. Another factor leading to divergence was perhaps the greater degree of continuity in the armed forces of the Eastern empire than of the Western, where barbarian armies, usually recruited on an ad hoc basis to fight a particular campaign, were prone to rapid turnover of personnel. Obviously, the corporate experience of any armed force, as of any business, would be reduced in such circumstances.

Now this is not to suggest that by this time "the east" had catapults and "the west" did not, for that is patently false. In any and every part of the old

Roman empire, prolonged peace brought with it redundancy of war material; walls fell into disrepair or were recycled into civilian buildings.[61] Despite Heather's sterling and partly successful attempt to absolve late Roman government and society from most of the criticisms that have been leveled against it, the primary sources still give me the distinct impression that at this time soldiers' wages were not the fiscal priority they should have been (so they began to help themselves to what was due). Money was instead poured into what Heather perceptively analogizes as "foreign aid." Demonstrably, the maintenance of unused ordnance and garrisons sooner or later appeared wasteful to some of those responsible for them. Justinian engaged in an extensive program of (re)fortification, to support his campaigns to re-acquire parts of the old empire, but when neighbors fought back, and war then broke out again in places not of Justinian's choosing, peoples and cities were sometimes caught unprepared. If in any area the previous peace had been continuous and lengthy, knowledge of war was generally as rusty or obsolete as were the weapons of war. So the people of Martyropolis, north of Amida in Mesopotamia, had nothing of value with which to defend themselves when besieged by the Persians in A.D. 531 (even though a garrison had been installed), while ten years earlier Petra, then recently refortified, *did* have machines with which to defend itself in a similar situation. Martyropolis, however, was blessed with "a clever man who devised counter-strategies to oppose the Persians' schemes," which included a wooden tower taller than the Persians' siege tower and a machine mounted atop it to drop a column onto the Persians' tower and smash it. That someone, somewhere was believed still capable of making stone-thrower catapults that could demolish walls is evident from the Petrans' countermeasures when building their new fortification towers: two of them were completely solid from the ground up, so that "they might never be shaken down by a ram or any other machine" (in the event one was brought down by sapping). The existence of powerful stone-throwing catapults is also attested by the recommendation of the Anonymous sixth-century author of the *Strategy* that walls should be at least three meters thick to withstand the stones thrown by them.[62]

New wars bring new weapons, and this was especially true it seems of the sixth century Persian Wars. Khosroes (Khusro) came to the throne in 531, and he was described by Prokopios as "strangely fond of innovations." A modular bridging kit involving hook-shaped irons that allowed the Persians to cross rivers at will seems to have been one such innovation. Khosroes too had defensive war machines significant enough that he could pretend for diplomatic reasons that timber that was actually intended for the construction of ships was instead to be used to set up machines on the walls of Petra.[63] It is perhaps not irrelevant that Justinian had driven those citizens with the most commitment to their intellectual health, and the most get-up-and-go in their bodies, out of the Roman empire in A.D. 529, when he forbade pagans to teach in schools. Some took refuge from this Christian-fanatic pogrom in Persia, where there was then toleration of different religious beliefs. Justinian apparently came to regret this quite quickly, for one of the clauses in the peace treaty between himself and Khosroes in 532 specifically allowed teachers who returned to the Roman fold to practice their religion in peace. There certainly were some very clever men around at this time; for example, Isidoros of Miletos worked on a revised edition of Arkhimedes' works, and Anthemios of Tralles was renowned for his mechanical models of natural phenomena. Both worked on building the Hagia Sophia.[64]

In Chapter 9 we considered the story about the Vitellians pushing their ballista onto a *testudo* in A.D. 69, in frustration at their inability to penetrate the structure with their missiles. By the Anonymous strategist's time, which was perhaps five hundred years later, there were a variety of methods for dealing with such a threat, and he even reports his predecessors' methods, though we do not know how far back these go of course. Historical methods in his time for dealing with a testudo included boiling water, or hot ashes, or powdered lime, or melted pitch poured from above and set alight, but most of the time, he says, they would bring large timbers onto the wall, fit them with metal tips like a sword point, and drop them, point first, on the tortoise. Internal evidence suggests that this anonymous sixth-century treatise on strategy was composed during the reign of Justinian (A.D. 527–565), and the military career of Belisarios was recent or contemporary. The author seems to have had practical military experience, perhaps even that of an engineer.[65]

He claims some new methods have been discovered by "us," and he proceeds to explain them (13.69–114). Some of these are at best rediscoveries that were known to Ammianus and Vegetius, and existed in alternative forms even earlier. Perhaps when he said "discovered by us," he meant "discovered by me in books." For example, the first of his "new methods" seems to be some kind of netting to trap incoming anti-personnel missiles of a type described by both those predecessors. At those points on the wall where attack is expected, special mats made of hair (especially goat's hair, as we learn from other sources), wool, or linen were hung loosely from the battlements to protect the men on the walls from the slings and arrows thrown by the attackers. In the absence of such mats use the citizens' bedding, says the Anonymous, or double-thickness cloaks says Vegetius. In earlier centuries animal skins, sails, curtains, and similar materials had been used for the same function.[66] Quite how such textiles were arranged to achieve the intended purpose is not explained; one assumes that they hung a little proud of the wall so that the defenders could pour or drop their munitions through the gap onto the tortoise below while being screened from missiles launched from anywhere except immediately below. The yielding, swinging matting absorbed the impact of incoming arrows, which passed through it only with difficulty (as noted by Vegetius). The barbarians had an equivalent sort of screening device made of goat's hair cloth ("Cilician"), according to Prokopios. A similar device is recommended as protection for the walls from stone-throwers: nets made of ropes a finger or more thick hung about a meter proud of the walls. These will yield and soften the impact of stones thrown by stone-throwing machines (πετροβόλοι μηχανὰὶ), observed the Anonymous. Recall the Tyrians' use of bags filled with seaweed, back in the fourth century B.C., and even earlier counter-devices (Chapter 4).[67] And compare the thick felt underwear described by DRB, specifically recommended by him to absorb impact, which plate armor cannot do.

The Anonymous then describes a machine that resembles a one-legged crane on a wheeled square platform. Its function was to drop a large stone on an enemy machine, and it was powered entirely by human muscle pulling and releasing ropes (13.122–135). One can see elements of the beam-sling in this structure, though here human muscle is being used to raise, rather than launch, the stone, and the lever principle is not apparently being exploited. Another machine based on the same sort of device (a one-legged crane) was the *tolleno*, the swing-beam. Here the cross-beam mounted at the top of the vertical pole had a light protected box at one end and ropes at the other. A few troops took cover in the box, and were lifted to the top of the enemy's wall by men pulling on the ropes on the other end of the cross beam (Vegetius 4.21). In the hands of skilled operators, this could be a relatively simple and quick method of lifting a small number of men to a modest height, which is what walls typically were by this time. Vitruvius advised his readers to leave one-legged cranes to the Greeks because it was so difficult to handle them, but that was a long time before Vegetius was writing, and anyway quite a lot of "Romans" were Greeks by now. Vegetius does not say so, but other people would have been required to swing the beam round and more, probably, to secure the upright pole with guy ropes, though wooden supports are not out of the question.[68]

Stone-throwing machines were a regular part of the armed forces at this time, for the Anonymous lists "stone-cutters, metalworkers, carpenters, and suchlike," as the technical people that formed a third part of a campaign force in his time. The other two groups were the combat troops and the suppliers of food, drink, and suchlike. Sharp-caster and stone-thrower catapults were still mounted on the prows of ships and were still being recommended for duty clearing the opposite bank of a river during an opposed landing. Interestingly, the Anonymous suggests pre-fabrication of boat components away from the front, each component being carefully labeled to indicate its place in the vessel, then transported to the river it is required to cross, where the pieces are reassembled and the joints caulked.[69] Our assumption is that catapults were disassembled into their modular components and reassembled regularly, and in the high Roman Empire, large stone buildings and structures such as the Pont du Gard were sometimes constructed like this, but this is the only reference known to me for application of the construction-by-numbers method to ships.

The *Miracles of Saint Demetrios*, in a section written by John Archbishop of Thessalonike (an eyewitness to the siege of Thessalonike by the Avars), records an intriguing story of technology transfer and development. One Bousas was a Byzantine soldier based at the fortress of Appiareia in Moesia Inferior, who was caught by the Avars while out hunting. When the townsfolk refused to ransom him back from his captors, he bought his own life by teaching his captors how to build a *helepolis*, a city-taker. We are not told any details of this machine, but there is absolutely no reason to suppose, as some historians of the Byzantine empire have done, that *helepolis* now meant catapult in general or stone-thrower in particular, for *petrobolos* still carried that meaning. A *helepolis* was first and foremost a siege tower, considered by many the ultimate siege machine—a mobile device that offered protection for its crew and height advantage to its soldiers, so that they could shoot down on the defenders, instead of being shot down by them. The heights in question were not very great in this period, for elephants were considered to be natural *helepoleis*, being tall enough to overtop many city walls built in the sixth century A.D. With Bousas' *helepolis* the Avars attacked and took Appiareia in (perhaps) 587.[70] They went on to take many Byzantine cities and fortresses, attacking Thessaloniki in 597 (or, if the date of the fall of Appiareia is wrong by a year or two, in 586) and here on the plain in front of the city they built *helepoleis* (siege towers), rams, and huge stone-throwers (πετροβόλους ὑπερμεγέθεις). The first catapults to be completed were in action on the third day. The haste with which the Avars built and —when the defenders of Thessalonike fought back employing defensive devices like grappling irons, the portcullis, and perhaps the wolf, — abandoned their new mechanical weapons, first rams and sheds, then stone-throwers, suggests, I think, real inexperience and unfamiliarity with such technologies, real ignorance of their capabilities and limitations. In fact, these particular machines may have been new to everyone, for here we have the earliest surviving description of the beam sling catapult, otherwise known as the one-armed traction stone-thrower or traction trebuchet. The earliest illustration of it is about a century later, on a wall painting in the ruins at Pendzhikent, 65 kilometers east of Samarkand (Map 6), of the late seventh or early eighth century.[71]

The beam was nailed to a thick cylinder, which formed a high horizontal axis that was supported by timbers akin to a roof structure (leading to its

Byzantine name Labdarea, after the letter L), the cylinder mounted across the top between two such frames, leaning slightly toward one another. The beam was fixed to the cylinder such that there was one longer arm and one shorter arm; naturally, since the cylinder could rotate, the beam came to rest with the long arm down. To that end was attached a sling. To the other end was attached a number of ropes (Figure 10.5). By pulling manually on the ropes and exploiting the lever principle operating through the unequal beam, a number of men could launch the sling payload into a high arc, in the same manner as an onager, but here powered by human muscle rather than by torsion spring. These machines tend to be tall, as the higher the cylindrical axis is raised, the longer can be the arm, and the faster the weight can be launched by the limited number of men that can pull quickly and simultaneously on the ropes (12–15 seems to have been normal). Given that both onager and beam-sling are mechanized versions of the staff sling, the development of the beam-sling may owe nothing to the onager but be a direct development from the manual staff sling, which was still in use among Roman troops. But if a Roman, whether Bousas or another, was helping the Avars to devise siege machines that they could build and use with the technologies and tools that they had, one should expect a certain amount of innovation as the mind of one culture was adapted to the body of another.[72] In due course, this machine acquired the name *kheiromangana*, the hand-mangonel, the *mangana* part of the name being derived ultimately, I think, from *monagkon*, the old Greek word for the one-arm catapult. *Kheiromangana* is an accurate and appropriate name for a manually powered one-armed catapult.[73]

The emperor Maurikios' sixth-century *Strategikon* is more concerned with food and water supplies, for besieged and besieger, than with artillery, and this reflected contemporary military theory that an army fights on its stomach, and that reducing cities by thirst or starvation was preferable to trying to storm them. But assault was still entertained and sometimes tried. Catapults were useful in almost all sieges, especially where it was difficult to get close to the walls, Maurikios observed. He recommended their use to throw incendiary bombs onto targets within the walls, as well as employing incendiary arrows. His anonymous contemporary advised that coastal city walls be built at least 18 meters from the water to be out of reach of ship-mounted sambukas, and not more than what equates to 62 meters from it, to ensure that any marine-landed attackers were constantly in range of missiles and stones thrown from the walls.[74] Thus we can deduce the *minimum* effective range of standard missile weapons in all conditions at this time was 62 meters.

The tenth-century treatise devoted to the subject of *Resisting a Siege* (commonly known by its Latin title, *De obsidione toleranda*) is similarly concerned with provisioning. When attack was unavoidable, then Nikephoros Ouranos recommended sapping, to undermine the walls, rather than the use

FIGURE 10.5 *Beam sling from an early twelfth-century manuscript. Otherwise known as the traction trebuchet or the traction catapult. This is a one-armed device. The arm is fixed, unbalanced, to a raised axle that can rotate freely. The sling is attached to the long side and a number of ropes to the short. Teams of men pull simultaneously on the ropes to launch the payload in a high arc. Human muscle power replaces the spring of a torsion catapult or the bent wood, horn and sinew of a tension catapult. (After Nicolle 1987)*

of complex or elaborate siege engines as described by "the ancients." Now scholars of this period have argued long and hard over the survival (or not) of torsion engines into the medieval period.[75] To take the story that far reaches beyond the scope of this book. However, I would make a methodological observation. The paucity of evidence for torsion machines in this period leads some to negate the value of that which exists, undermining such references by suggesting that they are literary *topoi* or repetition of classical sources describing things from a lost world. This method is, frankly, lazy and dangerously liable to self-fulfilment; there is a point at which source criticism becomes hypercritical. If there is also evidence that one person can make a difference (as there is with Bousas in late sixth century Moesia, for example), then it should be acknowledged as a rarity, not dismissed as false.

### How Technologies Are Lost

Skills that are not maintained in use are lost, observed Vegetius (3.10), and we all know the truth of it from our own lives. He went on to claim that it is possible to revive lost skills from books, but none of his examples of the revival of military skill demonstrably involve re-education from books, and may have resulted from reinvention or independent rediscovery. Besides, on their own, written sources are rather inadequate communicators of any kind of technical knowledge;[76] we have the classic modern experience of the instructions that came with relatively simple pieces of flat-pack furniture, and that were almost universally regarded as incomprehensible (they are now generally better). In any case, most people in the late empire appear to have been illiterate, including some of the elite and occasionally even the emperor or his alternative. In the worst cases, a stencil was made so that the illiterate incumbent could write "read" on all documents requiring signing off by the emperor.[77] Declining literacy meant that as time went on there was a catastrophic loss of knowledge also, for each new generation could learn only that fraction of the sum total of human knowledge that was known to their family and people they knew and could talk to; they could not access other ideas available only by means of written communication. Further, other types of knowledge, not communicated through writing, had also been lost over the centuries to A.D. 600. For example, Prokopios advised Belisarios to use the ancient device of trumpet signals to communicate with the men during the din of battle, instead of shouting, which was ineffective in the circumstances. He adds that although the old Roman armies used to know different trumpet signals for advance and retreat (and other commands too), "such skill has become obsolete through ignorance and it is now impossible to issue both commands by one trumpet," so he advised using the bronze cavalry trumpet for advance and the wood and leather infantry horn for retreat. Prokopios had apparently read his Vegetius, or perhaps even Polyainos (who offers lots of examples of ordinary and sneaky use of a variety of trumpet signals), and was offering a dumbed-down version for present needs.[78] The idea of using trumpets to send messages could be resurrected from books; the sounds could not, and so an adaptation arose. Belisarios apparently adopted the suggestion.

Prokopios' admiration for the workmanship of the past and recognition of the inability of his generation to equal it is apparent in his story of Belisarios' efforts to take the town of Auximum back from the Goths. Belisarios sent five Isaurians "who were skilled builders" into a cistern outside the walls "with mattocks and other implements suitable for cutting stone" [*sic*] to destroy it and thus cut off the enemy's water supply. Once inside, the men were protected from defensive missiles by the roof of the structure, which withstood "everything thrown at it." So the defenders sallied out, and the Romans came forward and battled around the cistern, to allow the saboteurs time to work. When the defenders finally withdrew and the assailants began to return to their camp, assuming the cistern was now destroyed, it transpired that the would-be saboteurs had been "unable to remove even a single pebble from the structure. For the old craftsmen, who strove above all for excellence in their work, had made this building to yield neither to time nor to men trying to destroy it" (6.27.19). Belisarios gave up trying to destroy it and ordered the troops to throw carcasses and toxic plants into it to pollute the water instead. There was good reason for the general preference of the age to try to reduce walled cities by starvation, rather than to try to take them by force.

People with curiosity and intelligence still existed, of course. For example, one of the Isaurians, while besieging the Goths in Naples, wanted to know how the aqueduct worked, so in order to explore it, climbed in through the break that Belisarios had made, and thereby discovered a hidden way into the city.[79] But all the shortcuts that books and education (qua apprenticeship, as well as reading) can provide were in short supply and many things had to be rediscovered empirically. When Totila had destroyed all the gates of Rome and Belisarios wanted to refit them, a lack of artisans in the city and in his army meant that he had to block up the openings with his best troops. And after a desperate fight for the gateways, they strewed *triboloi* around to assist against the next attack, employing the sort of techniques hitherto used for field camps, here to defend the eternal city. For Prokopios, innovation is a downsized ram, a simpler, lighter, more flexible version that could be carried by forty men over rough ground, and the same men, supporting both the ram and a light protective cover on their shoulders, simultaneously swung it to hit the wall. It goes without saying that such a ram would be effective only against a wall equally simple and light, and that those ancient cities whose classical walls had been maintained were more than able to withstand assault in this period. Indeed, when Prokopios says that the elephant may be described as "an engine for the capture of cities"—since at this time men mounted on elephants' backs were able to shoot down on defenders in towers on some city walls—he is clearly referring to walls that are only a couple of meters high, and towers little higher.[80]

Indisputably, siege warfare did not change its essential nature until the advent of effective cannon in the mid-1400s, or later. Old and new devices occurred side by side, performance varying with the skills and experience of those who built and operated them, and with the qualities and quantities of the materials used in the construction. Of course there were developments and continuities. Some continuities are obvious; others are suspect. When the anonymous tenth-century author of an untitled work on tactics (possibly Nikephoros Ouranos) writes about fighting for the walls he says it involves the same old technology we have been considering in earlier chapters: "battering rams, tortoises, stone-throwers (*petroboloi*), ropes, wooden towers and ladders, a mound, other city-taking devices and engines." He adds that "you will find these matters discussed knowledgably and fully, beautifully and usefully, in the books of the ancients," for which reason, he says, his writing on the same would be superfluous and he will be brief (27.6–12). He at least apparently detected no major differences in the essence of siege warfare between his own time and that of earlier authors, though unfortunately he does not say which authors he had in mind. When advising on the setting up of a besieger's camp, he says that "the inner rampart should be about two bowshots or more away from the wall, that is, far enough that sharps shot from a bow (τὰ ἀπὸ τόξου πεμπόμενα βέλη) will not reach it, and neither will the missiles from the stone-throwers (τὰ ἀπὸ τῶν πετροβόλων ἀκοντίσματα)".[81] A variety of interpretations of this passage are possible, one of which is that this distinguishes between the performances of the tension bow catapult shooting sharps, and the stone-thrower (powered by something else) which outranges it, but not by a huge amount.

A variety of old and new terms seem to indicate developments within lineages, for example Leo's late ninth/early tenth century *Taktika* (which modernizes Maurice's *Strategikon* of the sixth century) called wagon-mounted artillery *eilaktia*.[82] Machines of this type are elsewhere described more fully as "stone-throwing *manganika* called *alakatia* which can also shoot fire-arrows."[83] Haldon suggested that they were probably similar to the Roman *carroballista*, except that they were now tension-powered, i.e., bow catapults, rather than torsion-powered. In other documents dating to around A.D. 950 we read about the *toxobolistra*, *kheirotoxobolistra*, *megalai toxobolistrai*, and *kheiromangana*, which literally mean the bow-ballista, the hand-bow-ballista, the great bow ballista, and the hand-mangonel. The last is surely the beam-sling, or traction trebuchet as it is sometimes known. The first three seem to be different caliber bow catapults: medium, small, and large. In another document of about the same date we are told about supplies and parts for *tetrareai* (which seems to be a common alternative for *petraria*), *labdaraiai*, and *manganika* which sometimes seems to substitute for *eilaktia*. Clearly, more research needs to be done in this area to sort out the story of the catapult after antiquity. Springalds appear in the mid-thirteenth century, and looking at it one can guess the outlines of its development from the in-swinger *kheiroballistra*: wooden frame replaces iron frame and levers, bastard string replaces slider, horsehair permanently replaces sinew, and the arms, claw, stock, and windlass (when used as retraction mechanism; screw retraction is new) stay essentially unchanged.

Future research may throw up evidence for or against such a guess. But Vegetius was still being read and edited and promoted in the early Renaissance; Christine de Pizan, at the beginning of the fifteenth century, "argued for a roughly equal division of artillery between trebuchets and guns in her redaction of Vegetius."[84] Leonardo da Vinci, 1452–1519, evidently did not think that catapults and springalds were obsolete when he designed some new ones in the late fifteenth century and early sixteenth.

Perhaps research on early Islamic weaponry will help to resolve the difficulties. Early and classical Islamic armies constructed and used torsion-powered siege engines, and further developed "beam-sling machines of the mangonel type."[85] France suggested that by the time of the first crusade, torsion "had almost totally disappeared in favour of lever action," but there were still a few people around capable of the construction of torsion weapons. "Ordericus regarded Robert of Bellême as unusual, and the reference by Abbo to very skilled men building the Danish machines used against Paris makes the same point."[86] Everybody agrees that tension catapults were cheaper and easier to produce and maintain than torsion, and that they were more reliable, but less powerful.[87] Nevertheless, medieval catapults could still be fearsome anti-personnel weapons. At Mons-en-Pévèle in 1304 "four or five men were run through" with a large springald bolt. "In his account of the siege of Paris, Abbo tells us that on one occasion a single ballista bolt killed two skilled workmen, while on another Abbot Edles of St-Germain killed seven Danes with a single shot, causing him to jest that his victims should be sent, like skewered meat, to the kitchen!"[88] The machine that shot this would have been analogous to a 5-cubiter catapult built to formula, akin to the engine that shot the Vani megabolt.

Catapults were finally rendered obsolete as weapons of war when gunpowder weapons were made effective and reliable. The key finding of Richardson's work on the ballistics of historical weapons was that gunpowder weapons had velocities not just faster but an order of magnitude faster than any weapon that had preceded them. The velocity of the projectile jumped from speeds of about 68–100 mph (30–45 mps) to speeds of 400–1100 mph (180–520 mps). Whereas the arrow, bolt, or slingshot kills, if it does, by making the victim bleed to death, the bullet can kill that way *or* by shock, through the tremendous force with which it hits the victim at these velocities. Richardson also included some penetration tests. He found that none of the arrows or bolts of the bows and crossbows he used could penetrate a 2-mm thick sheet of steel. A medieval arquebus with a 50-grain charge firing a half-inch lead ball could easily pierce it, however; with a 65-grain charge it would penetrate two 2-mm sheets clamped together, and with a 90-grain charge it almost pierced three 2-mm sheets clamped together.

But this device, developed more than a millennium after the catapult was invented, belongs to another world and begins another story in man's unceasing efforts to dominate other men. Our tale ends here.

*Appendix One*

# The Calibration Formulae and Components

THE CALIBRATION FORMULAE APPLY only to torsion catapults; bow catapults had no equivalent, as far as we know.

The key parameter for building a torsion catapult is given by Ktesibios (aka Heron) in *Belopoiika* 113 (palintone) and 114 (euthytone), Philon *Belopoiika* 51 (palintone) and 54–55 (euthytone), and Vitruvius 10.10.1 (euthytone only).[1] The key parameter is ***the diameter of the spring hole***, here abbreviated D, which serves as an inflexible substitute for the diameter of the flexible spring.

TABLE A.1.

| EUTHYTONE/ EASY-SPRING | PALINTONE/ OPPOSING-SPRING |
|---|---|
| *Divide length of sharp by 9 (any units)*<br>*That number in those units = ideal diameter of spring hole* | *Weigh shot in drachmai**<br>*Find cube root of that number*<br>*Add one tenth of result*<br>*That number in daktyls = ideal diameter of spring hole* |
| E.g. arrow 3 span (= 36 daktyls) long | E.g. stone of 3000 drachmai (= 30 minas, or half a talent) |
| A ninth = 4 daktyls<br>Make the spring hole 4 daktyls | Cube root = 14 1/2<br>Plus a tenth = 16<br>Make the spring hole 16 daktyls<br>If (as usual) product has no whole cube root, take nearest whole number and then add a tenth of that |

*Ktesibios specifies the weight in minas, and therefore needs to multiply the weight of shot by 100 to turn it into drachmai. Philon has it more simply in units, *Bel.* 51.16. See also Marsden 2.58 n. 40.

These formulas tell the engineer what diameter to make the spring hole in order to shoot effectively the caliber of missile his employer wishes or has to employ. The dimensions of all other components of the catapult are based on a number of spring diameters, which are given for a selection of different catapult designs by Philon and Vitruvius.[2]

COTTERELL AND KAMMINGA STUDIED the mechanics of the palintone and assert that the ancient formula is mechanically accurate; it did in fact "give the correct size of torsion spring" for the load.[3] They also observe that, if a catapult were built to formula, the maximum range would be independent of the mass of the shot. This means that all catapults built to formula would have essentially the same range, regardless of caliber; unlike the situation with modern guns, big catapults throwing big stones would not shoot further than smaller catapults of the same design throwing proportionately smaller stones. It does not mean that the same catapult would throw shots of different masses the same distance.

If one catapult threw both light and heavy shot, lighter shot would be thrown further than heavier shot, by and large, although the relationship between weight and discharge velocity (and hence range) is not simple and the effects not obvious. Reinschmidt found that the kinetic energy at discharge is essentially constant above the formulaic shot weight, so that heavier shot has greater momentum (thus more impact on the target) but reduced range. But the kinetic energy is not constant at below-formulaic weight; if the shot is significantly below weight it may accelerate so fast that much of the catapult's energy is left in the springs and arms, which therefore crash forcefully into the stanchions. In these circumstances, the efficiency of the catapult and the range of the shot improve if the arms are not drawn back so far—but there is still energy left in the arms so they still crash into the stanchions.[4]

Shot was practically impossible to make precisely to a certain dimension never mind to a certain weight, so the reality was even more complex than theoretical studies suggest. Philon at one point even recommended throwing large *non-spherical* stones in the path of siege towers, as they were harder for the enemy to clear away (*Poliorketika* C 64, = 95.38); the shape would affect drag, of course. Stones of the same weight may vary in dimensions, and stones of the same dimensions may vary in weight, depending on the stone density, which is naturally variable even for stones from the same quarry. For example, one ball from Masada that weighs one mina is larger than others there that weigh two, but they were all made of Judean limestone or dolomite.[5] Thus when archaeologists talk of shot "caliber" they are talking about *roughly* spherical-shaped balls of varying weights and sizes *grouped* into ranges, rather than balls sharing a precise shape and number as such.[6] The same situation, though less pronounced, obtains with sharps.[7] It is in this slightly paradoxical sense then that these machines are "calibrated."

A brief digression is in order to deal with the issue of inscribed marks on some ballista stones. Lawrence suggested that the actual weight was marked on the stone sometimes so that the artilleryman could make appropriate allowances when aiming (1979: 45). This might convince if the said markings were accurate and consistent, but since they are not, it is not. For example, the 30 inscribed stone shot from Calahorra (ancient Calagurris, birthplace of Quintilian, on the Ebro in the Rioja region of Spain) demonstrate categorically that there is no direct correlation between the numeral inscribed and the weight (or the diameter) of the shot. The numeral XX, for example, is found on stones weighing between 4.5 and 10 kg., three stones each weighing 4.5 kg are marked with X, XIIII, and XX, and three stones weighing 14 kg are marked with VX (an interesting notation in itself, apparently reading right to left), XIIII, and XXXXII. One of the stones even has a Celtiberian symbol on it rather than a Roman numeral. And nearly 300 from the site have no inscription at all. Likewise, shot from Wallsend (ancient Segedunum, Hadrian's Wall's-end) includes X and IV on 70 mm diameter, 390–400 g. weight balls, while a XX measures 60 mm and weighs 280 g.[8] There is no correlation between the numbers inscribed and the weight or diameter of the stones. If it appears to be otherwise on one or a few individual stones (as for example some of the 353 from Rhodes, Laurenzi 1964: 148), it is mere coincidence.

Because the velocity of the shot as it left the catapult was, by and large, independent of the mass of the shot, all catapults of the same design, whatever caliber, could have been placed at the same distance from the target to concentrate firepower on one stretch of wall. The bearers of small caliber catapults would not have had to approach closer to hit the target. This insight from engineering is consistent with the literary sources, where varieties of caliber weapons are mounted on the same siege tower to attack the same target.

The euthytone catapult has the same basic mechanics as the palintone, although the ancients did not obviously realize it. Their calibration formula for the euthytone, that the spring hole diameter should be one ninth the length of the sharp, fails to mention the weight of the sharp. Clearly, sharps of the same length could, in theory, have grossly different weights, and the velocity and range of the sharp depends entirely in the first case and mostly in the second upon its weight, not its length. The euthytone formula, therefore, could not actually have achieved what it was designed to achieve, and what the ancients thought it achieved, namely, to take the intuition and luck out of catapult construction. If the euthytone formula worked for any particular shot, then it did so by accident, not by design. But as we saw when discussing the mechanics of the palintone, which are the mechanics of the euthytone as well, the machine is tolerant of weight variation within limits, and if most sharps of a certain length fell within those limits, then the formula would "work." Generally, increased shaft length would associate with increased shaft diameter and increased size and weight of the head, so the scaling would hold,[9] within limits.

Sharp heads could vary very significantly in percentage terms at the lower end of the scale. For example, two found at Qasr Ibrim that look almost identical and differ only half a millimetre in length (90 versus 90.5 mm) differ in weight by 50 percent (10.25 g versus 16 g),[10] but relative to the mechanics of the catapult and the power of the springs, such differences would usually be negligible if the weights were so small in absolute terms. In field trials of a reconstructed kheiroballistra, three bolts of 65 g and one of 80 g shot with the same drawback, on the same machine, and at the same elevation, all had the same range of 206 meters (225 yards).[11] With the development of the Dura bolt profile (Figure 5.1) the relationship of missile length to spring hole diameter perhaps changed, since this sort of bolt is more massive.

Wilkins argued that the length of a sharp with a regular cylindrical shaft probably corresponded to a Dura bolt of two-thirds that length (2000: 94). Reinschmidt guessed (pers. comm.) that someone would have realized that the shaft of the arrow is superfluous when one is not drawing back a hand-bow, and the development of the Dura profile perhaps marks a step on the road to that realization. Such a realization should have cast doubt on the calibration formula for euthytones.

The weight tolerances of any particular euthytone catapult were probably discovered in practice. Perhaps there was recognition of the inadequacy of the euthytone formula, particularly for heavy sharps, in the occasional use of palintones, with alternative slider and bowstring fitted, for sharp-casting rather than stone-throwing. Alternatively, that may simply reflect an economy of artillery provision. Certain sharps were known to require special treatment; for example, a fire-brand that had a relatively large and heavy iron cage head had to be shot from a relatively slack-stringed or long-armed catapult to discharge at relatively slow velocity so that the flame would not be extinguished.[12]

The spring-hole diameter, once found, was used as the unit for the rest of the construction. Philon and Vitruvius (but not Ktesibios) supply lists of some dimensions of other parts, expressed in spring-hole diameters, for both palintone and euthytone catapults. So using the formulae and the lists, the proportions and ratios between parts for engines of any caliber can be found. However, neither Philon nor Vitruvius supply *all* required measurements. Sometimes only one or other of them supplies the dimensions of a particular part. How useful then was one of these treatises on its own to an ancient mechanic? Is there enough information in one of them on key parts of the catapult to build one reliably? Can experience compensate for the missing data? Do the missing data belong to those parts whose dimensions are more or less defined by other parts for which one does have data? Or do they belong to parts for which there were large tolerances? Which are the critical elements of an ancient catapult? My intention in the rest of this appendix is to find answers to these and similar questions. At the end, I give a table with the key dimensions of popular caliber weapons.

## The Key Elements

### The Spring

The most important element in a torsion engine is the spring. It is shorter on the euthytone than on the palintone. On discharge of the weapon, the strain energy put into the springs on drawing the bowstring back is converted to the kinetic energy of the missile. The springs are the source of the machine's power. All other dimensions are derived from the spring size. The spring itself, however, is not actually part of the structure as such; it is a consumable that wears out, and a variable, since it might be made of two different materials—sinew or hair. The *space* for it in the engine is therefore the key datum.

The width of that space is defined by the spring-hole through which the spring-cords must fit, and the length of that space is defined by the distance between the levers top and bottom over which the spring-cords must pass. Both Philon and Vitruvius make clear that the width of the spring-hole was the unit for the modular construction of the rest of the catapult: all other measurements were expressed in terms of this unit. But neither author gives the *length* of the spring as such. To the modern mind, it is astonishing that none of the ancient authors bothers to tell his reader directly and clearly the ideal ratio of spring length to spring width.[13] The nearest anyone comes to it is Philon, when he sums the heights of five items on a palintone[14] and says "thus the total length of the spring when added up, excepting the levers, is 9 [spring-hole] diameters" (53.16–17). The length of the spring thus appears to be a consequence of construction, if not an afterthought in presentation. Philon's sum also explicitly ignores two more elements, (vi) the dimensions of the levers top and (vii) bottom, for which his recommended sizes are given a little later (in 53.23–24). But the height of the lever above each washer (top and bottom) is variable, because the lever may sit partially embedded, or not, in the washer, and the washer may sit partially embedded, or not, in the frame.[15] Therefore, Philon's excepting of the levers from his sum was reasonable from a practical perspective.

In fact, the third dimension is an often-overlooked aspect of ancient catapults, especially when playing the game of fitting artillery to fortifications. The height of the palintone's springs is 9D, plus an allowance for the levers. This means the height of the springs of the 3-talent catapult, for example, is 16′4″ (5 m), plus levers top and bottom. One needs to consider *the height of the ceiling*, as well as the floor plan, in reconstructions of artillery emplacements. Perhaps because our authors do not clearly state the height of the springs it is often not appreciated that sharp-casters took up much less room vertically. The springs on the euthytone were probably a little over half the height of a palintone, the frame being about 5 1/2 D high, so that even the 5-cubiter, launching sharps over 7 foot long, was only 4′ 7 1/2″ (1.410 m) high.

What are seen as failings in the ancients' presentation of figures used to be taken as evidence that, despite appearances, these treatises are not really practical manuals at all, but theoretical exhibitions. Landels contested this (1978: 100), as did Marsden, whose explanation was that, although the length of spring *appears* to have been ignored, "if we build the frame, washers, and levers in strict accordance with the list of dimensions, the springs which fit the frames will automatically turn out to be of precisely the right height in relation to their diameter" (1969: 33). Concerns about circularity of argument aside, this would be more convincing if we had clear figures from Philon for the dimensions of the frame, and did not have to find that by summing heights of stanchions and thicknesses of cross-pieces.

I have another explanation to offer for at least some of the shortcomings in ancient instruction manuals. We expect a simple ratio of spring length to width because we assume that Philon is talking about "a" "standard" catapult, which has "a" set of dimensions. But perhaps this was not the case. Philon refers to a good deal of variation in dimensional details—and the variety multiplies when we add the descriptions of catapults by Ktesibios, Biton, Vitruvius, the DRB, and Vegetius. Marsden thought that this variety was diachronic (different dimensions belonging to models of different times) but it may in fact have been synchronic (different dimensions being contemporaneous), at least in some cases. The same thing may apply to the washers: is the variety of washer type and technology a sign of change over time (as usually supposed), or is it a sign that there were simultaneously a variety of solutions to the same problem? Is it a sign of the variety that once existed in all areas of catapult design, but which *survives* only for the washers because (for the smaller caliber weapons at least) they were made of bronze, whereas most of the rest of the machine was usually made of wood, which does not survive? Why should a building in Epiros (Ephyra, variously interpreted as a fortified farmhouse or an Entrance to Hades), which was built in the fourth century B.C. and destroyed in 167 B.C., have seven different types of washer, belonging to six different caliber machines, varying in size from a hand-held specimen to something with a stock probably about 8 feet long? Was it because it had one new catapult, with the latest design of washer, periodically through its existence, or because at least seven different catapults were shooting Romans from here when the building was torched? To my mind, the latter is by far the more plausible explanation of the evidence. If so, then there might not have been *one* ratio of spring length to width for Philon to report. On the particular palintone that he chose to describe, it was 9D, plus levers. On other palintone designs, it may have been different. As pointed out above, the euthytone obeys the same laws of mechanics, and its springs were 5 1/2 D (on one variety of euthytone, anyway).

I think that what we are pleased to call the calibration "formulae" were treated by catapult-builders as guidelines rather than as rules. Philon continues, "they said that this size in particular (a spring 9D plus projecting height of the levers long) is fit in practice and seems to be neither too short nor too overstretched, but to have a medium and established format. Those with longer springs have long range and are easy to draw, but their blow is weak and ineffective, while those with shorter springs are hard to draw and cannot shoot very far, and the arms of these engines are often distressed" (*Bel.* 53.16–23). It is obvious from this that in Philon's time there were working palintone catapults with springs more than 9D long, and others with springs less than 9D long. Since all machines built to formula would have the same range and discharge velocity,[16] varying the spring volume (i.e., *not* building to Philonian ratios) was presumably the method by which they built relatively long range or relatively high impact weapons. In fact, it has recently been demonstrated that it is possible to extend the range considerably (at the expense of weakened impact, as Philon observes), but not to increase the kinetic energy of the missile, even at the expense of reduced range. The palintone that Philon describes has precisely the ratios required to optimize range *and* impact.[17] Further, as Philon observed and is confirmed by computer simulation, damage to the arms is increased by departing far from the formula in either direction; the arms of a catapult built to formula should not crash into the stanchions with any speed and hence force, whereas those of a catapult with nonformulaic proportions do.[18]

By Vitruvius's time, variety amongst euthytones was distinguished by name. If the frame was wider than high then the catapult was called short-sprung, and if higher than wide, long-sprung, and he adds that the former should have longer arms, to compensate for its relatively strong short springs, and the latter shorter arms to compensate for its relatively weak long springs (10.10.6).[19] The archaeological evidence supports a looser interpretation of the formulae too. Vicente et al. offer detailed comparison between the dimensions of the well preserved metal catapult components from Caminreal and those given for the same parts by Philon and Vitruvius; rarely do the numbers coincide.[20] Differences of 10–20 percent are common, and overall the Caminreal catapult was more compact than that which results from following the figures of Philon or Vitruvius; its washers (and thus springs) were significantly larger than one would expect in a frame of its size. Further, not all of the variation that occurs in the size of washers found to date can be accounted for as differences in caliber (Vicente et al. 1997: 181).

The catapult-builder should have used sinew or hair rope that had a diameter one-third of the diameter of the spring-hole, according to all the manuscripts of Philon (*Bel.* 54.23). Marsden thought this must be wrong and corrected it to one-sixth of the diameter of the hole.[21] The spring rope was not used in its manufactured state. It was stretched, or pre-tensioned, as it was being loaded into the catapult, by means of a machine called a stretcher.[22] This device stretched one length of rope at a time, until it was two-thirds its natural thickness (Ktesibios 108.3–4 Heron). So, following Marsden's correction of the Philon manuscripts, suppose the hole diameter was 6 inches, the rope for the spring should be 1 inch diameter, and that rope should be stretched until it was two-thirds of an inch in diameter. So it should be possible to get eight cords through such a hole (6 each reduced by a third, permits 2 more cords of three already-reduced thirds).[23] Ideally, each cord of the spring was tensioned to the same degree. The cords should have been as taut as the strings of a musical instrument (which was strung with the same material, sinew cord). The precise degree of tension was tested as each cord was loaded and stretched, by the tone emitted when it was plucked by the musical mechanic, or by callipers if the mechanic was tone deaf. Plucking the cords would have worked better, since rope would naturally vary slightly in thickness along its length, but the vibration produced by one length would be consistent. This method was also preferred when (literally) tuning the two springs of a two-armed engine so that both arms would be propelled by the same amount of force, else the projectile would not fly straight. The musical method at least would not work if the cords were packed so tightly that they could not vibrate.[24] This perhaps explains Philon's comments about the gauge of sinew rope being 1/6 D (which is usually overlooked in discussions and reconstructions).

Sinew fibers have a crimped or corrugated molecular structure (which perhaps explains why fresh sinew is so sticky), and this pre-stretching would have taken the crimp out of the fibers. As a result, the performance of the springs when the catapult was drawn would have been much better, consisting only of the elastic extension of already straightened fibers. For the springs of the catapult need to be elastic, but they must not stretch too easily. Nor must the material used be stretched out of shape easily. Ideally, it should spring back to its initial size and shape when the stress is removed, that is to say, it should be resilient. In technical terms, the Young's modulus of the material used must not be too low. Ideally one wants the arms to stop dead when they return to their rest position, so there is no recoil, then all kinetic energy is transferred to the missile and none is left in the arms or springs. To achieve that requires that the springs are as taut and as strong as possible. Otherwise, some of the force put into the springs on drawing the catapult is wasted on stretching the springs in the other direction (recoil) or lost in a collision with the stanchions.

Sinew has excellent properties as a store of energy. In fact, weight for weight it was better than any other material available to the ancients, and better than many materials available today, including spring steel.[25] They were familiar with its properties from the composite bow and other applications—the ancients did not waste a material that had evident strength, and was one of the inedible parts of the carcass. At religious sacrifice, which was the principal context in which ancients ate meat, the worshipers shared the meat, and the gods got the fat and bones. Sinew is generally extracted from the meat rather than the bone, though its function is to connect the one to the other, of course. In dismembering the carcass, the thick wedge of tendon at the bone is cut, and the sinews are traced back into the muscle until they give out. The lengths are typically short, though not obviously shorter than individual fibers of wool or human hair (women's was probably preferred because longer). No ancient source survives on the making of sinew rope, but the fibers were probably treated in a similar way to the manufacture of rope made of hemp, that is to say, after being dried, beaten, and separated, they were twisted into threads and the threads into rope.[26]

Hair had the advantage that it was a secondary product, that is, something that could be provided by living animals, so they did not need to be slaughtered in order to obtain it. This feature was exploited during some sieges when the women of the town were the animals in question. Further, hair did not need extensive pre-treatment before spinning. Philon and Heron both say that spring ropes, whether sinew or hair, were soaked in olive oil, and rubbed with grease or fat, prior to and after it was made into rope. I have no idea what this might do to the mechanical properties of sinew or hair. I expect it kept them supple, protected them from moisture, and stopped them from drying out and becoming inflexible, which sinew is apt to do very quickly.

Reinschmidt's computer simulation of a palintone built to Marsden's dimensions produced a discharge velocity of 221 feet per second, which equals about 67 meters per second. Baatz's estimate was 72 mps.[27] Both are hypothetical figures that involve guesses regarding various aspects of construction and performance for which no evidence survives (such as the elastic properties of the sinew, diameter of bowstring, amount of friction, etc.). They should be treated as *guides* to performance. Field performance of a reconstruction built exactly to the same specification would produce a range of values, depending on variables such as the quality of construction, the age and tautness of the springs, and even the weather. Wilkins and colleagues achieved almost 36 meters per second (118 ft/s) with their reconstruction of a 3-span, and 47 meters per second (154 ft/s) with their kheiroballistra.[28] Clearly, there is some difference between the performance of a simulated and a reconstructed catapult. Much of this is attributable to the properties of modern polyester rope, which performs poorly compared to sinew (as Wilkins understands very well), but which has served for spring material in the absence of sinew rope (2000: 99).

### The Arms

Both Philon and Vitruvius recommend arms 7D long for a euthytone and 6D long for a palintone, and about 1/2 D wide and about 1/2 D thick at the heel (thick) end (the arm tapers in width from heel to tip). Arms longer than this were easier to pull back but had weaker discharge; arms shorter than this were harder to withdraw but gave greater impact at short range (Philon 53.26–29). Evidently, 6D to 7D were recommended lengths for an "all-round" performance catapult, but there were catapults with arms longer or shorter than this for specific functions or requirements.

The arms contributed a little power to the performance of the catapult, insofar as they would have bent somewhat during the drawing of the weapon, and that strain energy would have been transferred to the projectile during discharge as they regained their initial shape. Torsion catapult arms were not like bow catapult limbs; they were not built to bend, but like all solids they would have bent under strain. How much or little would depend on the materials of their construction and the amount of strain imposed in drawing the weapon. The arms contribute their strain energy as the second stage of accelerating the projectile, after the bowstring and before the springs.[29]

The arm should have been inserted in the spring so that its heel was close to but not touching the heel-pad on the center stanchion. If the arm pressed onto the heel-pad in the brace position, it would be less efficient than an engine whose arms did not, because more kinetic energy of the arm would be transferred to the stanchion instead of to the projectile at the end of the discharge. Such a badly arranged machine would damage itself with every shot.

### The Bowstring/Slingstring

The bowstring was always made of sinew, not hair, nor, apparently, gut. The bowstring was made of the toughest sinew available, because all the force of the machine went through it during drawing of the catapult. However, unlike a hand bow, a torsion catapult's bowstring does not (or *should* not) hold the weapon in the brace position (strung but not drawn), and it is not (or should not be) taut when the catapult is at rest. When at rest, there is very little tension in the whole machine other than that in the springs—and their tension at this point should be only vertical, from being stretched taut between the levers. Consequently, stringing the catapult, in the sense of fitting the bowstring, was straightforward.

Ktesibios discusses the bowstring in some detail (*Bel*. 110.9–112.3). He recommends use of the most well stretched sinew available for this part (hair works well for the springs, he says (112.4), but does not recommend it here). He notes that this single cord does a lot of work and takes the force of the discharge. He comments that bowstrings are plaited in two ways: the euthytone catapult's bowstring is round in cross-section, since it must fit into the notch of a sharp, and the two-pronged claw by which it is retracted and loosed must fit over the bowstring on either side of the notch. The palintone catapult's bowstring is plaited flat in cross-section, like a belt, with loops at the end to go over the ends of the arms. At midpoint, where it meets the claw, is a "sort of ring, plaited from the same sinews" (as the band). A single claw, not a two-pronged claw, fits into this ring. This sort of bowstring is not of the same width its whole length, but widest at the centre, where it is "straight" or "upstanding" (ὀρθός), in order to hold and propel the stone efficiently. Ktesibios warns of the consequences of placing the slingstring too high or too low relative to the shot, and notes that generally the bowstring is set higher above the slider on a palintone than it is on a euthytone.

Vitruvius says nothing about the bowstring per se. Philon says very little, and on fitting it to the catapult next to nothing. His most informative comments come right at the end of his treatise, where he reports that Ktesibios "stretched tight the bowstring and adjusted the sling" (τὴν τοξῖτιν ἐντείνας καὶ τὴν σφενδόνην καταρτίσας 78.15–16). This implies a certain level of tension, but it could not have been more than a man could apply or else there must have existed some device for stringing catapults about which we are not told. Stringing Odysseus' hand bow was a matter for celebration in literature, so stringing a torsion catapult might have warranted a mention too, if it was difficult. Ktesibios at one point implies that the bowstring was loose when the catapult was prepared but at rest. He said (*Bel.* 82.11–83.2 Heron) they fitted the bowstring to the arms by "making loops, and securing these by pins AB and GD to prevent the string slipping off." Philon advises the reader to "fasten on the bowstring as usual; stretch and release this two or three times, without pulling the trigger but letting it go gently and gradually, until the springs slacken" (*Bel.* 66).

The bowstring itself contributed energy to the power of the catapult, insofar as it was stretched slightly during drawing, and the conversion of that strain energy into kinetic energy of the projectile would have been the first force transformation when the weapon was discharged. Thereafter, once the arms began to rotate under the torque in the springs, they would pull on the bowstring, which would accelerate the projectile until it lost contact.[30] Bowstring mass and elasticity reduce efficiency, since the string is in motion when the projectile is released, retaining some kinetic energy instead of passing it all to the missile.[31] Therefore the sinews used for this component were probably thoroughly stretched (confirming Ktesibios' advice), and the property of sinew that was being exploited here was simply its strength-to-weight ratio, rather than its resilience.

Whatever force the catapult is required to store, the bowstring must be strong enough not to break under that strain, and it may have to absorb some of the recoil too. Philon says that the arms wear badly on short-spring cata-

FIGURE A.1 *Stonebow slingstring. (Author sketch after Payne-Gallwey)*

pults, which are hard to draw back (*Bel*. 53.25–26); the fact that he does not mention bowstring breakages in this context suggests that they were adequate to the task. The same is true of 68.34–69.1, where he observes that the forces involved in drawing the bow often break the arms; again he makes no mention of the bowstring. The same is true of his discussion of the air-sprung catapult, 78.17–20. Whatever it was made of, and however it was fitted, the bowstring never seems to have caused any problems. Payne-Gallwey expected a well-made crossbow bowstring, preserved by rubbing it with beeswax, to last for twenty years in frequent use.[32] Wax would have been available to the ancients, of course, but applications of olive oil, grease, and fat are specifically mentioned by the ancients (above, under springs).

The length of the bowstring is given as 2.1 times the length of one arm (Philon *Bel*. 54.1–2). The width of the slider is given as fractionally over 1D; the string therefore was almost long enough, without being stretched, to span tip of arm to tip of arm across the stock even if the arms were laid perpendicular to the case, which they were not. Therefore the bowstring was long enough to reach across without much, if any, stretching. Consequently, when the weapon was discharged, the bowstring would not become taut and slow the motion of the arms before they returned to their brace position.

Ktesibios stated that "it is necessary to string the bowstring so as to keep the arms a little way off the side-stanchions, to avoid their crashing into them" (*Bel*. 102.1–2 Heron), implying that the bowstring should be sufficiently short physically to prevent the arms reaching the side-stanchions. If the catapult was built to formula and loaded with a projectile of approximately the right caliber, the springs themselves should have stopped the arms recoiling beyond their brace position. But no doubt this was not always the case in practice, and Ktesibios' advice was pragmatic. It was better to sacrifice the bowstring than have the stanchions damaged.

Slingstrings, that is, bowstrings for stone-thrower catapults, may have been single or double, we do not know. If double, they probably resembled that used on stonebows (Figure A.1).

### The Stanchions

If the stanchions are inadequate relative to the strength of the springs, there is the danger that, when the springs are stressed, the frame will collapse vertically, in the process exploding splinters and fragments of wood in the vicinity. They were usually iron-plated to strengthen them and prevent this happening. In the *kheiroballistra* the wood is dispensed with, and the frame is cast in metal, the so-called field-frame or *kambestrion*.

As mentioned above, if the catapult is well built to formula, the stanchions should not be exposed to damaging recoil from the arms. "When shot weight is the design value W that appears in the equation $D = 1.1W^{1/3}$, the proportions as given by Philon result in a machine that transfers almost all its strain energy to the shot, with very low (virtually zero) rotational velocity in the arms as they recoil back onto the stanchions. That is, this machine is very efficient in achieving maximal range for the energy stored, and does not try to demolish itself with every discharge."[33] If the arms crash forcefully into the stanchions, it is a sign that there is some shortcoming in the manufacture, maintenance, or loading of the weapon, or something has not been built to the recommended proportions. Reinschmidt found in a sensitivity analysis on his simulation catapult that performance degraded when any of the parameters were changed. For example, if the arms were made lighter the discharge velocity was reduced and the arms crashed into the stanchions more forcefully than they did with heavier arms. "The [ancient formulaic] device has been tuned so that all parts work together, with the result that changing any one component makes the performance worse. This would hardly be remarkable in a modern machine; the interesting thing is that the Greeks could do this so long ago" (Unpublished: 18).

There is more variation with the stanchions than any other component. First, the euthytone is shorter than the palintone. Second, Philon and Vitruvius give different recommended dimensions for the stanchions. I reproduce here the Philonian dimensions given by Marsden and the Vitruvian dimensions given by Wilkins.[34] Readers should be aware that the Vitruvian figures are represented in the manuscripts by a variety of symbols, some very prone to misreading, and their values probably unknown to the medieval scribes who copied them. Errors are common in such circumstances. There are, for example, four completely different symbols for 1/4 in the various Vitruvian MSS, and three different symbols for 6/16, one of which resembled the letter F and could easily be misread as F, E, or Γ. Wilkins has a more thorough examination of the matter than Marsden.

*Euthytone Catapult*

*Side stanchions*:

Type 1. Height: 3 1/2D, depth: 1 1/2D, width: 5/8D (Philon 55.5–7)

Type 2. Height: 4D, depth: ?, width: 5/8D (Vitruvius 10.10.2).

*Center stanchions*:

Type 1. Philon's machine with two centre stanchions: height: 3 1/2D, width: 3/8D, depth: 1 1/2D (55.7–9), i.e. almost identical to the side stanchion, just 1/4D thinner.
Type 2. Vitruvius' machine with one central stanchion: height: ?, width: 1 3/4D, depth: 1D (10.10.2). An aperture is cut in the middle of this for the slider to come through.
The Caminreal catapult, dated to 80–72 B.C., apparently has shorter stanchions (side and center) than either of these recommendations, at just 3D (Vincente et al. 180 fig. 22). This is probably because the spring-holes of this machine were enlarged after calibration, as Philon says sometimes happened, so that the figure we take for D is exaggerated. (The holes in this machine were so enlarged that the plating nails through the hole-carrier interfered with the springs and had to be bent.) The intended caliber of the weapon may therefore perhaps be found more reliably from the stanchions or cross-piece than from the enlarged spring-hole. For example, assume the stanchions of the Caminreal catapult were 3 1/2D (type 1 above); then this catapult, with stanchions 248 mm high, was an over-sprung version of the very common 3-span (36 daktyl) euthytone, rather than a short-stanchioned version of an odd-caliber 2 1/2-foot or even 3 1/3-span (40 daktyl) weapon. As the excavators of this catapult observed, there may be a variety of reasons for the discrepancies from formulaic norms, including adaptation to local circumstances and needs (Vicente et al. 1997: 181), but over-springing—that is, enlarging the spring-hole after construction—was common according to Philon.

*Palintone Catapult*

Palintones were almost twice as tall as euthytones.
*Side and center stanchions:*
Type 1. Height: 5 1/2, width: 1 7/12, depth: 1/2 + 1/6 – 1/16 (Philon 53.13–15)
Type 2. Height: 5 3/16, width: ?, depth: 11/18 (Vitruvius 10.11.5)

*Kheiroballistra*

The kheiroballistra had iron frames, connected together with iron struts. The stock was still wooden, and connected to the struts. There is much debate about the structure and functioning of this type of catapult, which was a Roman invention and for which we have only Heron's brief description, together with (currently) archaeological remains from Orşova and Gornea (Romania), Lyon (France), Sala (Morocco), and (perhaps) sculptural depictions on Trajan's column. The equivalent to the stanchions is the cylinders that hold apart the washers, which held the springs. Heron gives dimensions for this design of catapult exclusively in terms of daktyls, not formulae. It is possible to find formulae now of course, but Heron does not supply such. For a machine with a spring-hole diameter of 2 1/3 daktyls,[35] Marsden reconstructed rather tall cylinders, 20 daktyls, but as he noted (1971: 222–223 n. 17), an alternative reading of the text here is 10 1/2 daktyls, and archaeological examples found since his work was published are more in keeping with the latter. The archaeological remains of this catapult and its larger cousins (same design but not hand weapons) are wide and squat in comparison with Marsden's reconstruction. Taking 10 1/2 daktyls as the correct figure for the cylinder height (= stanchions), Heron's *kheiroballistra*, with pseudo-stanchions of about 4D, is closer to the euthytone than the palintone.

### The Case

The case carries the slider, along which the missile is shot. Both Philon and Vitruvius give us clear figures for the length of the case; 16D for a euthytone in Philon, 19D for a palintone in Philon, and 19D for both euthytone and palintone in Vitruvius, though his slider is 16D (10.10.4).

### The Hole-Carrier

Philon gives us clear figures for the size of the euthytone hole-carrier: 6 1/2D long by 2D wide in the middle tapering to 1 1/2D wide at the edges, by 1D thick (55.3–5). Vitruvius does not; we need to sum components to arrive at a total of 6D for length (10.10.2). The hole-carrier of the palintone is rhomboidal and not easily described in terms of length and width. Vitruvius gives clear dimensions for the first step of the construction, namely 2 3/4D by 2 1/2D (10.11.4).

### The Woodwork

Most of a catapult was made of wood. Most of that wood had to withstand significant stress. The arms of a torsion catapult were particularly prone to breakage. Since the catapult developed from the bow, it is very likely that the same principles were followed in the selection of wood for the construction of the bow for a catapult. Our sources do not elaborate on the selection and treatment of the wood for the war machines they describe; they occasionally talk of the sourcing of timber, but that invariably seems to be for earthworks, sapping, towers, and other temporary structures. I imagine that few catapults were built from scratch on site. Timber was used extensively in ancient society and knowledge of its various properties, and how best to exploit them, was, dare I suggest, better than it is today, at least among the laity. The processes for the aging of timber and the manufacturing of wooden bows probably did not differ much if at all from those of today, so comparative evidence from modern times can be used to fill this gap.

*Discussion*

Combining both Philon's and Vitruvius' lists, we lack dimensions for only a few parts, such as the tenons of the stanchions, which a competent mechanic can estimate, and whose estimates are unlikely to be significantly different from those arrived at by ancient mechanics following the same procedure. Philon's and Vitruvius' figures are usually, but not always, very similar. The differences appear to arise from different versions of the same basic design, as for example with the length of the case, which seems to have grown in Roman times, perhaps to counterbalance increasing weight of metal plating at the front.

It is possible to apply these formula and dimensions backward as well as forward of course, so that if a mechanic was asked to build the largest catapult that would fit, e.g., a certain tower room, then he could begin with the dimensions of the room, deduct the space needed for ergonomic reasons, and end with the appropriate caliber weapon. There is a tendency in the modern literature to assume that maximizing caliber was the principal desideratum, but towers could serve many purposes and the idea that such towers served as the equivalent of heavy gun emplacements has little support in the primary sources, which put large caliber weapons on the ground.

Philon and Vitruvius offer handy lists of ready-calculated dimensions for popular calibers. Philon's list for a stone-thrower is as follows.

> For a 10 mina shot, the spring-hole diameter is 11 daktyls;
> for 15 mina, 12 and a half and a quarter (i.e., 3/4) daktyls; for 20 mina, 14 3/4 daktyls; for 30 mina, 15 3/4 daktyls; for 50 mina, 19 3/4 daktyls; for 1 talent, 21 daktyls; for 2 1/2 talents, 25 daktyls; for 3 talents, 27 daktyls (Philon *Belopoiika* 51)

Vitruvius' equivalent list is in *De Architectura* 10.11.3. At first sight, this looks like a really useful resource. But herein lies the rub: numbers were peculiarly liable to distortion in transmission, as they were copied from one handwritten version to another, using a system where numbers are represented by letters or symbols that meant nothing to the average ancient or modern scholar, and the figures in Philon and Vitruvius have been distorted. Specifically, in Philon's list the third figure, 14 3/4 daktyls, is almost a daktyl larger than it should be; the fifth and sixth are also a daktyl larger than they should be. The seventh figure of 25 daktyls is incorrect for 2 1/2-talent shot, but it would be right for 2-talent shot, so perhaps the 1/2 is the error here. Consequently, this resource is not quite as useful as it might appear.

It was obviously better if the mechanic knew the formulae and ratios and could perform the computations himself. On the other hand, calculating cube roots is not easy, and the individual attempting to work out the numbers for himself might make mistakes too. As Reinschmidt pointed out (pers. comm.), an organized designer would only need to calculate the root once for each caliber machine he built, so would have no need of fancy methods or tools. He also believes that they would round off to the nearest integer, so a simple table of the cubes of integers would do the job. The variety of methods that the Greeks developed to find cube roots by geometry is more significant for the history of mathematics than of catapult construction. Both Ktesibios and Philon explain one method, and a number of mathematicians of course do so. Eratosthenes invented and set up in public in Alexandria a device similar to a three-dimensional slide rule to perform the calculations mechanically, though how widely that came into use is not known. Ktesibios lived before him so could not have known it, but Philon lived after Eratosthenes, and spent time in Alexandria, so in principle could have known his method. But he uses Ktesibios' method, with insignificant variations.

At the end of the day, these formulae were found by empirical means as *ideal* proportions for given calibers, and they were probably offered by the writers of the technical manuals as guidelines for effective all-round catapults. If the sort of errors we see in the lists of ready-reckoned dimensions crept in during antiquity, then the formulae would have failed, partially at least, in their aims, for using them would have resulted in unreliably performing catapults, just as not using them did, although not *as* unreliably, probably. For the errors are relatively small and these designs are robust and cope well with off-caliber projectiles—much better, for sure, than modern firearms do.

The benefits of standardization are recognized in these formulae, and the importance of that should not be underestimated. They represent standards, of materials and design, and together with the scaling law—the importance of which Philon clearly recognized in his emphasis upon it—they offer catapult builders the ways and means to construct reliable, safe, and mechanically efficient catapults to any size likely to be required in practice. In a world with almost as many different systems of measures simultaneously in use as there were states, it also avoided problems inherent in dimensions given in, say, "feet," which varied in length from place to place.

### The Tables of Dimensions

The largest caliber weapons of which traces have been found to date belonged to catapults that were at the small end of each table below. The largest diameter washer so far found is 160 mm (Hatra), which dates from the third century A.D. This could have belonged to a slightly over-sprung 3-cubit euthytone casting sharps c. 55 inches or 1.4 m long, or a slightly under-sprung 5-mina palintone throwing loads about 5 lb. or 2 kg in weight. The catapult remains are usually interpreted as those of a stone-thrower,[36] but the height of the frame tells against that in my opinion: it has a spring length-to-width ratio of 6 1/2:1, which is much closer to a formulaic euthytone than a palintone. Further, the surviving stories about the Hatrans' defense of their town revolve around sharps, particularly incendiaries, rather than stones (Cassius Dio 76.11). The next largest caliber weapon known, with a 136 mm spring-hole diameter (Ephyra 3), could have belonged to a slightly over-sprung 5-span euthytone casting sharps c. 45 inches or 1.1 m long, or an over-sprung 2-mina palintone throwing loads about 2+ lb. or 1 kg in weight. While these are the largest washers found, they correspond to the smallest caliber shot currently identified as catapult shot. 5-mina shot (about 2 kg) has been found at Rhodes, Pergamon, and Tel Dor (Israel), and there are six specimens of 3-mina shot (1.3 kg) from Tel Dor. The smallest yet entertained as catapult shot known to me is that of the smallest Vitruvian caliber, 2-libra (650 g, about 2 lb.), which has been identified among the finds from Shan Castle and Burnswark in Dumfrieshire. The stones weighing 340 g (about 1 Roman pound, libra) and 170 g (about 1/2 libra) found in the same locations are considered ammunition for staff slingers or hand-throwers, not catapults.[37]

There appears, therefore, to be a significant mismatch between the remains of ammunition and of catapults, the former being large and very large, the latter being small and very small. This no doubt has something to do with the indestructibility and identity of large stone balls, and the fact that washers for large caliber weapons were made out of wood (see Ktesibios 96.9–98.4 Heron) so have decomposed. But we should not perhaps expect to find much trace of catapults of whatever caliber in the archaeological record, for even broken ones would ordinarily, in peace or war, have been restored if working or repairable, or decommissioned and recycled for scrap metal if not. Recall the Rhodians' recycling of Demetrios' siege engines, and consider the sixty or so military daggers surviving, of tens of thousands once produced.[38] Mid-caliber shot was probably taken by the victors as part of the booty, along with doors, roof-tiles, and sundry other things that are not now considered as "movables." Moving large caliber shot would have presented the ancients with severe logistical problems, given the rather primitive nature of ancient land transport, where even the porter's trolley and wheelbarrow were unknown. But it may also be the case that at least some *glandes*, currently assumed to be slingshot, were in fact catapult shot of a caliber best considered as bullets, shot from weapons best considered as rifles. I would identify *glandes* in the British Museum weighing 439 g and 500 g as catapult shot, of a notional 1 mina caliber; so too *glandes* weighing c. 160 g, 105 g, and smaller calibers (see Chapter 5).

Throughout the book, including in the tables below, I have used the same equivalences as Marsden to convert ancient to modern measures: 1 daktyl =

19.3 mm = 0.76 inch. Not all scholars use the same conversion rates; Schramm, for example, equated 1 daktyl with 18.48 mm (1928: 220). Identifications of calibers with specific catapult finds are for indicative purpose only, showing only that these specimens are approximately of this caliber. In most cases too little survives to be able to identify the remains as definitely coming from a palintone or definitely from a euthytone.

When consulting the tables, remember that the notion of caliber with regard to catapults is not the same as the notion of caliber with regard to modern firearms. Catapults are much more robust with and accommodating toward off-caliber loads. They worked *best* when throwing loads of the caliber for which they were designed, but they could throw other loads, heavier or lighter, almost as well.

Note especially that Ktesibios instructs his reader to round to the nearest whole number when calculating cube roots; Philon does the same, and recommends adding or taking a bit from the tenth to reflect whether the nearest whole number was bigger or smaller, respectively, than the "true" number (discussed in Chapter 7). I have rounded the figures in the table to whole daktyls for length and width, and to within 1/4 inch or 1/4 ounce and 5 g or 5 mm in conversions to modern measures, *except* for the washers, which have nearest mm conversions to aid comparison with surviving artifacts. The ancients did not share our conception of precision and these figures will do. Rounding was a characteristic of ancient computations, as it remains today in most quarters outside laboratories and cash registers.[39] We are dealing with an environment where so-called "40-mina" stones could in practice vary between at least 16 and 20 kg. One can get lulled into a false sense of security by numbers arrived at by calculation and given to *n* decimal points. It was all a bit more rough and ready in the real world of antiquity where shot was cast or chiseled, sharps were wrought, and wood was hewn.

## Table A.2 Some Ancient Catapult Dimensions

Philonian Palintone *(usually a stone-thrower but can cast sharps)*
*Spring height = 9D*

| Caliber | Shot weight (*approx.* weight modern lbs. and kg) see n. 37) | Spring diameter D<br>D = in daktyls $1.1^3$ √ shot weight in drachmas | Stock length (19D) | Width[40] (6D x 2.1) |
|---|---|---|---|---|
| 4-drachmai | 1/2 oz. = 15 g[41] | 1 3/4 dactyls<br>= 34 mm = 1 1/4˝<br>*(Ephyra 6 and Elginhaugh)* | 645 mm<br>= 2´ 2˝ | 430 mm<br>= 1´ 5˝ |
| 6-drachmai | 3/4 oz. = 20 g<br>*"Via" and trident glandes in BM* | Almost 2 daktyls<br>= 38 mm = 1 1/2˝<br>*(Bath)* | 720 mm<br>= 2´ 6 1/2˝ | 480 mm<br>= 1´ 7˝ |
| 8-drachmai | 1 oz. = 35 g<br>*"scorpion" glandes in BM* | Near 2 1/4 daktyls<br>= 42 mm = 1 3/4˝<br>*(Xanten, Mahdia 3, Volubilis 1, 2)* | 805 mm<br>= 2´ 8˝ | 535 mm<br>= 1´9˝ |
| 16-drachmai | 2 oz. = 70 g<br>*70 g slingshot from Thebes inscribed EPAM (inondas)*[42] | 2 3/4 daktyls<br>= 53 mm = 2˝ | 1005 mm<br>= 3´ 3 1/2˝ | 670 mm<br>= 2´ 2 1/4˝ |
| 1/4 libra<br>(c. 18 dr.) | 2 1/4 oz. = 80 g | Smidgen more than the last = 56 mm<br>= 2 1/2˝ *(Ephyra 7)* | 1065 mm<br>= 3´ 6˝ | 705 mm<br>= 2´ 4˝ |
| 24 drachmai | 3 oz. = 105 g<br>*108 g glans in BM* | Near 3 1/4 daktyls<br>= 61 mm = 2 1/2˝<br>*(Ephyra 4 and Pergamon)* | 1170 mm<br>= 3´ 10˝ | 775 mm<br>= 2´ 6 1/2˝ |
| 1/2-libra<br>(c. 37 dr.) | 5 oz. = 165 gr.<br>*155 g "Bak" glans in BM* | Over 3 1/2 daktyls<br>= 71 mm = 2 3/4˝<br>*(Cremona 1; Mahdia 2)* | 1350 mm<br>= 4´ 5˝ | 895 mm<br>= 2´ 11 1/4˝ |
| 1-libra<br>(Roman lb, c. 75 dr.) | 10 oz. = 330 g<br>*(I suspect shot of this weight range is misidentified as something else)* | Over 4 1/2 daktyls<br>= 90 mm = 3 1/2˝<br>*(Cremona 2; Mahdia 1)* | 1,710 mm<br>= 5´ 7 1/2˝ | 1,135 mm<br>= 3´ 8 3/4˝ |

| CALIBER | SHOT WEIGHT (*approx.* weight modern lbs. and kg) see n. 37) | SPRING DIAMETER D D = in daktyls $1.1\sqrt[3]{\text{shot weight in drachmas}}$ | STOCK LENGTH (19D) | WIDTH[40] (6D x 2.1) |
|---|---|---|---|---|
| 1-mina (100 dr.) | 1 lb. = 1/2 kg (approx.); *two large glandes in the BM, 439 g and 500 g* | Over 5 daktyls = 99 mm = 4″ | 1,880 mm = 6´ 2″ | 1,245 mm = 4´ 1″ |
| 2-libra | c. 700 g (*e.g. Masada x 16*) | Near 6 dactyls = 113 mm = 4 1/2″ | 2,145 mm = 7´ 1/2″ | 1,425 mm = 4´ 8″ |
| 2-mina | 2 lb. = 1 kg (*e.g. Bar Hill x 24*) | Near 6 1/2 daktyls = 124 mm = 5″ (*Sounion*) | 2,355 mm = 7´ 8 3/4″ | 1,560 mm = 5´ 1 1/2″ |
| 5-mina | 5 lb = 2 kg (*e.g. Bar Hill x 9*) | 8 3/4 daktyls = 170 mm = 6 3/4″ | 3,225 mm = 10´ 7″ | 2,140 mm = 7´ 1/4″ |
| 10-mina | 10 lb = 4 1/2 kg; (*e.g. Rhodes x 46*)[43] | 11 daktyls = 212 mm = 8 1/4″ | 4,035 mm = 13´ | 2,675 mm = 8´ 9 1/4″ |
| 20-mina | 20 lbs = 8 3/4 kg (*e.g., Pergamon x 67*) | Over 13 3/4 daktyls = 267 mm = 10 1/2″ | 5,075 mm = 16´ 7 3/4″ | 3,365 mm = 11´ 1/2″ |
| Half-talent (= 30 mina = 40 libra) | 30 lb = 13 kg; (*e.g. Pergamon x 118*)[44] | Over 15 3/4 daktyls = 306 mm = 1´ | 5,815 mm = 19´[45] | 3,855 mm = 12´ 6 1/2″[46] |
| Talent = 60 mina = 80 libra | 60 lb = 26 kg (*e.g. Pergamon x 353*) | 20 daktyls = 382 mm = 1´ 3″ | 7,315 mm = 24´[47] | 4,865 mm = 16´ |
| 2-talent = 120 mina = 160 libra | 120 lbs = 52 kg (*e.g. Pergamon x 33*) | Near 25 1/4 dactyls = 486 mm = 1´ 7 1/4″ | 9,235 mm = 30´ 3 1/2″ | 6,125 mm = 20´ 1 1/4″ |
| 3-talent = 180 mina = 240 libra | 180 lbs = 78 kg (*Vani*) | Over 28 1/4 daktyls = 556 mm = 1´10″ | 10,565 mm = 34´ 8″ | 7,005 mm = 22´ 9 3/4″ |

Philonian Euthytone *(always a sharp-caster)*
*Spring height = 5 1/2D*

| Caliber | Sharp Length | Spring diameter D<br>D = in daktyls<br>1/9th sharp length | Case (Stock) length<br>(16D according to Philon) | Total Width<br>(7D x 2.1) |
|---|---|---|---|---|
| 1 span<br>= 12 daktyls | 232 mm<br>= 9 1/4″ | 1 1/3 daktyls<br>= 26 mm = 1″ | 416 mm<br>= 1′ 4 1/2″ | 382 mm<br>= 1′ 3″ |
| 1 foot<br>= 16 daktyls | 308 mm<br>= 12″ | 1 3/4 daktyls<br>= 34 mm = 1 1/4″<br>*(Ephyra 6; Elginhaugh)* | 544 mm<br>= 1′ 9 1/2″ | 500 mm<br>= 1′ 7 3/4″ |
| 1 1/2 span<br>= 18 daktyls | 348 mm<br>= 1′ 1 3/4″ | 2 daktyls<br>= 39 mm = 1 1/2″<br>*(Xanten)* | 625 mm<br>= 2′ 1/2″ | 575 mm<br>= 1′ 10 3/4″ |
| 1 cubit<br>(= Dura bolt, kestros) =<br>2 span =<br>24 daktyls | 465 mm<br>= 1′ 6 1/4″ | 2 2/3 daktyls<br>= 50 mm<br>= 2″ | 800 mm<br>= 2′ 7″[48] | 735 mm<br>= 2′ 5″ |
| 1 1/2 cubit<br>= 3 span<br>= 36 daktyls | 695 mm<br>= 2′ 3 1/2″ | 4 daktyls<br>= 77 mm<br>= 3″ *(Cremona 1; Ephyra 3; Sala)* | 1230 mm<br>= 4′ 1/2″[49] | 1130 mm<br>= 3′ 8 1/2″ |
| 2 cubit<br>= 4 span<br>= 48 daktyls | 925 mm<br>= 3′ 1/2″ | 5 1/3 daktyls<br>= 100 mm<br>= 4″ *(Azaila 3?)* | 1645 mm<br>= 5′ 4 1/2″[50] | 1505 mm<br>= 4′11″ |
| 5 span<br>= 60 daktyls | 1,158 mm<br>= 3′ 9 1/2″ | 6 2/3 daktyls<br>= 130 mm<br>= 5″ *(Ephyra 3; Sounion)* | 2045 mm<br>= 6′ 8 1/2″ | 1,870 mm<br>= 6′ 2″ |
| 3 cubit<br>= 72 daktyls | 1,390 mm<br>= 4′ 8 3/4″ | 8 daktyls<br>= 154 mm<br>= 6″ *(Hatra)* | 2465 mm<br>= 8′ 1″ | 2265 mm<br>= 7′ 5 1/4″ |
| 4 cubit<br>= 96 daktyls | 1855 mm<br>= 6′1″ | 10 2/3 daktyls<br>= 205 mm<br>= 8″ | 3275 mm<br>= 10′ 9 1/4″ | 3005 mm<br>= 9′ 10 1/2″ |
| 5 cubit[51]<br>= 120 daktyls | 2,315 mm<br>= 7′7″ | 13 1/3 daktyls<br>= 255 mm<br>= 10″ | 4115 mm<br>= 13′ 6″ | 3775 mm<br>12′ 4 1/2″ |

At first sight, these dimensions may appear to give machines that when viewed from above are too "square," and first impressions may be right, but only just. These figures are perhaps *very slightly* too square because the formula adopted for the width, which is the length of the bowstring, is a smidgen generous.

The formula for the length of the stock, 16D for a euthytone, 19D for a palintone, makes no allowance for the winding mechanism at the rear end of it. However, in the *Poliorketika*, in an undamaged section of text (A71), Philon says explicitly that a 1-talent stone-thrower has a case just 12 cubits[52] long—six feet shorter than the figure he gives in his *Belopoiika*—which with 4 cubit[53] handspikes *together* add up to about 19D.

> The case of a talent stone-thrower is 12 cubits, the handspike 4 cubits, so there will be no room for those who turn the winch to stand. (Philon *Poliorketika* A 71 [= 85.1-4])

The inconsistency between the two Philon passages is irreconcilable, but it is arbitrary to ignore the *Poliorketika*'s evidence in favor of the *Belopoiika*'s. By taking 19D for the palintone, the figures in the table could stand for stock length plus handspike, if one prefers the dimensions offered by the *Poliorketika*.

We have all got used to seeing elongated diagrams (especially Marsden's), in which the slider is invariably pushed forward, extending the apparent length, and the arms are partially drawn, reducing the apparent width. The problem has been exacerbated recently by a "working length = 30D" factoid. If you follow the instructions given in the *Belopoiika* texts, the dimensions that result are those given in the tables above. Note that the figures for the half-talent palintone are the same as Marsden calculated and reported clearly and concisely (*Historical Development* p. 145, give or take inches, which can be explained by rounding). His figures for the length of the small euthytones are slightly longer than mine because he followed Vitruvius, who gave 19D, whereas I have followed Philon wherever possible (because his account is fuller and clearer), and he says the euthytone length is 16D.

THE FINAL TABLE IS A PURELY SPECULATIVE attempt to estimate the sizes for the all-metal imperial Roman kheiroballistra. This is based on the assumption that the formulae it obeyed did not depart more than necessary from what we call the Philonian standard, since the basic mechanics are the same, and the Romans, like the Greeks, would have been trying to optimize performance on range and impact. The Orşova kheiroballistra shows that the frames could share some exact correspondences with Philonian recommendations (spring height of five times spring width).[54]

TABLE A.3 SIZE ESTIMATES FOR ALL-METAL ROMAN KHEIROBALLISTRA

| NAME | SPRING DIAMETER (d) mm | HYPOTHETICAL SHARP LENGTH (9 d) mm | HYPOTHETICAL STOCK LENGTH (16 d) mm | KAMBESTRION HEIGHT BETWEEN LEVERS (rings + cylinders)[55] mm | PHILONIAN EQUIVALENT |
|---|---|---|---|---|---|
| Elenova | 67 | c. 603 | c. 1,072 | c. 365 (5.4d) | 2-foot |
| Gornea 1,3 | 54 | c. 485 | c. 865 | c. 185 and 195 (3.4d, 3.6d) | 2-span |
| Gornea 2 | 59 | c. 530 | c. 945 | c. 195 (3.3d) | |
| Lyon | 75 | c. 675 | c. 1,200 | c. 300 (4d) | 3-span |
| Orsova | 79 | c. 710 | c. 1,265 | c. 410 (5.1d) | 3-span |
| Sala | 80 | c. 720 | c. 1,280 | c. 410 (5.1d) | 3-span |

The Orsova catapult is the only one with *kamarion* surviving, the arched metal strut that helped hold the frames apart (the *klimakion* or ladder performed most of that task). The width between springs can therefore be estimated for this specimen, though corrosion of the metal and uncertainty about precisely how the struts connected to the frames means that they must remain rough estimates. The Philonian formula for a euthytone catapult with D = 79 mm would give the Orsova catapult a width of about 1160 mm (2.1 times the arm-length, which is 7D). The actual distance between the inner edges of the frames of the Orsova catapult was about 1200 mm. It seems to me, therefore, that the exaggerated width of the kheiroballistra supports the in-swinging arm theory. This particular catapult has approximately the spring height we would expect by Philonian formula, but the smaller caliber Gornea catapults are significantly shorter (about 30 percent). This perhaps reflects an anonymous Roman engineer's appreciation of one or more of the following facts. (i) Catapult mechanics turns on missile weight above all else; (ii) missile weight is absolutely low with most sharps, so that (iii) less spring did the job as well if not better, because a typical sharp would be *so* far below a catapult's true caliber that, in order to avoid damaging it simply by shooting it, the arms could not be withdrawn to full extent. An apparently underpowered machine would work better in these circumstances.

*Appendix Two*

# Known Catapult Remains

THE WASHERS ARE FREQUENTLY the only part of ancient catapults that survive. This is probably because they were made of a lead-rich bronze,* which survives in the ground better than the other materials of which a catapult was typically constructed, viz. wood, iron, and sinew. However, since the washer diameter is the key parameter for construction of the weapon, this is actually the most important piece of the whole machine, and therefore independently enables confident estimates of the size of the weapon from which it came. In a few cases some of the metal plating of the frame, or the metal frame, or some part of it such as a field-frame, survives.

| LOCATION | FRAME MATERIAL | DATE | WASHER qty and avg. diam. (mm) | NOTES |
|---|---|---|---|---|
| Ampurias (Spain)[1] | Wood | c.100 B.C. | 4 x 81 | |
| Auerberg (Germany)[2] | Wood | c. A.D. 75 | 1 x 88 | Molds for casting washers, + 1 miscast thrown away |
| Azaila (Spain) 1[3] (house) | Wood | c. 80 B.C. | 1 x 94 | |
| Azaila 2 (temple) | Wood | c. 80 B.C. | [c. 94 mm] | Estimate from frame |
| Azaila 3 (counterplate) | Wood | c. 80 B.C. | [c.100 mm] | Estimate from counterplate 122 mm |
| Bath (GB)[4] | Wood | c. A.D. 100 | 1 x 38 | |
| Caminreal (Spain)[5] | Wood | c.75 B.C. | 4 x 84 | |
| Cremona (Italy) 1[6] | Wood | A.D. 69 | 4 x 73 | Average; 71-74 mm |
| Cremona 2 | Wood | A.D. 69 | 4 x 89 | |
| Elenova[7] | Metal | c.A.D. 250? | 2 x 67 | Kambestrion, welded, with crank? |
| Elginhaugh (GB) | Wood | c. A.D. 90 | 1 x 35 | Linear ratchet too |
| Ephyra (Greece) 1[8] | Wood | 167 B.C. | 2 x 84 | |
| Ephyra 2 | Wood | 167 B.C. | 3 x 83 | |
| Ephyra 3 | Wood | 167 B.C. | 4 x 136 | |

*Analysis of the Caminreal washers showed an average composition of 58 percent lead, nearly 38 percent copper, and nearly 4 percent tin, various impurities making up the remaining few percent. Vicente et al. Anexo 1, p. 195.

*(continued from previous page)*

| Location | Frame material | Date | Washer qty and avg. diam. (mm) | Notes |
|---|---|---|---|---|
| Ephyra 4 | Wood | 167 B.C. | 4 x 60 | |
| Ephyra 5 | Wood | 167 B.C. | 2 x 75 | |
| Ephyra 6 | Wood | 167 B.C. | 1 x 34 | |
| Ephyra 7 | Wood | 167 B.C. | 2 x 56 | |
| Gornea (Romania) 1[9] | Metal | c. A.D. 380 | 2 x 54 | Kambestrion |
| Gornea 2 | Metal | c. A.D. 380 | 2 x 59 | Kambestrion |
| Gornea 3 | Metal | c. A.D. 380 | 2 x 54 | Kambestrion |
| Hatra (Iraq) 1[10] | Wood | A.D. 241 | 3 x 160 | |
| Hatra 2 | Wood | A.D. 241 | — | Corners, rollers from a cata pult smaller than Hatra 1 |
| Herlheim[11] | Wood | 2d c. A.D.? | 1 x c.80 | |
| Lyon (France)[12] | Metal | A.D. 197 | 2 x 75 | Kambestrion |
| Mahdia (Tunisia) 1[13] | Wood | c. 60 B.C. | 2 x 94 | |
| Mahdia 2 | Wood | c. 60 B.C. | 2 x 72 | |
| Mahdia 3 | Wood | c. 60 B.C. | 2 x 45 | |
| Orsova (Romania)[14] | Metal | c. A.D. 380 | 2 x 79 | Kambestrion |
| Pergamon (Turkey)[15] | Wood | 2d c. B.C. | 1 x 60 | Lost; also mystery bracings found |
| Pityous (Georgia)[16] | Wood | 4th c. A.D. | 1 x 84 | Washer shape most resembles Hatra |
| Sala[17] | Metal | 4th c. A.D. | c. 80 | Kambestrion, cast in one piece |
| Sounion (Greece)[18] | Wood | c. 260 B.C. | 130 | Lost |
| Tanais (Ukraine)[19] | | c. 50 B.C.? | | Not yet published |
| Volubilis (Morocco) 1 | Wood? | 2d-3d c. A.D. | 1 x 41 | This and the next may have come from one machine |
| Volubilis 2[20] | Wood? | 2d-3d c. A.D.. | 1 x 44 | |
| Xanten (Germany)[21] | Wood | 1st c. A.D. | 2 x 43<br>2 x 44 | |
| Zeugma (Meydani) | Wood | 3d c. A.D.? | | Not yet published. |

I. R. Scott identified about sixty surviving Roman daggers, observing that "this is all that remains of several tens of thousands of weapons that must have been in use at any one time" (1985: 160). It is reasonable to assume that more efforts would have been made to recover a catapult than to recover a dagger, so that those unrecovered, which thus entered the archaeological record, the rediscovered subset of which are listed here, represent a smaller proportion of all catapults made than the number of surviving daggers represent the number of daggers made.

Recent reconstructions of catapults based on the archaeological finds at Caminreal, Cremona, Emporion, Mahdia, Lyon, Sala, Orşova, Gornea, and Hatra can (apparently) be found in F. Russo (2004) *L'artiglieria delle legioni romance*, Rome (*non vide*). Reconstructions of catapults described in the technical treatises can be found in Russo 2002, which on plate 8 includes a kind of bracing that is reminiscent of the mysterious Pergamon bracings (Figure 7.6 above).

# Notes

## Chapter One

1. N. Hammond, *Classical Review* 42 (1992) 376.

2. Bill Griffith, pers. com. The sling pouch is published with illustration in R. Birley, *Vindolanda Research Reports NS IV.1: The Weapons*, 1996: 11–14.

3. See K. Rosman et al., "Lead from Carthaginian and Roman Spanish mines isotopically identified in Greenland ice dated from 600 BC to 300 AD," *Environmental Science and Technology* 31 (1997) 3413–3416. Roman pollution is also visible in other environmental archives; see, e.g., M. E. Kylander et al., "Refining the pre-industrial atrmospheric Pb isotope evolution curve in Europe using an 8000 year old peat core from NW Spain," *Earth and Planetary Science Letters* 240 (2005) 467–485.

4. A. V. A. J. Bosman, "Pouring lead in the pouring rain: Making lead slingshot under battle conditions," *JRMES* 6 (1995) 99–103. Of over 500 lead *glandes* found at Velsen 1 in Holland, only one of five types was made in a proper mold. The rest had been poured into sand. Some had then been hammered at one end, producing the characteristic acorn shape; others were left just as they set. Type 1 retained the shape of the fingertip and nail that made the mould. Elsewhere wet clay was sometimes used to make the mold, and it was then easy to inscribe, which resulted in shot with letters or design in relief, Fougères in Daremberg et Saglio II.2.1608–1611.

5. A total of 101 inscribed bullets were found. D. M. Robinson, *Excavations at Olynthus X Metal and Miscellaneous Finds,* 1941, 418–443.

6. Timon's: J.-Y. Empereur, "Collection Paul Canellopoulos (XVII)," *Bulletin de correspondance hellénique* 105 (1981) 537–568 at 555 (a mold for making six such slingshots at a time). Sokrates: G. Manganaro, "Onomastica greca su anelli, pesi da telaio e glandes in Sicilia," *ZPE* 133 (2000) 123–134 at 129, illustrated plate 11, figures 19–21. Another example of ἐποίησῃ (apparently referring to the manufacturer of the shot) is in SEG 27 (1977) 966; see also SEG 31 (1981) 1607. Sertorius: F. Beltrán, "La pietas de Sertorio," *Gerion* 8 (1990) 211–226, esp. 225. World War II: W. C. McDermott, "Glandes plumbeae," *Classical Journal* 38 (1942) 35–37.

7. This was found in the Nile Delta. It was, perhaps, issued by Ptolemy Keraunos (the epithet means thunderbolt).

8. E.g., [Caesar] *Spanish War* 13, 18 (an even more unlikely tale). On the other hand, in his detailed account of Sulla's siege of Athens (*Mithridatic Wars* 5.31), Appian explains that the Romans began to take an interest in the incoming slingshot because of its continuity (compare gravel thrown at a windowpane); the treacherous missiles giving advance warning of the defenders' plans were then passed on to Sulla, who took counter-measures.

9. T. Richardson, "The ballistics of the sling," *Royal Armouries Yearbook* 3: 44. See, e.g., the 61 lead glandes recently found at Miletos and published by P. Weiß in *Archäologischer Anzeiger* 1997: 143–153, and the 11 ceramic glandes found at Montagna di Marzo in Sicily and published by G. Manganaro *Sikelika* 1999: 27–33.

10. An Attic mina weighed 437 grams; a modern pound is 454 grams. Balearics: Diodoros Siculus *Library of History* 19.109.1–2.

11. "Ballistic testing of historical weapons," *Royal Armouries Yearbook* 3: 50–52, 51 table 1.

12. Onasander: *Strategikos* 17. Cotta: Caesar *Gallic War* 5.35. Armor: Arrian *Taktika* 15. Teeth: Plutarch *Cato the Elder* 14, where this is said to have happened to Antiokhos at Thermopylai in 191 B.C. Collapse: *Jewish War* 5.13.3.

13. Thucydides: 6.43. For further discussion, see H. H. Scullard, *The Elephant in the Greek and Roman World*, 1974, 195–197.

14. Ron Rammage's article on http://slinging.org/27.html; accessed 10 March 2005. G. Alsatian's article at the same address adds that after 100 hours of practice one's accuracy can still be quite low (only 10 percent of shots hitting a large folio-size target at 25 meters). Note also his comments on rogue missiles coming down to left or right of his throwing position.

15. Richardson, "Ballistics" fin.

16. Selinous: Diodoros Siculus (hereafter DS) 1.54.7, Gaza: Arrian *Anabasis* 4.2.3, Syrakuse: Livy 24.34, Tyre: DS 17.42. 7, Utica: DS 20.54.4, Same: Livy 38.29.6 (an unusual type of sling, if Livy has understood his source), and Volandum: Tacitus *Annals* 13.38–39.

17. W. K. Pritchett, *The Greek State at War* Part 5, 1991, 33 suggested one-pound slingers; Vegetius 2.23 says all soldiers were trained to throw one-pound stones by hand "because this was a readier method, not requiring a sling." His comment does not preclude the use of a sling for such stones, of course.

18. See, e.g., Caesar *Civil Wars* 2.4.

19. Dion. Halik. *Roman Antiquities* 20.1.7. Polyainos *Stratagems* 1.39.2. Philon: *Poliorketica* A 79–80, C 51, and D 44 (=85.36–41, 94.36–40, and 100.6–11 Th). Garlan has a wonderful MS illustration of the minefield clearance, *Recherches de poliorcétique grecque*, Paris 1974: 399 fig. 70. Shallow water: Polybios 38.19. Against chariots: Vegetius 3.24. See Herodian *History of the Empire After Marcus* 4.15.2–3 for a particularly graphic description of their effects.

20. Maurikios *Strategikon* 12 B 6, 12 C; Anon. *Strategy* 29.25–32.

21. *Strategy* 38.21–27; 17.16–19.

22. Greeks: the Akhaians, Aitolians, Akarnanians, Argives, Elians, Kretans, Malians, Rhodians, Sicilians, and Thessalians; non-Greeks: the Balearic islanders, Kurds, Libyans, Persians, Syrians, and Thracians. Pritchett GSW 5: 54, with further discussion and references. The Aitolians were also attributed with the invention of the weapon, wrongly, since it goes back to the Stone Age.

23. Xenophon *Kyropaideia* 7.4.15; skill: see, e.g., Plato *Laches* 193b or *Laws* 7.794c.

24. Water-worn pebbles: Vegetius 4.8. Spanish Civil War: M. Korfmann, "The sling as a weapon," *Scientific American* 229 (1973) 34–42 at 41.

25. Tyrtaios fr. 11 W Sage 63.

26. *Iliad* xi 265, vii 264, vii 268; Philon *Poliorketika* C 25; Khosroes: Prokopios *Wars* 2.8.28.

27. Plutarch *Life of Aristeides* 19.1. Lazika: Prokopios 8.11.64. Kyropolis is modern Ura-Tyube, between Samarkand and Tashkent.

28. Here he was quoting Homer *Iliad* xi 265, 541.

29. Transient cortical blindness and temporary dysphasia in modern medical terms, J. Lascaratos, "The wounding of Alexander the Great in Cyropolis (329 BC): The first reported case of the syndrome of transient cortical blindness?" *History of Ophthalmology* 42.3 (1997) 283–287.

30. Livy 38.21.

31. Chinese army: G. Rausing, *The Bow: Some Notes on Its Origin and Development*, 1997: 119 (1st edition 1967). World War II and Vietnam: V. Hurley, *Arrows Against Steel: The History of the Bow*, 1975. Murder: Eriksson, Georén & Öström, "Work-place homicide by bow and arrow," *Journal of Forensic Science* 45 (2000) 911–916.

32. Philon *Poliorketika* B 53 = 90.14–24 Th.

33. See H. Soar, *The Crooked Stick*, 2004.

34. Skythian bow, Rausing 1997: 141–142. Kretans, A. M. Snodgrass, *Early Greek Armour and Weapons from the End of the Bronze Age to 600 BC*, 1964: 156, 202.

35. The best treatment of bow construction in antiquity is J. C. Coulston, "Roman archery equipment," in M. Bishop (ed.), *The Production and Distribution of Roman Military Equipment*, BAR S275, Oxford (1985) 220–348, at 248–259.

36. Coulston 1985: 284.

37. Trials: Massey 1994. Philip: DS 16.34.5, Theopompos F 52, Justin 7.6.13–16. Alexander: Arrian 3.30.11 (fibula), Plutarch *Alexander* 45.5 and *Fortune* 9 (quoted above) (tibia).

38. The spoon of Diokles supposedly "found" is a fake, but C. Salazar believes that it is a copy of a real one, *The Treatment of War Wounds in Greco-Roman Antiquity* 2000: 49.

39. Celsus *De medicina* 7.5, Loeb translation with my modifications. The treatment for poison was described in 5.27.

40. E.g., Euippos, spearhead in his jaw for six years, *Epidaurian Miracle Inscriptions* a 95–97; Gorgias of Herakleia, suppurating arrowhead in the lung for eighteen months, ibid. b 55–60; Antikrates of Knidos, spearhead behind his eyes, ibid. b 63–68. By Ammianus' time, the mid-fourth century A.D., some military units had their own medics, e.g., 16.6.2. Note that this man was later appointed head of the night patrol to protect public buildings; medics were still, or again, part-timers.

41. S. T. James and J. H. Taylor, "Parts of Roman artillery projectiles from Qasr Ibrim, Egypt," *Saalburg Jahrbuch* 47 (1994) 93–98.

42. M. I. Rostovtzeff et al., *Excavations at Dura-Europos, Preliminary Report of 6th Season of Work Oct 32 to March* 33, 1936: 456. V. Foley, G. Palmer, and W. Soedel, "The crossbow," *Scientific American* 252.1 (1985) 80–86, have analysis of the Dura bolt aerodynamics and comparison with other projectile types. For the full catalogue of finds, complete and incomplete, see S. James, *Excavations at Dura-Europos 1928–1937, Final Report VII: The Arms and Armour and Other Military Equipment*, 2004: 216–230.

43. *History of Alexander* 7.7.16.

44. Illustrated in R. Wheeler, *Maiden Castle* 1943: plate 53 D, and on display in the local museum.

45. S. Vogel, *Cats' Paws and Catapults: Mechanical Worlds of Nature and People*, 1998: 33.

46. Polybios 27.11.1–7; Livy 42.65.9; Suda s.v. *kestros*. Walbank, *Commentary on Polybios* s.v. thought this must be too long for a sling-delivered missile, and suggested two palms; he is followed by Griffiths (1989), who has the only modern discussion of this weapon known to me, at page 260. Further military innovations at this time are suggested by a fragment of Diodoros Siculus, who says that the Roman commander, the consul L. Aemilius Paullus, invented many new devices in the Third Macedonian War (30.20). This makes sense because, typically, new weapons on one side prompt new defenses amongst those people's enemies—if they fail to develop new defenses then they lose. That, after all, is the raison d'être for weapons development.

47. *SEG* 35.581. No one was immune from friendly fire incidents: Mithridates, King of Pontus and scourge of Rome, was once rammed accidentally by an allied ship from Khios, Appian *Mithridatic Wars* 12.25. He never forgave them, apparently.

48 There is some debate about the length of shaft on these missiles, and the method of delivery, but thrown like a *kestrosphendone* replicas reached 80 meters, whereas when hand-thrown they managed only 20–30 meters; see J. Bennett, "Plumbatae from Pitsunda (Pityus) Georgia," *Journal of Roman Military Equipment Studies* 2 (1991) 59–63.

49 Inadequate: Dion. Halik. 20.2.5–6. The devices like cranes that were mounted upon these wagons, however, did work, and already had a long history, going back certainly to c. 360 B.C., and possibly as far as 427. Masts are mentioned by Aineias Taktikos *How to Survive Under Siege* 32. 1, written c. 360; Timotheus reputedly used them at Torone in 364, and Nikias may have used them against Minoa (an island off Nisaia, Megara) in 427 (Thuc. 3.51) and Syrakuse in 415 (Thuc. 7.25), as A. H. Jackson suggested *JHS* 97 (1977) 200.

50. *IG* II2 1487. Kestros occurs on line 94.

51. Pritchett *GSW* vol 5: 38. There are at least forty surviving inscriptions that mention a *kestrophulax* officer, *IG* II2 1993, 2097, 2099, 2102, 2105, for example. At the beginning of the second century A.D. they seem briefly to have been called *kestrophoroi*, kestros-bearers, *IG* II2 2021, unless this refers specifically to hand-thrown or sling-delivered versions of the missile.

## Chapter Two

1. The two readings of *quadriremes* in the Loeb text and translation (here and in 42.2 below) were suggested by Wesseling; the manuscript has *triremes* and quinquiremes. Some scholars, e.g., B. Caven *Dionysius I* (1990) 96 think the MS reading should be retained, leaving quadriremes out of the picture; I agree. There is no good reason to change the text. A trireme is a "three-er"; a quinquereme is a "five-er."

2. καταπέλται παντοîοι κὰι τῶν ᾿άλλων βελῶν πολύς τις ἀριθμός 14.43.3.

3. Vim and vigor: DS 14.43.1. Most of the problems with rival claims and alternative histories of the origin of the catapult were first identified and the arguments formulated by Marsden, *Greek and Roman Artillery* vol. 1, 1969: 48–56 and Garlan, *Recherches de*

*poliorcétique grecque* 1974: 164–166.

4. Also known (in Kings 2.15.1–7) as Azariah.

5. Josephus *Jewish War* 3.7.

6. The Right Reverend J. B. Taylor, 1 and 2 Chronicles, in *The Lion Handbook to the Bible*, 3rd ed., *Israel's History,* 1999: 308.

7. Marsden, 1969: 52–53; Garlan, 1974: 164–165, and his review pp. 19–153, over 130 pages, of hundreds of sieges before Motya, at which there is *no sign* of artillery.

8. The inscription is dated 363 B.C.

9. F. G. Maier and V. Karageorgis, *Paphos*, 1984.

10. "The city walls of Phokaia," *Revue des Études Anciennes* 96 (1994) 77–110; the stone is discussed briefly on p. 90, and there are four photographs of it, 23 and 25–27. The date is discussed pp. 90–92.

11. "L'apparition des premiers engines balistiques dans le monde grec et hellénisé: un état de la question," *Revue des Études Anciennes* 102 (2000) 5–26. Kingsley's notion that the invention of the catapult should be shifted back vaguely into the 420s because he would identify as one two people (probably in fact, three) called Zopyros (a very common name in antiquity) warrants as little discussion as he offers of the history of the catapult. It is a great pity that Campbell repeated this nonsense, apparently in a late edit to the text of his book, whose title at least was not changed to suit, *Greek and Roman Artillery 399 B.C.–A.D. 363*.

12. Aineias Taktikos 32.5–6. Philon *Poliorketika* C 8–10, = 91.25–38 Th. See Garlan 1974: 378 figure 64 for illustration of how it might have worked.

13. Tacitus *Histories* 2.22. Thucydides *Peloponnesian War* 2.76.

14. Lysias *Oration* 33. Isokrates *Epistle 1 (Dionysios)*; see also his letter *To Philip 1* 11, and *To Arkhidamas* 19.

15. Inscriptions in translation conveniently in Rhodes and Osborne, 10, 33, and 34.

16. P. K. Hoch, "Emigrés in science and technology transfer," *Physical Technology* 17 (1986) 225–229, quote from 229.

17. On slavery in general the best book is O. Patterson, *Slavery and Social Death,* 1982. On slavery in ancient Greece see Y. Garlan 1988, N. Fisher 1993, and K. Bradley and P. Cartledge (eds.), *The Cambridge World History of Slavery* vol. 1, forthcoming.

18. For example, in the medical context, an outstanding healer called Demokedes was lured from Aigina to Athens and from Athens to Samos for higher (large) sums each time: the Aiginetans paid him 6,000 drachmai, the Athenians 10,000 drachmai, and the Samians 12,000 drachmai. The story is told in Herodotos *Histories* 3.131. 6,000 drachmai was a huge sum, equivalent to thirty-three years' worth of jurors' wages.

19. The first European handguns were also rather cumbersome, requiring some support at the front end, which was provided initially by a hook (hence the name *Hackenbüchse*) to steady the weapon on the battlement. An early Spanish version, known as the *escopeta*, was about 5 feet long and weighed about 30 lbs. (13 1/2 kg), and a larger variant called an *espingarda* used a fork to support the barrel. Ponting 2005: 116.

20. Coulston 1985: 259 and nn. 203–205.

21. Vicente et al. 1997: 186.

22. DS 14.34; 14.78.5.

23. Dionysios' offer of high wages to free artisans to encourage them to move to Syrakuse may have its closest analogy in the recruitment of doctors. We know that some cities offered large retainers to healers to move to their city (see n. 18 above). The fee served to bring a skilled person to a specific place, so that the people who lived there *could* employ that person should they so wish, and if they could afford to do so. The publicly paid fee was a golden handshake rather than a salary; the healer still charged his or her patients fees. So, in Dionysios' recruitment of weapons-makers, some of his effort might have been directed simply to trying to increase the number of artisans in Syrakuse that the self-arming among his forces could employ.

24. Model to follow: DS 14.41.4. Most significant causal factor: 14.42.1. A. Pacey, *Technology in World Civilization,* 1991, develops the notion of innovation by stimulation.

25. Caven 1990: 95. Ship-sheds: DS 14.42.5. Hamilton-Smith and Corlett, "Some possible technical features of a Salamis *trieres*," in S. Fisher (ed.), *Innovation in Shipping and Trade*, 1989, 9–36, at 10.

26. Supplies: DS 14.47.7. Mole: DS 14.48.3. The section demolished was apparently at about 550 meters from the mainland and 1,200 meters from the island. At least, the survey undertaken in the mid-1950s and early 1960s could not follow the causeway at this point: "it disappears at a depth below sea-level of about one metre; so the modern rise of water-level and consequent change of shore line cannot be supposed to account for its disappearance," E. Macnamara, *Motya* 1964: 88. Alternatively, the breach could have been made in the section between 400 meters and 565 meters from the island; that section is very dilapidated and has two narrow breaches, ibid., 101. Macnamara seems unnecessarily diffident about identifying the breach. There is no sign of any other construction crossing the lagoon, and many arrowheads were found around the north gate (to which the causeway leads) and the stone there shows signs of having been burnt, so at least one of Dionysios' "moles" was probably repair of the demolished section.

27. The symptoms as described by Diodoros 14.71 are consistent with smallpox, according to H. Zinsser *Rats, Lice and History* 124–127, but Grmek asserts that the Greeks did not know smallpox in Hippokrates' time (the Hippocratic Corpus was written over centuries, up until about the second century) *Diseases in the Ancient Greek World* 1991: 89.

28. Snodgrass 1964b: 127. My colleague David Gill kindly drew my attention to and provided me with a copy of this publication. More recently the Italians have undertaken study of the fortifications of Motya; for a report of their findings see A. Ciasca, "Fortificazioni di Mozia (Sicilia): dati tecnici e proposta preliminare di periodizzazione' in Leriche and Tréziny (1986): 221–227.

29. Tenth-century treatise: Anon *Tactics* 31.14–23. New type of arrowhead: Snodgrass 1964b: 128, J. Whittaker, *Motya* 1921: 340. Forged iron copies: Snodgrass 1964b: 130.

30 Hammond *History of Greece* 3rd ed. 1986: 480.

## Chapter Three

1. Likewise, in the history of science the giants seem to achieve everything. "A critical

history of technology would show how little any of the inventions of the eighteenth century were the work of one single individual," K. Marx *Capital* 1.392 n. 2, Everyman edition, cited by Thompson 1952: 81 n. 2. All inventions have a history and predecessors, as all scientific giants stand on the shoulders of others.

2. See A. W. McNicoll, *Hellenistic Fortifications from the Aegean to the Euphrates*, with revisions and an additional chapter by N. P. Milner, 1997:73, although he does not consider the continued effectiveness of *gastraphetai* as well in such circumstances.

3. Bomlitz: O. Mußmann, "Gunpowder production in the electorate and the kingdom of Hanover," in Buchanan (ed.), *Gunpowder: The History of an International Technology* (1996) 329–350. Waltham Abbey: P. Everson and W. Cocroft, "The royal gunpowder factory at Waltham Abbey: the field archaeology of gunpowder manufacture," in Buchanan ed. (1996) 377–394 at 384.

4. *Belopoiika* 81.1–2.

5. For example, shot-firing in mines was tried in Schemnitz in 1627 and was demonstrably spread from there by miners who were expert shotfirers to Freiberg, Nasafjäll (Sweden), Ecton (England), and many other places, Wild, 1996: 209. Similarly, staff from the Ballincollig gunpowder mill near Cork appear on the payroll of the DuPont mills in Delaware, Kelleher 1996: 364. In 1633, Chinese gun-founders taught the Portuguese how to make cast-iron guns, Ponting 2005: 204. Occasionally the vector may be the craftsmen of a particular culture: in the sixth century A.D. people known as the Sabiri were allies to the Romans as well as the Persians, and after those serving the Romans employed a type of ram new to both at a battle between them, the Persian Mermoeroes commanded those Sabiri who worked for him to build the same type, Prokopios 8.14.4–5, with 8.11.27–31. For general discussion of technology transfer, see Chapter 8. For transfer of a technique from one *medium* to another, see, e.g., Lapatin 1997, who argues that Pheidias in the fifth century B.C. applied bronze-working techniques to ivory-working techniques in order to create his huge chryselephantine statues.

6. Plutarch *Moralia* 191e and 219a. Ariosto surely had this in mind when he transferred the sentiment to the gun and wrote "through thee is martial glory lost" (1530s).

7. *Wujing zongyao* trans. Needham 1994: 122; I have changed the word order.

8. See, for example, DS 14.10, 62, 70; 15.47; Xenophon *Hellenika* 7.1.28–32.

9. Xenophon *Hellenika* 7.1.28–32. In the same way, towers later built for bow catapults could not, or could not easily, be refurbished to carry torsion catapults, but had to be rebuilt. Winter describes the problem as "the difficulty, not to say impossibility, of retro-fitting towers of the pre-torsion era with torsion artillery, especially stone-throwers," 1997: 261.

10. H. Lauter, "Some remarks on fortified settlements in the Attic countryside" in van de Maele and Fossey 1992: 80. It is principally for the reason that it is completely overlooked by a ridge 150 m away, but also because it has thin and high walls, that I would date the building of the tower at Isium (Myra E., Lykia) to the early (non-torsion) artillery period. Ober would seem to think it belongs here too, citing other similarities with "first generation" artillery towers, 1987: 601 n. 68. Isium has embrasures to all sides of chambers just under 5 m square on the second and third floors. It could probably have housed

catapults up to the size of a two- or three-cubit sharp-caster or 10-mina bow stone-thrower (see Appendix 1, "The Calibrating Formulae") in each of them.

11. A. W. McNicoll, *Hellenistic Fortifications from the Aegean to the Euphrates*, 1997, gives a full and excellent discussion of the issues in chapter 1.

12. J.-P. Adam, *L'architecture militaire grecque*, 1982, has good photographs and diagrams of the Messenian fortifications, and plates 145 and 146 are especially good for the embrasures, showing the small holes to either side which are thought to have allowed rods to pass through the walls to open wooden shutters mounted on the outside.

13. Philon *Poliorketika* A 41, = 82.50–83.3 Th. Lawrence *Greek Aims* 1979: 423, Fossey (1992) "The development of some defensive networks in eastern central Greece during the classical period" at 114–117 and 122–13l, and McInerney, *Folds of Parnassos* Appendix 3, pp. 340–354, argue that the seventeen other towers of Phokis that are built in a similar style but survive in a poorer state of repair are a system of defense, and should all be dated to the Third Sacred War or earlier (356–346) and not after Khaironeia (338). This point has been made in a number of other cases: it is argued, for example, with respect to the fortifications of Attike, Aitolia, Boiotia, and Megara.

14. This is the same place that Marsden calls Eleutherai, 1969: 136 n.2, and Adam sometimes calls by its modern Greek name Yphtokastro, 1992: 14 with plan on p. 15 and photo of walls with towers plate 3. Close-up photo in Adam 1982: 216 pl. 252. There are diagrams of most of the towers here discussed in Marsden 1969: 155–163.

15. Adam calls this Oinoi; he has excellent photos and diagram in 1982: 72–73, plates 97 and 98 (the latter showing the interior with holes for floor joists to three different floors) and figure 37a.

16. "Le réseau mégarien de défense territoriale contre l'Attique à l'époque classique (Ve et IV s. av. J-C)" in Maele and Fossey eds. 93–107, at 98–99.

17. Panakton invoice 1992–300, published by Mark Munn in "The first excavations at Panakton on the Attic-Boiotian frontier," *Boiotia Antiqua* 6, ed. J. M. Fossey (1996) 47–58 at 52.

18. J. Ober, "Early artillery towers," *American Journal of Archaeology* 91 (1987) 569–604. F. E. Winter, "The use of artillery in fourth-century and Hellenistic towers," *Échos du Monde Classique/Classical Views* 16 (1997) 247–292 at 258 agrees that Aigosthena is more likely to have held bow catapults than torsion, because of the arrangement of windows and size of the chamber.

19. Fossey: 1992: 116, 123. Khalkis in Aitolia, near Naupaktos, 1969: 129 n. 4. Poseidonia/Paestum, 1969: 133. Marsden would date the older part of the tower to 330–300 B.C. Its construction might fit with the Latin or Samnite Wars. There are photos of the Paestum tower (which has been lived in and the roof maintained) in Adam 1982: 246–247, plates 279 and 280.

20. Ober 1992: 148–149. Pinara is on the coast of Turkey, east of Halikarnassos; there is a photo of this relief in Adam 1982: 38 figure 7. Winter 1992: 192–193.

21. Inscription: R&O 43. One would choose as a proxenos either one of the most prominent citizens in his home state, or one who is always available at home in the foreign state. Someone who would be abroad for months at a time would be of no use to a

Boiotian in trouble in Karthage, so this man is unlikely to have been a Karthaginian merchant, as some have supposed.

22. The principal difference between the ancient and the modern analogues is that in antiquity the proxenos was a citizen of the state where he lived and represented foreigners' interests; today he or she is a foreigner where he or she lives and a citizen of the state he or she represents.

23. G. Glotz, *Melanges N Iorga* 331–339. Roesch *REG* 97 (1984) 45–60. Adam: 1992 at 22–27, with figure 1.16 and plate 8. Alexander quote: Curtius 7.7.16. Kiernan: 2004: 27.

24. Ober 1992: 166.

25. Plutarch *Sayings of Kings and Commanders* Timotheos 2, = *Moralia* 187c. The reckless general in question was Khares apparently, Plut. *Life of Pelopidas* 2 (278d).

26. 15 (*Liberty of the Rhodians*) 9.

27. *RE* 6A.2 (1937) 1329. Diodoros Siculus does not mention the first siege and does not mention artillery at the second siege, 16.21. But according to him Epaminondas of Thebes had something to do with the Social War (15.79.1), and so did Mausolos, satrap of Karia (16.7.3). If the story is connected with the 350s rather than the 360s, then these people are potential vectors for the catapult too.

28. *IG* II2 1422.8–9; usually dated 371/370, is dated 363/362 by Cole 1981. They were apparently still there in storage in the 350s (*IG* II2 120.36–37), and they and many more (1,800; Garlan 1,900 [1974: 216]), were still there in 320: *IG* II2 1469B.77–80. Ancient temples were stuffed with war material taken as booty; the presence of catapult bits is not in itself evidence that a temple deposit was a functional military store.

29. L. La Follette, "The Chalkotheke on the Athenian Acropolis," *Hesperia* 55 (1986) 75–87.

30. Inscriptions: R&O 33, 34. Cole 1981 suggested they may have been obtained in 373 as part of the war booty when the Athenians caught some of Dionysios' triremes (DS 15.47.7). This assumes that *gastraphetai* were carried on warships, which they were not at Motya, although that was some years earlier. See also Marsden 1969: 65–66.

31. 8.8.8, cited by D. Whitehead, *Aineias the Tactician* How to Survive Under Siege, translated with an historical commentary and introduction, 2002 on §32.

32. *Mechanics* 11, translated in Irby-Massie and Keyser 2002: 164.

33. About 150 years later, around 200 B.C., Philon recommended that holes be made right through the wall from the inside in order to employ counter-rams against siege towers and similar devices, *Poliorketika* C 14–17.

34. *Hellenika* 7.5.18–27. Diodoros 15.84–89 tells the story of the battle more briefly.

35. J. McInerney, *The Folds of Parnassos* (1999), is a history of Phokis and its people.

36. Diodoros 16.34.5. Strabo *Geography* Book 7 fr. 22.

37. *Strategemata* 2.38.2. See also Hammond 1992: 376.

38. Marsden 1969: 59.

39. The staff sling and monagkon had a leather sling pouch, the hand sling one made of flax or hair, according to Vegetius 3.14. A relatively large (75 mm wide), relatively very

thick, molded leather sling pouch was apparently found by C. van Driel at Vindolanda, whence other artillery (and leather) finds have been made; Griffiths 1994. It has not yet been properly published, its whereabouts are unknown, and Dr. van Driel has not responded constructively to requests for information on it.

40. N. Hammond *History of Greece* 1986: 296.

41. Atheniaos Mekhanikos 10W; Vitruvius 10.13.3.

42. V. Nutton *Ancient Medicine* 2004: 60.

43. *IG* II2 3234. The word is καταπελταφέτας, one who discharges a catapult, not one who makes it.

44. Pergamon is first mentioned in Greek sources in 401 B.C.; it becomes great when the Attalid kings, successors of Alexander, make it their capital from 302. Abydos, whence came the Isidoros who built an enormous bow-catapult of about forty pounds caliber that is described by Biton, is about eighty miles north of Pergamon. Isidoros was said to have built his monster *gastraphetes* in Thessalonike, which was not founded until 316/315. So bow catapults, sometimes thought of as "pre"-torsion catapults, were clearly still being built *at least* twenty-five years after the date at which we think the torsion catapult was invented. Non-torsion stone-throwers of Isidoros' type would have certain advantages in some circumstances, as Marsden recognized (1971: 83 n. 22), and there is an argument for putting Isidoros much later, in the early second century, c. 190 B.C.

45. There are some inscriptions with translation and commentary (including a good overview of the *ephebeia* at this time) in R&O 89, dated to 332 B.C.

46. Briareos: Timokles *fr.* 12.4–5, Edmonds 2.610–611. Inscription: *IG* II2 1627 B 328–41. The bits reappear in later inventories into the 320s. Text and a different translation in Marsden 1969: 57 with discussion on 58. A text and English translation of a larger part of another very similar inscription, which gives a good idea of the whole and of the type, is R&O 100.

47. Marsden 1969: 60.

48. See R. Payne-Gallwey, *The Book of the Crossbow* (1903) 129–130.

49. ξύλα σεσωληνισμένα ἰσομεγέθη τῶν σαγιττῶν ἐν σχήματι καλάμου κατὰ τὸ μέσον ἐσχισμένου

Cod. Ambros. Gr. B 119 sup. (139), quoted in G. T. Dennis, "Flies, mice, and the Byzantine crossbow," *Byzantine and Modern Greek Studies* 7 (1981) 2 n. 8. "Mice" in ancient sources can mean any rodent, since they were not normally distinguished from one another.

50. Bürgerbibliothek, Berne, MS cod. 120/II, 40, reproduced in D. Nicolle, "Armes et armures dans les épopées des croisades," *Les Épopées de la Croisade* (Stuttgart, 1987) 17–34, at 34 figure 51.

51. Paul: 6.88.2. DS 31.38. Musket: Dennis 1981: 5.

52. Artilleryman: J. C. Balty, "Apamea in Syria in the second and third centuries AD," *JRS* 78 (1988) 91–104. The same nomenclature problem bedevils the history of weapons in China, and we are no better, using "gun" for a huge variety of devices shooting anything from potatoes to shells and "missile" for an equally diverse range.

53. J. Scarborough, "Nicander's toxicology II," *Pharmacy in History* 21 (1979) 3–34 at 17.

54. The only thing remotely resembling a tube on an onager is the washer, and the normal term for the washer is **χοίνικις**, *khoinikis*.

55. Catapult building came under the remit of architects as late as Vitruvius' time (Fleury 1994: 22–26), so application of the same terms in the two domains is to be expected.

56. Ktesibios 78.3, "withdrawal-rest" of Marsden's translation. It is the curved bar at the back of the machine on which the catapultist pushed with torso.

57. *IG* II2 1467 B column 2, 48–56. Again conveniently in Marsden 1969: 56–57. Not to be confused with *IG* II2 1422, mentioned above.

58. Siege: Diodoros Siculus 17.24.6. Lexicographer: Tzetzes *Khiliades* 11.381 line 608.

59. Athenians on Persian side: DS 17.25.6. Tower: 17.26.6. Arrian: *Anabasis* 1.22.2, 1.23.2. Neoptolemos: DS 17.24.4–25.5 and Arrian 1.20.10.

60. For example, all Hornblower (*Mausolus* 1982: 7, 256, 273) has to say about Magnesia on the Maeander is that its gardeners were mentioned in a letter from Persian King to satrap, and it was the refuge (and, we might add, penultimate resting place) of Themistokles, architect of the Greek victory over the Persians at the battle of Salamis, who was forced into exile by his ungrateful compatriots.

61. Priene's fortifications, thought by McNicoll to date to shortly after Alexander's passing through in 334, include in the surviving towers one embrasure that McNicoll thought suitably sized for "only a small arrow-shooter." Perhaps small sharp-casters, belly-bow descendants of Dionysios' original design, were standard by this time. The walls of Kolophon, probably built between 313 and 305, may have been designed for similar machines. Priene: McNicoll 1997: 52 and plate 23. Kolophon: McNicoll 1997: 70 with n. 176, and 68 for the date.

62. Little towns: Polybios 1.53.11. Ephebes: Ma, 2000: 347, on an early third-century inscription from Koressos. Ma's argument is that during the centuries of "big war" between Alexander's successors, "small war" between small communities carried on much as it had done in the classical period, and with this I wholly concur. However, I disagree with his view that "developments in artillery technology and fortifications would favour the defenders" (2000: 353) for reasons that will become clear in the next chapter. Small war armaments were obsolete when faced with big war technologies.

63. As noticed by A. Wilkins, 2000: 100, his translation. The reference is to Arrian's *Taktika* 43.1.

64. Elginhaugh: Hanson, forthcoming. Reliefs: Baatz 1994 and 1999. Pliny: *NH* 8.26.

65. Van Wees 2004: 239, citing Herodotos 7.102.1.

## Chapter Four

1. "**ΠΑΛΙΝΤΟΝΟΝ** and **ΕΥΘΥΤΟΝΟΝ**," *Classical Quarterly* 14 (1920) 82–86, at 85. Garlan 1974 unfortunately has a typographic error in his reference to this paper, giving *CJ* for *CQ*.

2. Flavio Russo, part of whose latest work arrived just before this book went to press, entertains a similar idea: see esp. pp. 53–54 in *L'artiglieria delle legioni romance*, 2004.

3. E.g., "la baliste au contraire possède deux demi-cadres," P. Fleury, *La mécanique de Vitruve*, 1993: 229 (contrasting the palintone ballista with the euthytone scorpion).

4. The difficulty of interpreting the word "palintonos" is indicated by the variety of proposals that have been made: Garlan, *Recherches*, 1974: 223 n. 2, summarizes seven suggestions. It naturally means something like "opposing spring" rather than the linguistically forced "bent-spring," as Barker pointed out. He preferred the translation "counter-sprung."

5. Marsden dismissed Barker's argument as "ingenious but mistaken" (1971: 23 n. 1) but failed to give any reason why he thought it was mistaken.

6. Lienhard 2000: 171–174 with examples, p. 198 for the example of the telegraph—early versions had one wire for each letter of the alphabet, instead of one wire and a code for each letter—and pp. 203–205 for Linotype versus the Paige compositor.

7. Some early guns, both in the Islamic world and the European, also launched both shot and "arrows." Islamic *midfa*: Ponting 2005: 82; European "quarrel-guns": Morin 2002: 59–60 and figure V-8 (surviving specimens of "gun-quarrels" from Damascus, now in Paris) and Ponting p. 108 figure 12.

8. κα[ταπάλτ]ας διπήχεις τριχοτ[όνους ἐ]ντελεῖς ||| κατα[πάλτας] διπήχεις τριχο–τό[νους οὐχ] ὑγιεῖς οὐδ' ἐντελε[ῖς |||] ἑτέρους δύο κατα[πάλτας τριχοτό[νους οὐχ ὑγιεῖ]ς οὐδ' ἐντελεῖς ...τρ[ιχ]οτό[νους]

9. Compare the search for reliable gunpowder: the Surveyor General of the Board of Ordnance from 1750 to 1782 thought that if Henry Strachey just kept on experimenting with different compositions of raw materials and durations of grinding "etcetera," he might "hit on the right method as others have done," quoted by Buchanan 1996: 244. As she explains (pp. 244–249), it is likely in this case that the firm producing the gunpowder in question had production (and quality control) methods that were more advanced than were those of the Board's preferred suppliers, and probably more reliable, at that time, than were the Board's methods for testing that quality.

10. The earliest evidence for the use of gunpowder as a propellant in a metal-barrel gun is 1128, Forage 1996: 511, yet popular tales would have the gun invented by one Niger ("Black") Berchtoldus in 1380, Kramer 1996: 55. It was hundreds of years more before this invention ousted longbows and crossbows as the range weapon of choice on the battlefield. Similarly, after the discovery of nitrocellulose in 1832–1833, it took fifty-two years of development, by a variety of European powers, to make a smokeless powder good enough to replace gunpowder, Vernidub 1996. (The smoke from gunpowder hindered visibility on the battlefield.)

11. For example, Lavoisier's colleague Le Tors died experimenting with chlorate explosives in 1788, Bret 1996: 265. King James II of Scotland was killed by one of his own guns when it exploded during the siege of Roxburgh in 1460, Ponting 2005: 115.

12. Equipment: DS 16.3.1. Max Weber elaborated the relationship between troops and their leader and the differences that result when troops are self-supplying; *Economy and Society*, "Non-legitimate domination," Part 2, Chapter 16, ii, "The occidental city," 8.

13. Polybios: 18.29.

14. Polyainos *Stratagemata* 4.2.10.

15. There are some very good pictures, plans, and descriptions on the Web, at the Perseus Web site, http://www.perseus.tufts.edu/.

16. The worry about fifth columnists is very evident in another author of this period, Aineias Taktikos, in his work *How to Survive Under Siege*.

17. Reported by Plutarch, *Sayings of Kings and Commanders* Philip 14 (*Moralia* 178b).

18. *Stratagems* 4.2.20.

19. Petroboloi: DS 20.48.3. Hand-thrown stones: DS 19.109.2.

20. The smallest yet found is one of the Ephyra washers which has a 34 mm spring hole diameter. It dates from before 167 B.C. when this fort or fortified farmhouse was destroyed; D. Baatz, "Hellenistische Katapulte aus Ephyra," *Mitt DAI Athen* 97 (1982) 211–233, reprinted in Baatz 1994. This corresponds to a one-foot caliber sharp-caster. Currently seven of around 40 mm caliber are known.

21. Sackcloth: DS 19.106.4. Aineias *How to Survive Under Siege* 32.3. Philon *Pol*. C 3, and D 10 and 17, respectively, = 91.3–9, 97.24–25, 98.7–8 Th.

22. See Arrian 2.18–24; DS 17.40–46; Curtius 4.2–4; see also (though catapults are not mentioned in these sources) Plutarch *Life of Alexander* 24–25; and Justin 11.10.10. Generally, DS and Curtius are better than Arrian on Tyrian countermeasures. Arrian, despite being the best source overall on Alexander, is sometimes too favorable to him, especially when Alexander made a mistake or got the worst of some engagement. Arrian has a tendency to launder such blemishes out of Alexander's career.

23. DS 17.46.3, Curtius 4.4.12. The first good evidence for wall-destroying stone-throwers is Demetrios Poliorketes' use in 307 against Salamis in Kypros, and then in 305/4 against Rhodes (DS 20.86–88; 20.95).

24. DS 18.51.1. P. T. Keyser, "The use of artillery by Philip II and Alexander the Great," *Ancient World* 25 (1994) 27–59 at 45.

25. France made this point for medieval history, *Western Warfare in the Age of the Crusades 1000–1300*, 1999.

26. Arrian 1.5.5–6.8, also DS 17.8.1 and Plutarch *Life of Alexander* 11.5.

27. 329 B.C., crossing the river Iaxartes in Sogdiana, also reported by Arrian 4.4.

28. Tacitus *Histories* 2.34. Cassius Dio 71.3.1. Further examples are cited in the following chapters.

## Chapter Five

1. For example, arbalet, arbalest, arbalist, arbalista, arbaliste, arblast, arbelaste, alablaste, alblast. R. Payne-Gallwey, *The Book of the Crossbow,* 1995 (1st ed., 1903) 3.

2. "Katapulte und mechanische Handwaffen des spätrömischen Heeres," *Journal of Roman Military Equipment Studies* 10 (1999) 5–19, esp. 12.

3. The Ephyra washers date from before 167 B.C. when this fort or fortified farmhouse or necromanteion was destroyed; D. Baatz, "Hellenistische Katapulte aus Ephyra," *Mitt DAI Athen* 97 (1982) 211–233, reprinted in Baatz, *Bauten und Katapulte des römischen Heeres*, (Stuttgart, 1994).

4. L. Allason-Jones, "Small objects," in W. S. Hanson, *Elginhaugh: A Flavian Fort and Its Annexe,* 2007. I wish to thank Professor Hanson for providing an advance copy of this publication.

5. Bath: 38 mm, Mahdia 3: 45 mm, Volubilis 1 and 2: 41 and 44 mm, Xanten: c. 40 mm. Bath: D. Baatz in B. Cunliffe, *The Temple of Sulis Minerva at Bath 2,* 1988, 7–9. Mahdia: Baatz, "Katapultteile aus dem Schiffswrack von Mahdia (Tunesien)," *Archäologischer Anzeiger* (1985) 679–691, reprinted in Baatz 1994. Volubilis: C. Boube-Piccot, *Les bronzes antiques du Maroc IV*, 1994, 188–197. Xanten: H.-J. Schalles, "Eine frühkaiserzeitliche Torsionswaffe aus der Kiesgrube Xanten-Wardt," in H. G. Horn et al., *Archäologie in Nordrhein-Westfalen,* 2005, 378–381. I owe the last reference to Dr. Jon Coulston. The internal washer diameter is not given in this interim report of a recent find, but the frame size is (220 x 260 mm), and if it was built approximately to the formulae given by Vitruvius, the washer diameter would be about 40 mm. Full publication of this important find will hopefully occur in the near future.

6. Initial publication, identifying it as catapult ammunition: M. I. Rostovtzeff, A. R. Bellinger, C. Hopkins, and C. B. Welles, *The Excavations at Dura-Europos: Preliminary Report of Sixth Season of Work 1932–33*, 1936, 455–456 and plate XXIV, 2, 3. Aerodynamic: Foley, Palmer, and Soedel, "The crossbow," *Scientific American* 252.1 (1985) 80–86 record the results of tests in a wind tunnel.

7. S. James, *Excavations at Dura-Europos 1928–1937: Final Report VII: The Arms and Armour, and Other Military Equipment*, 2004.

8. M. Vickers, *Oxford Classical Dictionary*, 3rd edition, s.v. "Measures."

9. Chapters 33, 34, 37–40.

10. W. Griffiths, "The hand-thrown stone," *Arbeia* 1 (1992) 1–11.

11. Saguntum: Livy 21.11.7–10; Capua: Livy 26.6.4–5.

12. Avaricum: Caesar *Gallic War* 7.25. See also 8.41.

13. Plutarch *Sayings of Kings and Commanders* Timotheos 2 = *Moralia* 187c.

14. Given the date, this would have been a tension (or bow) catapult, a *gastraphetes*.

15. E.g., Tacitus *Histories* 4.29; Amm. 19.1.7; Prok. 5.21.14–18, 5.23.9–12. Caesar *Civil Wars* 3.22 offers another probable example: Milo, when talking with the praetor Q. Pedius during the siege of Cosa, was struck and killed by a stone from the wall. While it is indisputable that slingshot could kill, for *this* to be slingshot requires a concatenation of three circumstances, each in itself unlikely: 1. two experienced men were standing close enough to the wall of a besieged town to be in effective range of slingshot; 2. the slinger was deadly accurate; 3. the slingshot hit somewhere on his body to kill him without penetration (see below).

16. [Caesar] *African War* 29. See, e.g., Livy 35.48, Tacitus *Histories* 1.79, *Annals* 3.43, Curtius 4.35, Ammianus 16.10.8; Prokopios 8.11.48 (the immobilized leader is in this case a Roman, Bessas).

17. Emperors in iron: e.g., Otho, Tacitus *Histories* 2.11. Trajan targeted: Cassius Dio 68.31.3.

18. E.g., Dittenberger *Sylloge Inscriptionum Graecarum*, 3rd ed., vol. 3, 958 ll. 23–32;

ἐξάγειν εἰς μελέτην ἀκοντισμοῦ καὶ τοξικῆς καὶ καταπαλταφεσιας τρὶς τοῦ μηνός. I owe this reference to Serafina Cuomo, who also cites 1061 in her forthcoming book on ancient technology.

19. *Poliorketika* 188.6: ὁι λιθοβόλοι μονάγκωνες οὕς τινες σφενδόνας καλοῦσιν

20. Livy 38.29.6: *glans . . . velut nervo missa exécutiatur*. The present hypothesis explains this statement by Livy, which has hitherto been found baffling. That σφενδόνη was a term with a variety of applications is suggested also by its use for a sort of sling employed to lift cargoes off ships; W. Wyse, "On the meaning of σφενδόνη in Aesch. *Ag.* 997," *Classical Review* 14 (1900) 5. "Sling" in English is the name for a variety of strap-like devices used in the lifting industry.

21. C. Foss, "A bullet of Tissaphernes," *Journal of Hellenic Studies* 95 (1975) 25–30. Another candidate for very early lead slingshot is one found at Miletos inscribed "Dromas Ainian," SEG 47. 1635. Dromas was commander of the Ainianes sometime after Kyros the Younger's bid for power in 401; Xen. *Anab.* 1.2.6. The current latest fixed point for *glandes* is a slingshot inscribed with "Legio II Italica," which indicates a date after A.D. 165; Griffiths 1989: 271.

22. There are five so inscribed among those recorded in CIL I.22 I 847–888, and one LEG XI. It is perhaps not coincidental that the fifteenth legion was famous for its "artillery."

23. W. Griffiths, "The sling and its place in the Roman Imperial Army," in C. van Driel-Murray (ed.), *Roman Military Equipment: The Sources of Evidence* BAR IS 476 (1989) 255–279, at 259 and 275 n. 22. He cites top weight of 140 g, but most museum specimens have not been weighed, and a medium sized one in the BM that I weighed was 155 g. Two other specimens were much larger, at 439 g and 500 g—the first being almost exactly a mina; it is on display at the museum.

24. Τινὰ δε καὶ κινούμενας διὰ γιγγλύμου τὰς ἀκίδας ἔχουσι συναγομένας, αἵτινες ἐν τῆι ἐξολκῆι ἐξαπλούμεναι κωλύουσιν ἐξέλκεσθαι τὸ βέλος, Paul *Epitome* 6.88.2. All these types are archaeologically attested. Extraction is pulling the projectile out by the route it entered; expulsion is pushing it through to the other side.

25. μεγέθει δέ, καθ᾽ὸ τὰ μέν ἐισιν μεγάλα ἄχρι τριῶν τὸ μῆκος δακτύλων, τὰ δὲ μικρὰ ὅσον δακτύλου, ἃ δὴ καὶ μυωτὰ καλοῦσι κατ᾽ Αἴγυπτον, τὰ δὲ τούτων μεταξύ. Paul was writing an epitome, and here as elsewhere we are probably dealing with excessive abbreviation of text.

26. 6.88.2–4. If a catapult's slider was similarly shaped on top (instead of being grooved) it could have shot tanged or socketed heads without any shaft at all.

27. C. Salazar, trans., "Getting the point: Paul of Aegina on arrow wounds," *Sudhoffs Archiv* 82 (1998) 170–187, at 183–184.

28. Salazar 1998: 178.

29. *Strategikos* 19.3.

30. Ingentum vim pilorum, velitarium hastarum, sagittarum glandesque et modicorum qui funda mitti possent lapidum paraverat, Livy 38.20.1–2.

31. For the formula see K. G. Sellier and B. P. Kneubuehl, *Wound Ballistics*, 1994, 218.

32. M. Belkin, "Wound ballistics," *Progress in Surgery* 16 (1978) 7–24.

33. A. Eriksson et al., "Work-place homicide by bow and arrow," *Journal of Forensic Sciences* 45 (2000) 911–916. Staff at the National Laboratory of Forensic Science in Sweden determined the velocity of arrows shot by the modern compound bow used in this crime to be 57–58 m/s, regardless of head type, but acknowledge that the performance of the bow varies with the ability of the person wielding it. They also note that most animals bagged in modern hunting with modern bows and arrows are shot at a range of 18 m or less, and only very few at a range of 35 m or more. On modern bow launch velocities see also B. Karger et al., "Experimental arrow wounds," *Journal of Trauma* 45 (1998) 495–501, citing 45 m/s for a longbow (the same as Richardson, next note) and 67 m/s for a modern compound bow.

34. T. Richardson, "Ballistic testing of historical weapons," *Royal Armouries Yearbook* 3 (1998) 50–56. These are the only results based on modern test equipment known to me. In "Schleudergeschosse aus Blei—eine waffentechnische Untersuchung," *Saalburg-Jahrbuch* 45 (1990) 59–67 (reprinted in Baatz 1994), D. Baatz assumed, working backward from maximum attested ranges for "slingshot," that a slinger could launch slingshot at 75 m/s, but these ranges must have been achieved by catapult. For comparison, Richardson found best launch velocities of 45 m/s for a longbow, 44 for a steel crossbow, 43 for a yew crossbow, 38 for a staff sling, 32 for a sling, and 19 m/s for a hand-thrown spear. (By contrast, a medieval handgun with a 50 g charge, of modern gunpowder, achieved 181 m/s. One sees immediately how gunpowder revolutionized range weapons—revolutionized *after* more than two centuries of development of that new technology. The first mention of guns in Europe dates to 11 February 1326. A survey of ordnance for Nuremburg's defense 104 years later revealed 501 handguns and 607 crossbows. Crossbows and firearms co-existed until gun and gunpowder technology was sufficiently advanced to oust the crossbow. Gunpowder, like sinew, deteriorated in a moist atmosphere. Changes to the ignition mechanism were needed before the gun really took over in the mid-sixteenth century. B. Hall, "Weapons of war and late medieval cities: technological innovation and tactical changes," in E. Bradford Smith and M. Wolfe (eds.), *Technology and Resources Use in Medieval Europe* [1997], 185–208. Centuries of development may similarly have been involved in the history of the catapult to bring it to the kind of performance implied in these Roman sources.)

35. Experiments made in 1907 and reported in Sellier 1994: 218.

36. E.g. Virgil *Aeneid* 9.589. Lucretius 6.177. Ovid *Metamorphoses* 2.727. Seneca *Natural Questions* 2.57.2. His nephew Lucan, *Civil War* 7.513.

37. A. V. A. J. Bosman, "Pouring lead in the pouring rain," *Journal of Roman Military Equipment Studies* 6 (1995) 99–105 at 99 and fig. 1. I owe this reference to Bill Griffiths.

38. Missiles in bodies: Livy 38.27. Pragmatic: Livy 38.44.9–50.3. Hannibal's book: Nepos *Life of Hannibal* 13.2.

39. E.g., G. H. Donaldson, "A reinterpretation of RIB 1912 from Birdoswald," *Britannia* 21 (1990) 207–214, esp. 210–213, and J. G. Crow, "A review of current research on the turrets and curtain of Hadrian's wall," *Britannia* 22 (1991) 51–63, esp. 53. The old argu-

ments against nonlegionary forces having catapults seem to be based on one or other of the (related) assumptions that (i) catapults were large caliber, or (ii) catapults were not very or at all mobile. E.g., "the majority of weapons found at *numerus* forts are spear and lance-heads, and *what at first sight appear to be ballista bolts*. Baatz is sceptical about the presence of artillery, and suggests that the finds represent hand-thrown spears. The same is probably true of the stones found in the Odenwald forts and watchtowers, and at the cohort fort of Oberscheidental. The stones *are too varied in size* and *irregularly shaped* for use as artillery ammunition, and *the towers of the numerus forts are too small* to support *ballistae*," P. Southern, "The numeri of the Roman imperial army," *Britannia* 20 (1989) 81–140 at 109, italics mine. Note the presence in the Peel Gap turret of "two piles of small, rounded pebbles . . . perhaps the remains of sacks of crude slingshots" and the "unexpected . . . ballista bolt head" (Crow 1991). On the irregularity of stone size and shape being no bar to interpretation as catapult ammunition, see Appendix 1: The Calibration Formulae, below. The Elginhaugh washer is another case for nonlegionary troops being armed with these weapons, since Hanson could find no legionary presence at the fort, which was built predominantly if not entirely for cavalry, Hanson (n. 4). Cavalrymen could use little catapults from horseback, as Arrian's force apparently did (*Taktika* 43.1), though these "machines" have been assumed hitherto to have been tension (bow) catapults armed with a thrust of the leg.

40. W. Griffiths, "Reconstructing Roman slings," *Arbeia* 3 (1994) 1–11 at 9. The Slinging.org website's safety advice includes the recommendation that a slinger of average height using a 36-inch sling "needs an open area of at least 27 feet diameter and height," at least when learning how to control it, http://slinging.org. Slings were really battlefield weapons, as Griffiths observes.

41. Illustrated in J-P. Adam, *L'architecture militaire greque*, 1982, fig. 73 and 279, 280, respectively.

42. B. Gourley and H. Williams, "The fortifications of Stymphalos," *Mouseion* Series III 5 (2005), forthcoming. I thank the authors for making this paper available to me before its publication, and the editor of the journal for permission to cite it in advance. Kassandros' general Apollonides captured the city in 315 B.C. and garrisoned it thereafter, refurbishing and refitting the fortifications in the process. See also *Archaeology in Greece 1997–8*: 28.

43. See, for example, E. Gil, I. Filloy, and A. Iriarte, "Late Roman military equipment from the city of Iruña/Veleia (Alava, Spain)," *Journal of Roman Military Equipment Studies* 11 (2000) 19–29 at 21.

44. This is shown most graphically by a couple of cases from a battle outside Rome reported by Prokopios: one Koutilas fought part of the battle with a javelin embedded in his head, while an Arzes fought on with an arrow extending through his face almost to his neck. The latter was extracted by expulsion that evening by a doctor called Theoktistos, and the victim survived. The former collapsed when the javelin was extracted forcefully, and died some time later. *Wars* 6.2.14–31. As in the latter case, medical intervention was perhaps not always for the best. Another in the same campaign, Trajan by name, ignored a barbed head buried in his face, and when Prokopios was writing, some eight years later,

his body was slowly expelling the arrowhead itself, and was expected to complete the job in due course, 6.5.24–27. Philon: *Poliorketika* C 72.

45. C. G. d'Ampmartinois, 1583, quoted in Payne-Gallwey 1995: 158.

## Chapter Six

1. And of the warship. For example, Antigonos had no navy when he began his fifteen-month long siege of Tyre in 315. It had not been an island since Alexander built the mole (and it has remained connected to the mainland ever since), but in order to take it he needed command of the sea, and so he commissioned ships from all areas under his control (which were many at this point in time). During the early days of this siege, he apparently took the cities of Joppa (coastal) and Gaza (inland) quickly with only a subset of his forces (DS 19.59.2 and 61.5). The fleet when assembled had no triremes (3-ers) apparently, but ninety *tetrereis* (4-ers, a warship better known in its Latin form, the quadrireme), ten *pentereis* (5-ers or quinquiremes), three *ennereis* (9-ers), ten *dekereis* (10-ers), and thirty ships without decks (19.62). The *tetrereis* and *pentereis* first appeared in Athenian naval lists only in 330 and 325, respectively, just ten years earlier (Morrison and Williams 1968: 3). The variety of ship types in Antigonus' fleet suggests that innovation and experimentation were still (or again) a feature of naval architecture in the last decades of the fourth century, as they had been in its first decades in Syrakuse, and, as in Syrakuse at the other end of the century, it is likely that this was accompanied by innovation and experimentation in other technical areas, too. For further information on these ancient warships, a good place to start is still L. Casson, *Ships and Seamanship in the Ancient World* 1971, esp. chapter 6 for the polyremes. The debates on ancient ships are many, various, and old. See, e.g., W. L. Rodgers, *Greek and Roman Naval Warfare* 1937 (reprinted 1964) Appendix 1 to chapter 9 (pp. 254–261). Rodgers' account depends to some extent on data on seventeenth-century ships, which method has been criticized, with some justification, by A. Tilley, *Seafaring on the Ancient Mediterranean* 2004, chapter 5. Most issues will not I think be settled until some warship wrecks are excavated and published (work is currently in progress).

2. Instability: Diodoros 20.40.7. Kleonymus: Polyainos 2.29. Polyainos' reference to the Troizen garrison commander as "a general of Krateros" enables us to date this to sometime between 324 B.C., when Alexander made Krateros Viceroy of Europe, and 320, when he died. More about Kleonymos can be found in Plutarch's *Life of Pyrrhus*. Pydna: DS 19.36 and 49.

3. Amphipolis: DS 19.50.7. Telesphoros: 19.74.2.

4. Trial and error: Ktesibios *Bel.* 112.9, and Philon *Bel.* 50.27–28. Calibration formulae: Ktesibios *Bel.* 113.1–4, Philon *Bel.* 50.9–13.

5. Philon *Belopoiika* 49. 12–50.4. Marsden.

6. Marsden 1971: 157 n. 8.

7. Polybios 4.56.1–5. Polybios was a careful historian, so this unusual terminology may be deliberate, signifying a different design of machine.

8. *Bel.* 72.10–73.5. Marsden.

9. McNicoll 1997: 47. For the period before, Pierre Ducrey showed how unusual were

successful assaults on walled cities, "Les fortifications grecques: rôle, function, efficacité," in Leriche and Tréziny (eds.), *La fortification dans l'histoire du monde grec* (1986) 133–142. Going back before that, Adcock summed up the situation with the remark that "in historic Greece down to the first phase of the Peloponnesian War there is no certain trace of a Greek city being stormed by Greeks," 1957: 57. A recent comprehensive treatment of the subject is in P. B. Kern, *Ancient Siege Warfare*, 1999.

10. Two decades would be a relatively short time for an invention to reach this point. For comparison, steam engines had been around for about seventy years, and some six hundred had been built, before Watt made the improvements that quadrupled their efficiency, J. Lienhard, *The Engines of our Ingenuity* 2000: 92.

11. E.g., the development of powder mills in Lower Saxony, Mußmann 1996: 330–331. Later in Britain Waltham Abbey was purchased just before the Napoleonic Wars, Everson and Cocroft 1996: 379; and Ballincollig during them, Kelleher 1996: 361; and both, together with Faversham purchased in 1759 (during the Seven Years War), were involved in then-cutting-edge gunpowder technology. An "experimental section" of the Royal Engineers was formed in 1916 "to test weapons produced by field companies at the front" and examine captured enemy equipment. The Royal Engineers' Museum (in Chatham) preserves a German bolt-action rifle that is simply a version scaled up for anti-tank duty in 1918, together with various devices developed by and for the troops in the trenches. The atom bomb is probably the most famous example of new technology developed during war.

12. "The act of invention," *Technology and Culture* 3 (1962) 490.

13. Ktesibios *Bel.* 93, Philon *Bel.* 52.20–53.7 (two methods), and Vitruvius *De architectura* 10.11.4.

14. Convergence: Vogel 1998: 36–37. Cars: B. S. Hall, "The corning of gunpowder and the development of firearms in the Renaissance," in Buchanan ed. 1996: 87–120 at 87.

15. Lysimakhos: DS 19.73.3–5. Philip: 19.74.6. Elephant in Britain: Polyainos 8.23.5. Gauls: Caesar *Gallic War* 2.12, and 30–31. These submissions by awe occurred in 57 B.C., shortly after the Romans arrived in the area. Antigonos' general: DS 19.78.2. Demetrios: DS 20.100.6.

16. Persikon: DS 20.27.1–2. Mounykhia: DS 20.45.6–7 Loeb with modifications.

17. DS 20.46.2.

18. DS 20.47.7–8. Moreover, some of Ptolemy's garrisons in the Delta were routinely equipped with sharp-casters and many types of missile by 306 (DS 20.75.4), and his river boats were "all equipped with missiles of every kind and the men to use them" (20.76.3).

19. Terrifying equipment: DS 20.48.1. Stone-throwers: 20.48.4. Helepolis: 20.48.7.

20. Agathokles: Diodoros Siculus 20.54.6–7 Loeb modified. Torture: 20.71.2. Deserters: 20.75.3.

21. Dispositions: DS 19.82–83. Antigonos: 20.109.1.

22. 2.28.1. Magas was probably either the brother of Ptolemy I or the father of Berenike, who married Ptolemy III.

23. Tunis: DS 20.17.5. Vegetius gives many ways to counter elephants in his time—

fourth century A.D.—which culminates with the advice to post relatively large two-horse *carroballistas* (carriage-mounted sharp-casters) behind the line to shoot them when they come in range, and to use broader and stronger iron heads than normal "to make larger wounds in large bodies," 3.24.

24. 1.40. He uses unusual terms for the type of missile shot by troops stationed in front of the ditch, υσσος, and the even more odd γρόσφος, which LSJ interpret, on the basis of Polybios' use of the terms, as referring to the pilum and "a kind of javelin," especially that used by the Velites. English lacks the same variety of terms for missiles that evidently proliferated in a bellicose world in which many kinds of missiles were developed and deployed and for which many names were coined. Insignificant town: 1.53.10–11.

25. Polybios 11.12.4 Loeb modified.

26. In wagons: Polybios 11.11.3. Philopoimen: 11.12.5. Firearms: Morin 2002: 61.

27. A formulaic three-span sharp-caster would have a stock length of four and a half feet. The number of ships in the text is probably corrupt; he could have had as many as 180 ships.

28. DS 20.49.4–51.2 Loeb modified.

29. Greatest weapons: DS 20.92.5. Energy: DS 20.92.2; see also 20.103.3.

30. Overwhelming machines: DS 20.103.7 Loeb modified.

31. DS 20. 91. 5–6, this *helepolis* was designed by Epimakhos of Athens.

32. Shake the walls: DS 20.95.4. Damaged curtain wall: 20.95.5. On the wall: 20.96.3.

33. On the other side of the Euphrates, however, fortifications were usually made of mudbrick, and walls constructed in this material were sometimes colossal. For example, it has been calculated that *just* the rampart of Aï Khanoum (about 300 km southwest of Samarkand in central Asia) required the production and laying of 10,000,000 mudbricks (each weighing between 40 and 45 kg), six million of which went into the curtains. The rampart was 6–8 meters thick. P. Leriche, "Fortification grecques: Bilan de la recherché au proche et moyen orient," 1986: 39–49, at 48–49.

34. A. McNicoll has a table of the ground area, in square meters, of 34 artillery towers, figure 155, in "Developments in techniques of siegecraft and fortification in the Greek world ca. 400–100 BC," in Leriche and Tréziny, eds. (1986) *La fortification dans l'histoire du monde grec* 305–313. Only 12 such towers (8 of them at the one site of Herakleia-on-Latmos) exceed 100 square meters.

35. Strabo (*Geography* 14.2.5) ranked Rhodes' *rebuilt* walls, together with her harbors and roads, as without equal in the world of the first century Mediterranean that he knew. This accolade does not refer to the (earlier) walls attacked by Demetrios, but the ones built to repair the damage he caused and perhaps in anticipation of a return match.

36. Smash ladders: *Poliorketika* C 39, = 94.1–6. Sheets of wicker: Philon *Pol.* D 51, 100.25–29 Th.

37. Eleven artisans: DS 20.93.5 Loeb modified. Priskos: Cassius Dio 75.11.2.

38. Abdaraxos, Agesistratos, Apollodoros of Damascus, Arkhimedes of Syrakuse, Athenaios Mekhanikos, Biton, Bromios, Geras of Khalkedon, Diades, Diognetos of Rhodes, Epikrates of Herakleia, Epimakhos of Athens, Loukios Iulios Euemeros, who was

an engine-maker in Rome (gravestone, *IG* XIV 1717), Harpalos who served with Xerxes, Hegetor of Byzantion, Heron of Alexandria, Kallias of Arados, Kharias, Kharon of Magnesia, Ktesibios, N. Magius of Cremona, Mamurra, Pephasmenos of Tyre, Philon of Byzantion, Poluidos of Thessaly, Poseidonios, Priskos of Byzantion, Sericus (*fl.* A.D. 370), the Theodoros "who was learned in the science of mechanics" of Prokopios 2.13.26 (possibly the same Theodoros who wrote to Proklos suggesting that "the universe is as it were a machine, the spheres interacting like certain (cog) wheels," prompting Proklos to write *On providence, fate and that which is within our power*), Tryphon of Alexandria, Vedennius Moderatus, Vegetius, Venetios, L. Vibullius Rufus, Vitruvius, and Zopyros of Tarentum. The Anons are the fourth-century A.D. author of *De rebus bellicis* and that of a Byzantine *Poliorketika*, and the engineer from Eretria of IG XII.9.259.

39. Philon *Pol.* B 49, = 89.46–51.

40. DS 20.99. Evidently they maintained their abilities in this department, for they were able to keep Mithridates out just as effectively over two hundred years later, in 88 B.C., Appian *Mithridatic Wars* 4.24–27.

41. About 110 feet; compare the Statue of Liberty, at 152 feet.

42. Perhaps we should call it second-best; Demetrios did, after all, lose this fight.

43. Pliny *NH* 34. 41. About the same time, i.e., 300 B.C., an ordinary bronze portrait statue at Athens cost half a talent (the raw material for it probably cost about a tenth of that); Rhodes and Osborne 266. The figure given for the cost of the colossus thus seems reasonable, at 600 times the cost of an ordinary bronze sculpture. When the remains of the Colossus were recycled in A.D. 653, legend has it that 980 camels were required to carry the pieces. A conservative estimate for the quantity of metal involved is 200 tons. For more on the statue see W. Hoepfner, *Der Koloss von Rhodos und die Bauten des Helios* 2003.

44. Dedication: Vit. 10.16.6–8.

45. Along with 100 talents of silver, freedom from custom dues, and a statue group; Polybios 5.88.6–8. DS adds that the silver was for the reconstruction of the city walls, 26.8.

46. Livy 24.34.13.

47. DS 26.15.

48. This is about 250 kg, and is almost certainly a confusion by Plutarch with the size of stones dropped on ships from beams (see below).

49. Caesar *Civil Wars* 2.9. Plutarch *Marcellus* 15.9; Polybios: 18.29. Philon: *Bel.* 53.18–30.

50. Polybios says this was achieved "by means of a rope running through a pulley," presumably connected to a trigger mechanism on the holding device, which would have been a Lewis bolt, tongs or suchlike, all of which are documented in Greek archaeology from the fifth century B.C.

51. See C. Rorres, and H. G. Harris, (2001) "A formidable war machine," online at http://math.nyu.edu/~crorres/Archimedes/Claw/harris/rorres_harris.pdf For video of this and other reconstructions of the claw http://math.nyu.edu/~crorres/Archimedes/Claw/Models.html.

52. Polybios *Histories* 8.3–7. Livy 24.34.9. Plutarch *Life of Marcellus* 15–17.

53. Philip V: DS 28.5. Antiokhis III: DS 28.12. Near Oropos: DS 29.1. On Euboia: 29.8. Treaty: DS 29.11.

54. O. Lordkipanidze, "The fortification of ancient Colchis," in Leriche and Tréziny 1986: 179–184 at 184 n. 14, with figure 224. The weight of this mega-sharp was obtained for me by Darejan Kacharava, to whom I am enormously grateful, on one of her annual visits to the site. Being iron, it has probably acquired weight through corrosion, but we are anyway looking at about a 3-talent caliber.

55. Vani II, 1976: 176 fig. 136.8. It weighs 900 g and is missing its point, which would have been pyramidal, on a round shaft. Thanks to Darejan Kacharava for this information.

56. Prokopios 2.29.18 for location; 8.4.4 for maintenance of garrison.

57. Epiros' fate: Polybios 32.5 and 30.15. Crushing of a nation: Plutarch *Life of Aemilius Paullus* 29.

58. Ephyra no. 4. The Pergamon washer is exactly the same caliber and very similar in design, and Gaitzsch consequently dates it to the same period, "Hellenistische Geschützteile," esp. 239.

59. DS 31.8.10. Plutarch *Life of Aemilius Paullus* 32.

60. No room on the walls: DS 22.10.7. Pyrrhus' army: 22.10.2. Lilybaion: 22.10.7. Myttistraton: 23.9.3.

61. Machines threw down wall: DS 23.18.4. Lilybaion: 24.1.2. Helpless: 24.1.4.

62. *Poliorketika* A2–7, 11, 29, 66. The same sort of advice was given by Alberti in the 1480s, in response to the emergence of effective cannon. P. Brun, "Les fortifications d'Hyllarima, Philon de Byzance et Pleisarchos," *Revue des Études Anciennes* 96 (1994) 193–204 dates developments in fortifications that are recommended by Philon to the turn of the fourth-third centuries B.C. (thus preceding him). This site is inland from Herakleia-on-Latmos, Priene, and Iasos.

63. This is conveniently depicted in Adam 1982: 110 figure 74, showing three palintones aiming out of hour-glass shaped embrasures and three euthytones set to shoot out of regularly shaped loopholes.

64. Two different examples of how the urge to install big stone-throwers distorts theories here: Winter (rightly) criticized Ober's typology of towers because it assumed that "towers incapable of accommodating large arrow-shooters and stone-throwers must pre-date the introduction of such weapons" (1992: 208). But then Winter himself proposed that a tower in Rhodes could have housed catapults that shot "two 1-talent, one 40-mina and four 10-mina balls, plus two 4-span and six 3-span bolts, i.e. 200 minas of shot and eight missiles" (1992: 199). Seven stone-throwers and eight sharp-casters might fit in the space available, but no literary source supports the idea that an arsenal might have almost as many stone-throwers as sharp-casters; normally the latter outnumber the former several (if not tens) to one.

65. People often seem to be surprised, not to say disappointed, when their structures turn out to be too small for anything bigger than small or medium caliber weapons.

66. Marsden rightly asked (1969: 145) how such a machine might have arrived in a room

that has no door and embrasures that are not much more than a meter square (at best).

67. McNicoll 1997: 155.

68. See Philon *Poliorketika* 95.49–51 and Polybios 26.30 for examples.

69. Built-in: *Poliorketika* A 25 = 81.34–37 Th. Largest located at foot of wall: A 32, 82.6–9. Ten-pounders: C 6, 91.15–19 Th. Emphasis mine.

70. This is well illustrated in Marsden 1969: 118–119 with figure 1, where it is shown that a catapult that could shoot 400 yards at a 45° angle will reach only 70 yards if shot horizontally from a 42-foot-high tower or 63 yards if shot horizontally from a 27-foot-high wall.

71. Philon *Poliorketika* C 63; McNicoll 1997: 105; see also Adam 1982: 111.

## Chapter Seven

1. The surname recognizes the oddity, as far as the Macedonians and Greeks were concerned, of sibling marriages. Ptolemy II only followed Egyptian custom and married his sister (Arsinoë II) after his marriage to his first wife, Arsinoë I, the daughter of another of Alexander's generals and mother of Ptolemy III, ended. The surname therefore must postdate the second marriage.

2. Modern opinion on Ktesibios follows Kenny 1932.

3. Degering 1905: 14–16; Marsden 1969: 3, 1971: 1–3.

4. Marsden 1971: 2. Cuomo 2002: 170.

5. Philon *Bel.* 51.18–20. Ktesibios *Bel.* 114.2–3.

6. The earliest known method of finding the cube root is Arkhytas', a contemporary of Dionysios I of Syracuse, c. 400 B.C. The second is that of Eudoxos, of about the same date or slightly later, and the third and fourth (which is a mechanical solution attributed to Plato) are by Menaikhmos, who was a student of Eudoxos and lived about 350. Then we have Ktesibios' solution (which is to large extent duplicated by Philon and Apollonios), sixth is Eratosthenes' solution, seventh that of Nikomedes, and eighth that of Diokles, Sporus and Pappus. All are discussed by Heath 1981: 1.244–270. The variety of methods developed suggests widespread if occasional desire to scale three-dimensional objects (not just catapults; also, e.g., buildings and sculptures from models) and testifies to the fragmentation of mathematical knowledge and education across space and time in antiquity.

7. Greek astronomers had discovered that the year was about 365 1/4 days long in the fifth century B.C., but such calendars were not adopted for civil purposes in the Greek city-states. Despite this order, it seems that the Canopus calendar was not actually implemented in Egypt either, at least in official and religious documents, and the Egyptians continued to use their 365-day calendar for roughly the next two centuries. However, a 365 1/4-day calendar was followed after Augustus imposed the Julian calendar on the whole of the Roman Empire, in 30 B.C., after defeating Antony and Kleopatra. The Julian calendar takes its name from Julius Caesar, because he sponsored a Greek mathematician (Sosigenes) to produce an accurate and workable calendar for the empire, which came into force in Italy in 45 B.C. About eighty years earlier, Hipparkhos had calculated the length of the year with such precision that he realized that calendars needed to follow

the same leap year rules that we follow today. That is, add a day every fourth year, unless the number of the year is divisible by 100, in which case do not add a day, except if the number is also divisible by 400, in which case put the extra day back in. But Sosigenes' calendar was simpler, and was, basically, the Egyptian calendar plus the extra day every fourth year. By omitting Hipparkhos' refinement, Sosigenes/Caesar created the small error that, after 1,600 years, required Pope Gregory to remove ten days in 1582 to realign the calendar with the sun and inaugurate the Gregorian calendar. By the time the British (and American colonies) accepted the Gregorian reform, another 170 years later, eleven days had to be dropped instead of ten.

8. Eutocius 88.3–96.27 Heiberg.

9. The Rosetta Stone, which enabled the decipherment of hieroglyphs, was written to commemorate his coronation on coming of age in 197 B.C.

10. A. W. Lawrence 1979: 71 prefers to date Philon to the 240s and the reign of Ptolemy III.

11. A. Chaniotis, "The divinity of Hellenistic rulers," in Erskine ed. 2003: 439.

12. Polybios 4.47.6. See also V. Gabrielsen, *The Naval Aristocracy of Hellenistic Rhodes* 1997.

13. In southern Italy, where Philip V had been campaigning to help the Greeks fight Roman expansion. This man, Herakleides by name, is described as an architect (Polybios 13.4.6; Polyainos 5.17.2), which literally means chief builder and sometimes means a mechanical engineer as well as or instead of a civil engineer. We met another son of Tarentum who was a catapult-builder in Chapter 2: Zopyros. It was perhaps as a result of this sabotage that the Rhodians became so unusually security conscious that Strabo felt moved to comment on it: "some of their naval stations were kept hidden and forbidden to most people, and death was the penalty for anyone caught spying on them or going inside them," 14.2.5 (653). The Ptolemies also developed a "secret" naval base, though by the time [Caesar] was writing it was clearly dysfunctional, if not derelict, *Alexandrian War* 13.

14. Poly. 16.2.10.

15. See, e.g., *Idyll* 17, for Ptolemy II.

16. "Leaning forward or backward"; this may be a reference to in-swingers.

17. The Caminreal specimen seems to be just such as catapult. With a spring-hole of internal diameter of 86 mm average, it was over-sprung for a 3-span.

18. Reinschmidt 1999: 18.

19. Fitting metal plates to support weakened wood sounds like an ancient mechanical analogue to a modern software patch.

20. Wells 1973, especially 254 and 256 n. 4.

21. "Tools and Weapons" in the Egyptian Collection at University College London, Warminster, p. 48 and pl. 58. These come from the Ptolemaic (i.e., Macedonian) and Roman levels.

22. J. J.-M. De Morgan, *Mémoires de la Delegation scientifique en Perse* vol. 1 (1900), 107 with figure 170, and 150–151 for the bows; bronze and iron arrowheads on 118 (fig. 215) and 122–123 (figs. 264 to 269), flint arrowheads on 194.

23. 2 Samuel 22.35: "He teacheth my hands to war; so that a bow of steel is broken by mine arms." These words are spoken by David in a victory song, which relates to a date before 900 B.C., but was written later. Job 20.24 (early OT but undatable): "He shall flee from the iron weapon, and the bow of steel shall strike him through."

24. Skythian bowl: Hermitage inv. no. ДН 1913.1/40, dated 400–375 B.C. Marcus column short bow: eg. E. Petersen *Die Marcus-Säule* (1896) Tafel 19, 21, 22, 35, 45, 46; one of them much more conveniently reproduced in Coulston 1985: 340 figure 24. There is also a very strange looking "bow" borne by a cavalryman in Petersen Tafel 66 scene 57. Comparative evidence: Rausing 1997: 157 note 11, 109, respectively.

25. De Morgan 1900: 151. Rausing 1997: 91 and 137.

26. Berlin Antikenmuseum F 2294. Realistic: J. D. Beazley, "Un realista greco" 1966. Cavalier 2000.

27. Wells 1973: 256 nn. 3, 5.

28. *Bel.* 76.27–77.4. Peddie notes that similar complaints were made about the Bren gun when it was introduced in the 1930s, yet it went on to serve throughout World War II and after, 1994: 95. He also notes (idem) that the Chinese repeating crossbow, which originated in antiquity, was last employed in 1895.

29. [Hyginus] *De munitionibus castrorum* 58. See, e.g., Ammianus 21.12.7, siege of Aquileia.

30. Lucan *Civil War* 3. 709–725.

31. DS 20. 97.1–2.

32. E. Schramm (1918) *Die antiken Geschütze der Saalburg* 62–66, esp. 64.

33. See Vicente et al. 1997: 195 Anexo 1.

34. Ricochet: A 8, = 80.6–11 Th. Bossing: A 29, = 81.47–82.2.

35. E.g., Asine, Lawrence plate 25.

36. *Pol.* A 73, = 85.8–9 Th.

37. Hall 1996: 101.

38. *Pol.* C 6, = 91.15–18 Th., and D 17, = 98.12–13, respectively. For dimensions see Appendix 1, The calibration formulae.

39. *Pol.* C 56, = 95.9–15. He imagines having to deal only with small craft because large craft should have been kept out by the submerged boons, artificial reefs, and suchlike that he recommended earlier.

40. Harbor defense: C 57, = 95.15–17. Fire bearers: C 58, = 95.17–20.

41. C 68–9, = 96.2–10 Th., and C 67–71, = 95.49–96.26 Th., respectively. He recommended using a 30-mina stone-thrower also against enemy siege towers.

42. *Pol.* A 64–5, = 84.24–34 Th.

43. A. Sleewsyk, "The lineage of the triacontor," in Tzalas (ed.) *Tropis VI* 1996: 517–527.

44. 55.18–56.8. One is reminded of the Pergamenes' apparent taste for 37 1/2-mina caliber shot, preferred there only by 1 talent shot. 353 x 1 talent balls found, 166 x 37 1/2-mina shot, 126 x 40-mina shot and 118 x 30-mina shot. Or are the 37 1/2-mina caliber stone balls really undersized 40-mina caliber shot (so Marsden, 1969: 81)?

45. 52.1–19; see Heath 1981: 262–264.

46. Some of these Gauls had entered Asia Minor in 278 B.C., and settled near the center of it after a decade or so raiding widely across the area. Hellenistic monarchs imported others as mercenaries. Celtic was still spoken here as late as the sixth century A.D.

47. This building was reconstructed by the American School at Athens and stands in the Agora, housing a small selection of finds from the area.

48. M. J. T. Lewis, "When was Biton?" *Mnemosyne* 52 (1999) 159–168, esp. 167. Ktesibios: opening of his treatise. In a similar manner, long continued peace during the High Roman Empire led to the moats around fortified cities being "choked with rubbish" and the walls in large part collapsed, Ammianus 29.6.11.

49. Salonae: Caesar *Civil Wars* 3.9. Thasos: Polyainos 8.67. Matrons: Vegetius 4.9. Bow: chapter 1.

50. Garrison: Polybios 16.24.1; see also 16.1.3. Eumenes II: Polybios 21.19–20. Livy 42.42.6. First strike: Polybios 32.16.

51. Thessalonike: Biton 49.1–2. Abydos: Polybios 16.29–34. Attalids: Polybios 21.1–24.

52. Ath. Mekh. 27.7; date about first century B.C. Mithridates' siege: Appian *Mithridatic Wars* 4.26–27.

53. Pergamon court dull: E. Kosmetatou, "The Attalids of Pergamon," in Erskine (ed.) 2003: 169. Fault: Cassius Dio 69.4.4. Ammianus observed that Hadrian shared with Valentinian the fault that, in order to appear to surpass all men in good qualities, they put down other people who possessed them; 30.8.10. Cassius Dio, before going on to tell the story of Apollodoros' fall, calls Hadrian jealous of "all who excelled in any respect," and says that he caused the downfall of many and utterly destroyed several, because he wanted to surpass everybody in everything, 69.3.3. See Cuomo (forthcoming) chapter 2 for discussion of other cases of "technician-warriors," including Demetrios Poliorketes and Appius Claudius.

54. Work on agriculture: Pliny *NH* 18.22. Galen: Kosmetatou, 2003: 165. Galen, A.D. 129–216, was one of the great doctors of antiquity. Beginning as repairer of gladiators, he wrote voluminously, performed dissection spectacles, became a medical celebrity, and attended three emperors: Marcus Aurelius, Commodus, and Septimius Severus. His views dominated medicine for the next 1,500 years.

55. W. Gaitzsch, "Hellenistische Geschützteile aus Pergamon," in C. Dobiat (ed.) *Festschrift für Otto-Herman Frey*, 1994: 235–242; Baatz consultations, p. 242 n. 26.

56. 47.8, 49.11, 50.3–7, καὶ ἐνδεδεμένος δὲ ἔστω ἑκάτερος σιδηραῖς λεπίσιν This again distinguishes his bow stone-throwers from his *gastraphetai*, which are not fitted with such braces.

57. This was an important subject in a kingdom with a building program at home and abroad, as Pergamon had.

58. Benjamin Franklin once made the same sort of point, in a letter written in 1776, wherein he pointed out that bows and arrows were not wisely laid aside, because "a man may shoot as truly with a bow as with a common musket. Secondly he can discharge four arrows in the time of charging and discharging one bullet. Thirdly his object is not taken

from his view by the smoke of his own side. Fourthly a flight of arrows, seen coming upon them, terrifies and disturbs the enemies' attention to their business. Fifthly an arrow striking any part of a man puts him *hors du combat* till it is extracted. Sixthly bows and arrows are more easily provided everywhere than muskets and ammunition." Quoted by T. Esper 1965: 382, who adds that Lee endorsed this when he recommended the long-bow over the flintlock musket for the British army in 1792.

59. Triggers were so common when Biton wrote that he did not bother to describe one for any of his four catapults. He assumed that the reader would know what a trigger looked like and how one was made.

60. Marsden, for example, links him with the stone-throwers used by Onomarkhos against Philip in 353, 1969: 59, 75.

61. G. Reger, "The economy," in Erskine (ed.) 2003: 339.

62. How well a particular machine was built, tuned, maintained, operated, and similar variable factors would also have some influence on performance, but their effects are relatively minor in most cases.

63. Aristotle: *Physics* 4.8 = 215a24–216a11. Vegetius: 3.14.

64. Foley, Palmer, and Sodel 1985: 84.

65. James 1994: 97 and n. 35.

66. For an alternative interpretation of these "foreshafts" (and other socketed and tanged heads) see Baatz, "Zur Geschützbewaffnung römischer Auxiliartruppen in der frühen und mittleren Kaiserzeit," *Bonner Jahrbuch* 166 (1966) 194–207, reprinted in Baatz 1994: 113–126 at 122–126.

67. Reinschmidt 1999: 3–4.

68. *Mechanics* 15, see also the notion of resistance in chapter 34. Chapters 32, 33, and 34 of the *Mechanics* have discussions on this topic that appear to predate the *Physics*.

69. For further reading on Philoponos' ideas, see R. Sorabji, *Matter, Space and Motion,* 1988, chapter 14.

## Chapter Eight

1. Inv. 88/946, discussed by A. R. Birley, "Vindolanda: notes on some new writing tablets," *Zeitschrift für papyrologie und epigraphy* 88 (1991) 87–102 at 90–92. Another tablet, 93/1398, refers to nine pounds of hair, which Bowman and Thomas suggest may have been for catapult springs (or saddle stuffing, or wigs), "New writing tablets from Vindolanda," *Britannia* 27 (1996) 299–328 at 300–304.

2. E. Evans, "Military architects and building design in Roman Britain," *Britannia* 25 (1994) 143–164. She adds that there may sometimes have been a number of subordinates or trainees allocated to work with him and perhaps succeed him, p. 148.

3. Literally, up-sprung and down-sprung. This must not be confused with what I refer to throughout the book as over-sprung and under-sprung, which refers, respectively, to a catapult with an enlarged spring-hole relative to some of its other dimensions, or undersized spring-hole relative to some of its other dimensions. *Anatonos* and *katatonos* are not attested as such in the surviving Greek treatises, but synonyms are; Philon talks of longer-spring (μακροτονώτερα) and shorter-spring (βραχυτονώτερα) types, *Bel.* 53.20–21.

4. Appian describes a fire-lance as a wooden shaft, two cubits long, covered with flax and smeared with pitch and sulfur, that was shot by catapult and fanned into flame by the draft as it flew though the air (*Illyrian Wars* 2.11). The head of such a missile seems to be preserved at Azaila. See M. Beltrán Lloris, *Arqueologia e historia de las ciudades antiques del Cabezo de Azaila*, Zaragoza 1976: 60 fig. 28. Museo Arqueológico Nacional no. 3034. At 85 mm long, it is approximately the same length as the Wroxeter example (82 mm), which has a damaged tang. The Dura Europos examples are longer and socketed, S. James (1983).

5. Fleury 1993: 253–254.

6. Marsden's "working length" added disproportionately long windlass handspikes (10D) set horizontally, to the stock length. There are all sorts of problems with Vitruvius' text, especially the numbers therein, and all editors consider the text corrupt on the length of the handspikes; to be specific, the word for handspike is missing and there is no numeral, just seven dots. Vitruvius *De Architectura* 10.10.3 and 5. Wilkins 2000: 82 gives manuscript readings and editorial suggestions. It is not clear how Marsden read X (10) from seven dots; he says nothing of relevance in the accompanying note (a typographic error perhaps).

7. The argument is dealt with in detail in Rihll 2006.

8. Marsden 1969: 21–22 (travel of straight arms), 1971: 230 (travel of curved arms), though his figures are more precise than our knowledge of these machines warrants. Schramm, for example, read the text differently and the arms on his reconstruction curve much less than Marsden's.

9. Marsden 1969: 204. Fleury 1993: 261.

10. Marsden's idea that the new washers allowed some 30 percent more spring-cord to be inserted in an elliptically shaped washer of nominally the same diameter is not supported by my calculations. He suggested that Philon's formulae would give a spring-cord weight of "just over 9 1/2 minai" for a three-span (1969: 206). I do not know how he arrived at that figure; I get 11.4 minai. Agesistratos' record-breaking three-span machine with 12 minai of spring-cord (Athenaios Mekhanikos 8.5–9 W) therefore had about 5 percent more spring-cord than a Philonian machine would do.

11. Wilkins 2003 figs. 12 and 17 and accompanying text. The Xanten catapult will shed light on this when fully published, since the corrosion surfaces apparently show great detail of the wooden parts that have since shrunk away; Schalles 2005.

12. Some of the remains of the plating of the catapult found in the temple there are identified as the battleplate, complete with small aperture for slider, and cutaway for wedge, Díez 2002: 296.

13. Flattened end: GR 61.10–24.35, weighs 30 g, excavated from a tomb in Kamiros on Rhodes. Thunderbolt: GR 1964.1–7.67, weighs 46 g, found in the Nile delta. Issued by Ptolemy Keraunos Highest quality: GR 68.1–10.61, weighs 32 g, also inscribed. I wish to thank Peter Higgs and other members of the British Museum's staff for assisting me in my examination of the material and supplying information on findspots.

14. It apparently occurs on the reverse of a single early electrum issue from Miletos, *if* this coin, which I found offered online by a coin dealer, is genuine.

15. 5.99.7: καταπελτῶν μὲν ἑκατὸν πεντήκοντα πετροβολικῶν δ᾽ ὀργάνων πέντε καὶ εἴκοσι

16. J. Lienhard, *The Engines of Our Ingenuity* 2000: 11–12.

17. This is probably the most appropriate place to say that I do not think that the "Casalbordino cylinder" is a windlass cover. The Imagines Italicae project does not think it is, either. Those who wish to pursue the issue should read Adam Morton's "Finding the corkscrew," *Studies in the History and Philosophy of Science* 37 (2006) 114–117.

18. 10,000 mules: Plutarch *Life of Sulla* 12. Arkhelaos: Appian *Mithridatic Wars* 5.30–32.

19. *Mithridatic Wars* 5.34: ὁ Σύλλας ἐκ καταπελτῶν, ἀνὰ εἴκοσιν ὁμοῦ μολυβδαίνας βαρυτάτας ἀφιέντων, ἔκτεινέ τε πολλοὺς, καὶ τὸν πύργον Ἀρχελάου κατέσεισε καὶ δυσάρμοστον ἐποίησεν . . . The heaviest lead shot known to me weighs 500 g and was found in Kephissia, Attike, British Museum GR 1865.7.20.106 (unpublished). It has a relatively long inscription but I cannot make out enough even to tell whether it is Greek or Latin.

20. Jones and Robinson, "The House of the Vestals (VI. i. 6–8, 24–26): its development and implications," in preparation, reported in advance of publication with the authors' kind permission. I also want to record my thanks to Jo Berry, who alerted me to the discovery of the shot in the insula behind the wall and provided this reference.

21. For an overview of recent archaeological work in the Iberian peninsula (though it does not mention the catapult finds), see S. Keay, "Recent archaeological work in Roman Iberia 1990–2002," *JRS* 93 (2003) 146–211.

22. Reported very briefly in Cabre 1925: 14.

23. Díez 2002. Díez is of the opinion that the counterplate belongs with the temple catapult (Azaila 2) and not the acropolis catapult (Azaila 1) because it appears in a photograph associated with the 1925 excavation. He notes that the acquisition records at the National Archaeological Museum are rather vague and there are over 200 fragments of iron and bronze preserved.

24. Díez 2002: 298.

25. Azaila seems jinxed: Díez 2002: 296 refers to a catapult at the ramp but his reference to "Beltran Lloris [*Arqueologia e historia de las ciudades antiques del Cabezo de Azaila*, Zaragoza] 1976: 181 fig 88.1.6" is wrong somewhere, since this page of this book concerns ceramics and the last figure in the book is fig. 82. Baatz' reference to the same book (in Baatz "Katapultfunde 1887–1985," 281) also contains a misprint—read "page 176" for "page 17." R. Sáez is working on the material from the site (pers. com.) and has a number of papers (in Spanish) in press.

26. Details in Díez (2002) 298 n.4.

27. Caesar *Gallic War* 7.41. Alesia: 7.81. Corfinium: *Civil Wars* 1.17. Ategua: *Spanish War* 13; the comment was that one of the six people thrown from the tower by a direct hit from said ballista was a slave *whom the townsfolk had charged with keeping an eye on it*. This has its closest parallel in the lookout posted to watch the biggest catapult at the siege of Jerusalem (Chapter 9).

28. 300 paces: Livy 44.35.17. Losses: Livy 44.35.22. Incidentally, we note that being placed in the wooden towers of a temporary fort constructed in the field, these must have been little catapults.

29. Dyrrhacium: Caesar *Civil Wars* 3.51, 55. Uzitta: [Caesar] *African War* 56.

30. Plutarch *Life of Sulla* 18. Caesar *Gallic War* 2.8, 8.14. Caesar *Civil Wars* 3.45.

31. Reputation: Livy 44.35.9. Philip: Livy 32.10.11.

32. Vehicles: Livy 42.53.4. "Everywhere": Livy 44.35.9.

33. Pots of snakes: Nepos *Life of Hannibal* 10.4–11.6. Hasdrubal's head: Livy 27.43–50. Pots of insects: Herodian 3.9.5. The last was clearly highly unusual since Herodian barely ever mentions catapults explicitly and unequivocally. One of those few cases is 7.8.10, where the "machines and war engines" taken on campaign against the Germans in A.D. 238 seem to have been partially responsible for his slow progress back to Rome following Gordian's coup.

34. Britain: Caesar *Gallic War* 4.25. Brundisium: Polyainos 8.24.7. Actium: Plutarch *Life of Antony* 66; Dio Cassius 50.31–35, see also 50.16.2, 18.5–6, 23.1–2, 29.1–3. A.D. 69: Tacitus *Histories* 2.14–15.

35. Massive sharps: Caesar, *Civil Wars* 2.2, Lucan *Civil War* 3.464–468. Pulverizing shot: Lucan 3.469–473.

36. Appian *Civil War* 5.118–119. A scythe on a pole was a regular piece of land-based military equipment. It was a military version of a reaping hook or tree pruning equipment, to cut at a distance.

37. E.g., Caesar *Civil Wars* 3.44–45 (Pompey), *Alexandrian War* 1 (Caesar's forces), 2 and 21 (Alexandrian army), 9 (Pompeian forces), 19 (naval forces, on the prows of ships); *African War* 20 (Caesar's camp at Ruspina, complete with smithies), 29 (at Leptis).

38. Nervii: Caesar *Gallic War* 5.42, 7.22.

39. Caracalla: Cassius Dio 79.3.3. Rhodian ships: Polybios 1.20.15–16; 1.59.8. Liburnian: Vegetius 4.33.

40. *Alexandrian War* 3. He also remarked on the Egyptians' originality, saying that they invented much on their own account.

41. Tacitus *Histories* 4.23, 28.

42. *Civil Wars* 3.10. Chief engineers—plural—were probably sequential rather than simultaneous.

43. Hanno: Polybios 1.74. Spaniards: Livy 39.30.5–6. Iapydes: Appian *Illyrian War* 4.19.

44. Negotiator: *CIL* XIII 6677. Codex Theodosius: 9.40.24. Cassiodoros: *Letters* 5.16.

45. Catapults among the spoil: e.g., Caesar *Civil Wars* 2.22. Hawaii: C. Ponting *Gunpowder* 2005: 200.

46. Pliny *NH* 18.22. However, some of this material found its way to Rome, for when the Numidian monarchy was abolished in 46 B.C., its library was confiscated by the first governor, one Sallust, the same who went on to write the *Jugurthine War* and the *Conspiracy of Cataline*.

47. Caesar: *Civil Wars* 1.44. Antalkidas: Plutarch *Sayings of Spartans, Agesilaos* 71 (=

Moralia 213f). The Thebans essentially brought Spartan history to an end at the battle of Leuktra in 371 B.C. This anecdote relates to a battle in 378.

48. Plutarch *Life of Antony* 38.

## Chapter Nine

1. M. C. Bishop, "The military *fabrica* and the production of arms in the early Principate," in Bishop (ed.), *The Production and Distribution of Roman Military Equipment* 1985: 1–42, esp. 11–13; J. Oldstein, "Manufacture and supply of the Roman army with bronze fittings," same volume, pp. 82–94, and J. Paddock, "Some changes in the manufacture and supply of Roman bronze helmets under the late republic and early empire," same volume, 142–159, all stress that, for a variety of different reasons related largely to date and location, the manufacture and maintenance of weapons and other military equipment was carried out by private individuals (civilians), as well as staff in forts and other military establishments. For an overview of production empire-wide, and a map of *fabricae* in the *Notitia Dignitatum*, see Bishop and Coulston, *Roman Military Equipment* 2nd ed. 2006, chapter 9.

2. Baatz, "Ein Katapult der Legio IV Macedonica aus Cremona," *Röm. Mitt.* 87 (1980) 283–299.

3. Wilkins 2000: esp. 88–89.

4. For example, in A.D. 214–215 Caracalla built "two large engines for the Armenian and Parthian wars, so constructed that he could take them apart and carry them in ships to Syria," Cassius Dio 78.18.1.

5. Tacitus *Histories* 3.19–20.

6. Tacitus *Histories* 3.23. The two heroes were run through immediately, presumably by the men manning the weapon who would, one imagine, be liable to their commander on a charge of negligence.

7. *nec restitui quivere impedientibus vehiculis tormentisque*, 3.25.

8. Tacitus *Histories* 2.82 and 84.

9. Tacitus *Histories* 4.30.

10. G. H. Donaldson, "*Tormenta, Auxilia* and *Ballistaria* in the environs of Hadrian's wall," *Archaeologia Aeliana* 17 (1989) 217–219. Tablets from Vindolanda of approximately the same time refer to orders for "100 pounds of sinew" and "9 pounds of hair"; A. K. Bowman and J. D. Thomas, "New writing tablets from Vindolanda," *Britannia* 27 (1996) 299–328 at 302–304, and A. R. Birley, "Vindolanda: notes on some new writing tablets," *ZPE* 88 (1991) 87–102 at 90–92, respectively.

11. Manufacture by the legions: Vegetius 2.11. The Tenth's reputation: Plutarch, *Life of Pompey* 69.

12. Josephus *Jewish War* 5.6.2–3. Hundreds of stone balls were recovered at the site. The largest ball recognized as a ballista shot (i.e., excluding those roughly shaped, for rolling down the hill) weighed 22 kg, which is a "light" talent. Of fifty complete shot, 40 percent were of approximately 2-mina caliber, and only 14 percent were over 10 mina. See Holley 1994. Two mina is about 2 lbs., or 1 kg. One of three European cannon of this sort of

caliber that was used by the Tai-Ping rebels to defend the gates of Kahding in 1862 is on display at the Royal Engineers' Museum, Chatham. Medium caliber "artillery" were not toys, neither in antiquity nor for the next 2,000 years, any more than rocket-propelled grenades are now. They are weapons for the destruction of targets bigger or tougher than individual men and smaller or more fragile than tanks and towers, and there are many such targets in war.

13. Lightness: Josephus *Jewish War* 3.7.23. Withdrawal: 3.245–246. Experience: 5.9.2. Titus hit: Cassius Dio 66.5.1.

14. For the major campaigns see K. Strobel, *Die Donaukriege Domitians* (1989) and *Untersuchungen zu den Dakerkriegen Traians* (1984).

15. Cassius Dio 67.7.4, 68.9. Wilkins entertains similar possibilities, *Roman Artillery* 41 on fig. 28.

16. Ashes: J. B. Campbell, *OCD*3 2003: 1543 s.v. "Trajan."

17. Orşova is ancient Dierna, in J. J. Wilkes, "The Roman Danube: an archaeological survey," *Journal of Roman Studies* 95 (2005) 124–225, code Ms 43. Gornea is not in Wilkes. Novae, on the opposite bank, is Wilkes Ms 23.

18. "Trajan's artillery: the archaeology of a Roman technological revolution," *Current World Archaeology* 3 (2003) 41–48.

19. This is part of the simplifying tendency we have noted before, to ignore variations leaving just a handful of types, and specifically, just those designs for which we have surviving literary descriptions.

20. "Pseudo-Heron's *cheiroballistra*: a(nother) reconstruction: 1. Theoretics," *Journal of Roman Military Equipment Studies* 11 (2000) 47–75 at 57.

21. In recent years, it was M. J. T. Lewis, according to White (1984: 218), who first came up with the idea that the wide *kheiroballistra*'s arms could be set on the inside, and that this would allow the arms to be pulled back much further, increasing range by as much as 70 percent, but he has not, as far as I know, published on it. He is, however, cited as the source for the article on "Trajan's artillery" in *Current World Archaeology* 3 (2003) 41–48. The late John Anstee pursued the idea (1998), and Aitor Iriarte is now the most vocal proponent of it. But the idea has a long pedigree: Prou imagined the palintone as an in-swinger back in 1877 (and, it should be added, the *kheiroballistra* as a bronze-spring sharp-caster); some of his diagrams are conveniently in Gille (1980) *Les mécaniciens grecs* 23 and 113.

22. Its springs seem to be a distinctly bigger caliber than the cart-mounted weapon of scene 40, as can be seen when the springs are compared with body parts of the legionaries who operate them: the spring of scene 66 is about the size of the legionary's calf, whereas the spring of scene 40 appears to be a couple of fingers' width, at most.

23. More realistically, we are dealing with a maximum arm travel of about 50° in the case of the out-swinger, and 100° or more in the case of an in-swinger. That is a huge difference, and it would have significant influence on performance.

24. See R. Bockius, "A Roman depiction of a warship equipped with two catapults?" in *Tropis VI* ed. H. Tzalas, 1996, 89–97. Unfortunately the image appears on a tombstone

for someone whose name has been lost and which is undatable by current methods.

25. K. D. White, *Greek and Roman Technology,* 217.

26. Milimete-type guns: Morin 2002: 59–60. Scorpion as name for operator of the weapon: J. Ch. Balty, "Apamea in Syria in the second and third centuries AD," *Journal of Roman Studies* 78 (1988) 91–104 at 101.

27. βέλη or κούφα παλτα, Arrian *Taktika* 43.1. Hadrian's level of interest is indicated by Cassius Dio 69.9: he visited all the garrisons and forts in one province after another, personally inspecting and investigating "absolutely everything," not just "weapons, machines, trenches, ramparts and palisades, but also everyone's private affairs," rank and file, and he drilled the men in various battle formations.

28. Ambleside 16, Birdoswald 1, Birrens 3, Chesterholm 14, Corbridge 9, and Housesteads 1. On the Antonine wall another 1 turned up at Bar Hill (together with 6 clay shot) and 2 at Cramond. The second largest find of lead shot comes from Windridge near London, where over 100 were found, dating to perhaps the first century A.D. Finally, four have been found at Charterhouse. Griffiths, "The sling and its place in the Roman Imperial Army" 1989, with table and map pp. 272–273.

29. Roman crossbows: Baatz, "Die römische Jagdarmbrust," *Archäologisches Korrespondenzblatt*, 21: 283–290 and A. MacGregor, "Two antler crossbow nuts and some notes on the early development of the crossbow," *Proceedings of the Society of Antiquaries of Scotland* 107 (1975–6) 317–321. Crossbow spanning devices: Payne-Gallwey 70–89, 121–125, 131–144. Springald: J. Liebel, *Springalds and Great Crossbows* trans. J. Vale, 1998, 14–16.

30. See sections 4, 19, and 25–26.

31. Pers. comm. On the undesirability of an elastic sling pouch, see Denny, "Bow and catapult internal dynamics," *European Journal of Physics* 24 (2003) 367–378, esp. 374–375 and 377. I owe knowledge of this sling pouch to Bill Griffiths, whose paper "Reconstructing Roman slings," *Arbeia Journal* 3 (1994) 1–11, has the only published drawing of it (by C. van Driel) known to me.

32. L. 3167. See R. Birley, *Vindolanda Research Reports New Series 4 Small Finds Fasc. 1 The Weapons* (1996) 11–14.

33. M. Sommer, *Hatra: Geschichte und Kultur einer Karawanenstadt im Römisch-Parthischen Mesopotamien* (2003) has an aerial view of the site (abb. 50) which shows well its defensive strength.

34. See also H. Lawson, *The Hatra Ballista* 2005: 29, 39–41. Arms 6D long would fit as in-swingers in this frame; 7D would be tight.

35. Baatz, "Recent Finds" 7, *Bauten und Katapulte* 230. Baatz considers the Hatra ballista similar to the 10-pounder of Vitruvius (throwing 3.27 kg balls), because (i) the spring volume is roughly the same (the Vitruvian machine had less than half a liter more); (ii) the washer height is almost identical; and (iii) the lever width is almost identical. More generally he thinks it is Roman in manufacture or inspiration because its dimensions "fit" Roman measures (8 foot long crosspieces, washers 6 feet apart, 3 feet high by 1 1/2 deep etc.) However, there are three points to note: (i) The frame is much too wide for the height as a Vitruvian stone-thrower. (ii) The washers are bigger than the recommended

size (and the spring-*diameter* was the key parameter to which the ancients built remember). (iii) The springs are significantly shorter than recommended for a 10-lb. stone-thrower. Wilkins meanwhile (2003: 69–70 and figures 54 and 55) compares the Hatran machine with the Vitruvian 5-pounder—half the caliber supposed by Baatz. This gives some idea of how significant is the discrepancy between the Hatran machine and a Vitruvian machine.

36. Dura Europos was founded on the banks of the Euphrates as a military colony by the Seleukids in about 300 B.C., occupied by the Parthians from about 100 B.C., and by the Romans from A.D. 165. The double name results from the change of ownership.

37. See M. I. Rostovtzeff et al., *Excavations at Dura-Europos (6th Season)* 1936, chapters 6 and 13.

38. Cotterell & Kamminga give the mechanics, 1990: 161–163.

39. Marsden 1969: 131.

40. McNicoll 1997: 5; Marsden 1969: 91.

41. Distance shooting: Cotterel and Kamminga 1990: 181. Modern maxima: Denny 2003: 368.

42. 1918/1980: 29.

43. 120 yards: Soar 2004: 111. Vegetius: 2.23. Weaker bow range: Cotterell and Kamminga 1990: 182. Tudor specimen: Soar 2004: 102.

44. *Inscriptiones Tyrae, Olbiae, Chersonesi Tauricae ali. loc.* 1.195.

45. 1994: 294–302. His specimens of each type weighed 40 g, which appears to be a reasonably common figure for ancient shot.

46. Article on http://Slinging.org/27.html. His figures are those in miles per hour and feet per second. I have converted this to meters per second for comparison with other authors' figures.

47. Richardson 1998b: 51 table 1.

48. T. Richardson (1998a), "The ballistics of the sling." A fuller version of this paper is on the web at www.slinging.org/25.html.

## Chapter 10

1. The conceit extended to a daughter, too, Constantia. Herodian's observation: 8.8.4.

2. *The Fall of the Roman Empire* 2005: 107.

3. See P. Allen, "The Justinianic plague," *Byzantium* 49 (1979) 5–20.

4. Ammianus 29.1.41–2.4. "Innumerable writings and many piles of volumes" burned publicly were on "the liberal arts and jurisprudence," he says, while private owners "burnt their entire libraries" to avoid being accused of and tortured for practicing "magic." The emperor Valens was as much a vandal as the illiterate hordes who threatened Rome from without. It started before this time. For example, in A.D. 200 Septimius Severus demolished the fortifications of Byzantion and removed "all the books that he could find" in "practically all the sanctuaries" of Palestine and Egypt, Cassius Dio 76.13.2. This may have been hostility to superstition, rather than learning as such, since he is also said to have accurately observed the variation of the sun's motion and the changing lengths of

day and night through summer and winter while he was in northern Britain, Dio 77.13.3. The same cannot be said of the actions of his son Antoninus (aka Caracalla), however, who in A.D. 212 ordered the burning of the books of the Aristotelians, and abolished their *sussitia* (dining clubs for common meals) in Alexandria and elsewhere, Cassius Dio 78.7.3. Then in A.D. 238 Praetorians loyal to Maximinus deliberately set fire to Rome when the population, hostile to the same, cut off the water supply to their camp. The result was that many rich houses—the principal repositories for books at this time—were looted and burned, some unfortunates were burned alive, and "the section of Rome that burned down was greater in extent than the entire area of any city elsewhere," Herodian 7.12.7.

5. For general historical accounts of these centuries, see P. Heather *The Fall of the Roman Empire* (2005) and T. Gregory, *A History of Byzantium* (2005).

6. A. Cameron, "The date of the Anonymous *De Rebus Bellicis*," in M. W. C. Hassall (ed.) *De Rebus Bellicis* BAR IS 63 (1979) 1–11.

7. T. Gregory, *A History of Byzantion*, 2005: 15.

8. See E. A. Thompson, *A Roman Reformer and Inventor*, 1952, for text, translation, and notes on the Anonymous *De rebus bellicis*.

9. There are three chapters on coinage, its current unreliability and the author's proposed reform of the mint; one chapter on the current corruption of provincial governors and his advice to appoint decent men to the post; and one chapter on current and proposed terms of military recruitment and service, after which the anonymous turns to military devices, which actually constitute two-thirds of the treatise. The Panopolis papyri portray the sort of corruption he may have had in mind; see e.g. 2.229–244, dated (in our terms) 16 February 300.

10. I follow Marsden for the interpretation of the last sentence.

11. The Bodleian MS was produced in 1436 for Pietro Donato, Bishop of Padua, copied from a borrowed ninth- or tenth-century MS in Speyer cathedral (of which almost nothing remains), which was itself perhaps a copy of an earlier Carolingian MS. The illustrations were made by Peronet Lamy. R. Ireland, *Anonymi Auctoris Libellus De Rebus Bellicis*, ed. trans. and commentary, published as Part 2 of M. W. C. Hassall (ed.) *De Rebus Bellicis* BAR IS 63 (1979) esp. "The transmission of the text," pp. 39–75.

12. M. Henig, "Late Antique book illustration and the Gallic Prefecture," in M. W. C. Hassall (ed.) *De Rebus Bellicis* BAR IS 63 (1979) 17–28 at 17.

13. Actually, other MS illustrations (P, M, M*) clearly show that his left foot is supposed to be on the ground while his right rests on the hub of the axle. Our illustrator has drawn the left leg a bit short.

14. J. Alexander observed that the majority of the DRB's illustrations are "landscape views from a normal view-point showing the machines in use by the soldiers . . . a realistic context for the objects is thus provided," "The illustrations of the Anonymous *De Rebus Bellicis*," in M. W. C. Hassall (ed.) *De Rebus Bellicis* BAR IS 63 (1979) 11–15 at 13.

15. Maurikios *Strategikon* 12 B 6. The crews of these ballistae, together with carpenters and metal workers, seem to have been regarded as a discrete unit, under a single officer.

16. Marsden: 1971: 242–243. Cranequin: See Viscount Dillon, "Military archery in the Middle Ages," chapter 7 in A. Lever, ed., *Archery*, 1894.

17. Vitruvius 9.8.5. There is a reconstruction of the device in M. J. T. Lewis, "Theoretical hydraulics, automata and water clocks," in O. Wikander, ed., *Handbook of Ancient Water Technology* 2000: 343–370 at 364.

18. By E. Gil, I. Filloy and A. Iriarte, "Late Roman military equipment from the city of Iruña/Veleia (Alava, Spain)," *Journal of Roman Military Equipment Studies* 11 (2000) 19–29. The part found is iron, semicircular, and 94 x 41 x 2 mm, and is half of an elevating device of exactly the type described for a dioptra; see Heron's treatise of that name, dating to the first century A.D., and M. Lewis, *Surveying Instruments of Greece and Rome* 2001: chapter 3.

19. The screw press had existed since the first century at least (a carbonized example survives at Herculaneum) and reappears certainly as a spanning mechanism in some designs of spingardae (Latin) / spingarda (Italian), springald (English) / Springolffen (Habsburg lands) / espringale (French). See J. Liebel, *Springalds and Great Crossbows* trans. J. Vale, 1998, 12–14 and Appendix 1.

20. All these MS illustrations are reproduced as black and white plates in Hassall, ed., 1979.

21. Museum of London acc. Nos. 20816 and 20815, weighing 27g and 23g, respectively. I wish to thank Penny Hall for showing and discussing with me the bolt heads and a variety of other items in the Museum's collection. The lead weight a little behind the head, rather than the head itself, is the trademark of *plumbatae*, and it is not obvious that the Burgh Castle head, for example, which lacks the lead weight, should be interpreted as a *plumbata* head rather than a large caliber ballista head.

22. H. Hodges, "The Anonymous in the Later Middle Ages," in M. W. C. Hassall (ed.) *De Rebus Bellicis* BAR IS 63 (1979) 119–126 at 121–124. Hodges takes a very different view of the functionality of such weapons in the times both of the DRB and of Leonardo.

23. See J. P. Wild, "Fourth century underwear with special reference to the Thoracomachus," in M. W. C. Hassall, ed., *De Rebus Bellicis* BAR IS 63 (1979) 105–110 at 106–107.

24. DRB 15. Anon. *Strategy* 16.14–27, 54–64.

25. B. Cotterell and J. Kamminga *The Mechanics of Pre-Industrial Technology* 1990: 180.

26. This was made of inflated skins which, having been dressed by a specific (Arabian) process, formed flattish air sacs instead of rounded ones. Each skin forms a module, with hook on one side and eye on the other, and to construct a bridge, a number of them are inflated (with bellows, according to the accompanying illustration), hooked together, and then the whole is rolled up. The inflated bridge is then released to unroll itself across the river, a little like a modern aircraft emergency chute I suppose, only much slower of course. It is secured with ropes above and below, and made safe underfoot by laying mats over the potentially slippery leather.

27. Generally, Marsden's views on late Roman developments no longer find favor with scholars of that period, e.g., P. Southern and K. R. Dixon, *The Late Roman Army* 1996:

158–160. R. P. Oliver, "A note on the de rebvs bellicis," *Classical Philology* 50 (1955) 113–118, took Thompson to task for his lack of understanding of the mechanics of artillery, and argued in 1955 that Prokopios' description resembles DRB's in all important particulars to such an extent that the drawing of the *ballista fulminalis* in the *De rebus bellicis* MS could serve as an illustration of the passage in Prokopios *Wars* 5.21.14–17, given below. Readers must decide for themselves; I do not agree. Before Oliver, E. Schramm (1918), S. Reinach (1922), and I. A. Richmond (1930) read DRB as referring to a steel bow. Marsden admitted that Prokopios' description could be of "the non-torsion *manuballista*" (1971: 247).

28. M. W. C. Hassall, "The Inventions," in Hassall (ed.) *De Rebus Bellicis* BAR IS 63 (1979) 77–95, at 82.

29. I. Kelso, "Artillery as a classicizing digression," *Historia* 52 (2003) 122–125. See also J. Matthews, *The Roman Empire of Ammianus Marcellinus* 1989 301–302. Marsden called all the late authors "certainly laymen, as far as artillery was concerned," 1971: 234. Their vagueness is the direct cause of the variety of interpretations now offered for their words.

30. Note this analogy with a mechanical saw. Ausonius, who was a contemporary of Ammianus in the fourth century A.D., refers to water-powered marble saws on the Mosel River (*Mosella* 362–364). A water-powered cam-operated marble saw is now attested over 1,900 miles away (> 3,000 km) as the crow flies, in a sixth- or seventh-century A.D. context in the old sanctuary of Artemis at Gerasa (mod. Jerash) in Israel, J. Seigne, "Une scierie méchanique au VIe siècle," *Archéologia* 385 (2002) 36–37, with preliminary plan of the installation and photograph of an abandoned column part sliced lengthwise into slabs. As Seigne observes p. 37, "la sophistication dans la simplicité de la machine de Jerash indique qu'elle ne correpondait pas à un prototype." Another has recently been discovered at Ephesos, and a relief depicting another one found at Hierapolis; papers at a conference on "Force hydraulique et machines à eau dans l'Antiquité romaine," Pont du Gard, 20–22 September 2006. On Ausonius and the Mosel region, see Heather 2005: 40–45.

31. C. W. Fornara, "Studies in Ammianus Marcellinus II," *Historia* 41 (1992) 420–438 esp. 421–427.

32. Julian and artillery: e.g. Ammianus 17.1.12; 18.7.6. His speed: Ammianus 21.9.6. Julian's *Mechanics* is no longer extant. As Gilliver observes, *The Roman Art of War* 1999: 177, there is no reason to dispute his authorship of a treatise on such a topic, as his interest in it is evident in Ammianus' account of his campaigns. Military units in the *Notitia*: e.g., *ballistarii seniores*, Oriental 7.8; *ballistarii Theodosiaci*, Or. 7.21; *ballistarii Dafnenses*, Or. 8.14; *ballistarii iuniores*, Or. 8.15. It also records two *fabricae ballistariae*, at Autun and Trier in Gaul, Occidental 9.33. For a summary diagram of the *Notitia*'s information (though it is not sufficiently detailed to indicate which are *ballistarii*) see Goldsworthy 2003: 204, and for an overview of army organization since Diocletian's reforms, Gregory 2005: 38–39 and Heather 2005: 64–67.

33. Crocodile: Ammianus 22.15.15. Persian infantry: 19.7.4. Issued to men in trenches: Maurikios *Strategikon* 12 B 21.

34. 4.29, Milner trans. modified. It is not coincidental that one of the ballistae in the *De rebus bellicis* is called the *fulminalis*, the lightning or thunderbolt ballista.

35. Factory: 31.6.2. Romans had the upper hand and Goth injuries: 31.15.13. Catapults shooting in all directions: 31.15.6. Novelty: 31.15.12.

36. State factory making mural tormenta at Amida: Ammianus 18.9.1. His involvement: 18.8.13.

37. Ammianus 19.2.7; tragularii: Vegetius 2.15.

38. Mobile mural artillery at Amida: Ammianus 19.5.6–7. Function of ballistae at Bezabde: Ammianus 20.11.20–22. Hasty removal of engines: 20.11.23.

39. Sniper: Ammianus 19.1.5–7. Bezabde: Ammianus 20.7.2.

40. Unigastos: Prokopios 6.27.13–14. Justinian's parsimony: 7.1.33. Belisarios' generosity: 7.1.8. According to Prokopios, Justinian's repeated and prolonged failure to pay the troops must be held as the principal cause of corruption, rebellion, treason, or suchlike, again and again through his history, e.g. 7.11.14; 7.12.7–8; 7.30.7–8; 7.36.7; 7.36.26; 8.26.6. Equally, in the west, ensuring that the troops were paid and were not forced to pillage the people they were supposed to be defending is a notable feature of many of King Theodoric's letters (actually written by Cassiodorus) to his lieutenants, e.g. 2.5; 3.38; 5.11; 5.13; 10.18.

41. For plans of Singara's fortifications and a summary history of the site see Oates 1968: 97–106.

42. Singara's armament and fall: Ammianus 20.6. Use of those munitions against Amida: 19.2.8.

43. Ammianus 19.5.1, repeated 19.7.2.

44. Ammianus 19.2.15 Loeb modified.

45. Constantius: Ammianus 21.6.6. Sericus: Ammianus 28.1.8. Valentinian: Ammianus 30.9.4.

46. Ammianus 19.6.5, 19.7.4. Sixth-century use: Maurikios *Strategikon* 12 B 21 and e.g. Anon *Strategy* 26.13–16.

47. This "measure" has had a particularly long life, going back to Prokopios at least; he says that each side of Hadrian's tomb is about a stone's throw in length, *Wars* 5.22.13.

48. T. Esper, "The replacement of the longbow by firearms," *Technology and Culture* 6 (1965) 382–393 at 390.

49. 4.22, and individual items appear elsewhere singly or together.

50. Less useful soldiers have onagers and ballistae: 3.3. Sinew the best material for springs: 4.9.

51. Dismemberment: Ammianus 24.4.28: During the siege of Ktesiphon an architect, whose name he forgot [the final insult!], was standing behind a scorpion [*sic*; read onager] when an insecurely-loaded stone was hurled backwards. The man's chest was crushed and he was thrown on his back, and his body was so damaged that parts of it could not be identified. Discharging empty: Ammianus 19.6.10.

52. E.g., J. Haldon, *Warfare, State and Society in the Byzantine World 565–1204*, 1999.

53. Biton, 62.9. Marsden 1971: 98 n. 55.

54. Specialist sharp-caster operators: 5.27.6. Unspecialised onager operators: 5.28.29.

Goths terrified: 5.22.24; 5.27.10. Goths unfamiliar with them: 5.27.15. Goths using Roman equipment: e.g., Ammianus 31.5.9; 31.6.3.

55. The wolf (*lupus*) was an iron framework resembling a portcullis but with spikes protruding from the intersections of the latticework, at right angles to the frame, rather like teeth sticking out of a jaw, which is presumably the origin of the name. The device was stood against the city wall, spikes pointing out, and the top edge was pushed to cause it to fall forward onto hapless assailants caught too close to the gates.

56. Belisarios' ladders: Prokopios 5.10.21. Vittigis' siege tower: 5.21.3.

57. Lemboi: 5.5.13–16. Mills: 5.19.8–28.

58. Goths copying: 7.19.16. Lembos: 7.19.18–19 and 7.19.3–4 for the target.

59. Impaling: 5.23.9–12. Vegetius' towers: 4.17. Vegetius' large ballistae: e.g., 4.18. First Gothic siege tower: 5.22.1–9, 5.24. Second Gothic siege tower: 6.12.2–13. End narrative: Prokopios 8.35.9.

60. Dio 77.10.3–5.

61. Redundancy: e.g., as the empire grew the cities of Italy felt their walls were not needed anymore and let them fall into disrepair, Herodian 8.2.4. Recycling: e.g. in the reign of Valens the walls of Chalcedon were used to build baths, Ammianus 31.1.4.

62. Martyropolis and Petra: Prokopios 1.21.8 and 2.17.15. Clever man: Malalas 18.66. Petrans' towers: Prokopios 2.17.22. Powerful stone-throwers: Anon. *Strategy* 12.3–6.

63 Fond of innovations: Prokopios 1.23.1. Bridging kit: Prokopios 2.21.21–22. Timber: 2.29.1.

64. Justinian's building program was the subject of another book by Prokopios. For the context of this program, see Gregory 2005: 128–135. For the cost, some idea can be gathered from Heather 2005: 400.

65. Dennis 1985: 3.

66. Anon. 13.72–88; Ammianus 20.11.8; 24.2.10; Vegetius 4.6. Earlier versions: e.g., Aineias Taktikos 32.1 and 9–10; Philon *Polior*. 95.35 C63.

67. Matting absorbs impact: Vegetius, 4.6. Cilician cloth: Prokopios 2.26.29–30. Anon. 13.115–120. Tyrians: Chapter 4.

68. See Lendle *Texts und Untersuchungen* 1983: 117–127.

69. Groups: Anon. 14.27–32. Opposed landing: 19.15–18. Prefabrication: 19.56–65.

70. Theophylaktos Simokatta *Historia* 2.16–17.

71. *Miracula S. Demetrius* ed. Lemerle, 148–154. Illustration: Nicolle's editorial note in text in Nicolle (ed.) 2002: 124, not illustrated, unfortunately. The earliest Chinese reference to the beam sling is centuries later, dating to 1044, and refers to engines going back to the last half of the tenth century. Loose modern usage sometimes labels the beam-sling a trebuchet, failing to distinguish between machines powered by traction and by gravity. Clear and accurate use of terminology is essential to avoid confusion. Similar problems, with similar cause, beset the history of technology in China; see J. Needham et al. *Science and Civilisation in China* 5.7 (1986) 11–12, 22, 276. On a related issue, it is commonly said that the stirrup was a Chinese invention and the Avars introduced it into Europe; e.g., A. Rona-Tas, *Hungarians and Europe in the Early Middle Ages* (1999) 103.

However, this was a reintroduction, for its predecessor, a simple leather loop, may have preceded it by several hundred years. Lynn White Jr. attributed the stirrup, qua big-toe stirrup, to the Indians in the second century B.C., "The act of invention" (1962) 489. Now the earliest stirrup-like device, a hook in which the rider's foot rests, hanging by a chain, is dated to c. 400 B.C. and comes from Kul Oba in Skythia; H. Nickel, "The mutual influence of Europe and Asia," in Nicolle (ed.) *A Companion to Medieval Arms and Armour* 2002: 107–125 at 108 and plate X 5.

72. There is a significant amount of evidence from recent history for the stimulus that such syntheses can provide; see, e.g., P. K. Hoch, "Emigrés in science and technology transfer," *Phys. Technology* 17 (1986) 225–229 for examples from the twentieth century. It might also be significant that the mechanics of the beam-sling are very similar to those of the shadoof, a very simple water-lifting device that goes back to classical times.

73. "Manual" is from *manus*, Latin for "hand," which corresponds to *kheir* in Greek.

74. Maurikios 10.1. Anon *Strategy* 11, 15–16. Compare the Chinese beam-sling, which seems to have had a range of about 85 yards; Needham and Yates 1994.

75. The debate is well summarized by Rogers, *Latin Siege Warfare in the Twelfth Century*, 1992 Appendix 3, and Bradbury, *The Medieval Siege* 1992: 250–256. I owe these references to John France.

76. J. M. Ziman "Ideas move around inside people," *The Fifth J. D. Bernal Lecture* Birkbeck College, London (1974) 5–6.

77. Prokopios *Secret History*, 6.15, talking of Justinus, A.D. 518–527, and Anonymous Valesianus, *History of King Theodoric*, 14.79, talking of his contemporary in the west Theodoric, King of Italy from A.D. 493–526. Justin appointed Prokopios to Belisarios' staff, so Prokopios was well placed to know details like this.

78. Prokopios' recommendation: 6.23.23–29. Vegetius: 2.22, 3.5. Polyainos: , e.g. 3.9.5; 4.3.26; 5.16.4.

79. Prokopios 5.9.11–17. The Goths later tried the same trick against Belisarios' forces when the latter had seized Rome, but he anticipated it and had them blocked, 6.9.1–11.

80. Gates: Prokopios 7.24.9–12. New style ram: 8.11.29–31. Elephants as siege towers: 8.14.35–36.

81. 21.81–84, G. Dennis ed. *Three Byzantine Military Treatises* 241–335.

82. This seems to be a variant of *elakatai/alakatai*, which Haldon thinks refers to the windlass needed to span the weapon (see LSJ II.4); alternatively it may mean "spindle-shaped," which is the basic meaning of the word (LSJ I).

83. Haldon1999: 136.

84. B. Hall *Weapons of War* p. 201.

85. Editor's note on J. Coulston's paper "Arms and armour of the late Roman army" in Nicolle (ed.), 2002: 23.

86. J. France *Victory* (1994) 49–50. Robert flourished circa 1092, and was described as an expert in siege machinery and builder of a great stone-thrower. The siege of Paris was A.D. 885–886. Abbo died A.D. 923. When the Avars attacked Constantinople in 626, they used *petrariae*, which the Whitbys translated without comment as stone-throwers,

*Chronicon Paschale* p. 174. However, the precise nature of this device is unclear. The Thessalonikans had *petrariae* when they were attacked by the Avars in the late sixth or early seventh century, and on that occasion sailors described as *empeiromagganous*, experienced with machines, helped the besieged operate their *petrariae* (Haldon 137). Sailors would be particularly experienced handling one-legged crane-like devices, which were used to load and unload vessels, as well as cross-beams and rigging generally. This would also explain why they are associated with large numbers of ropes; see sources cited by Liebel 1998: 3 (though he thinks they were ballistae).

87. E.g., Chevedden (1995) 163, Haldon (1999) 137.

88. Mons-en-Pévèle: Liebel (1998) 18–19. Siege of Paris: France *Victory* (1994) 49.

## Appendix 1

1. Vitruvius does not give a formula for the palintone, apparently to avoid inflicting geometry and multiplication on his audience (10.11.1–2), and gives a list of spring-hole dimensions for popular calibers instead (10.11.3).

2. Ktesibios does not give dimensions of components in spring-hole diameters or even standard measures for most components; I take this as further evidence that this treatise comes from the third century B.C. and is Ktesibios' text, not the first century A.D. and Heron's.

3. 1990: 190–192.

4. Reinschmidt Unpublished: 20–21.

5. Holley 1994: 355 (stone), 358–359 (weight versus diameter). His Appendix B, p. 364, giving a more or less uninterrupted progression of weights from 0.6 to 5 kg with rarely more than a pair weighing "the same" (to one decimal point) illustrates the general point about variability as well as anything could.

6. The balls found at Goritsa, for example, were grouped into calibers of 6–9 kg (nominal 20 mina), 9–12 kg, 12–15 kg (nominal 30 mina), 21–24 kg (nominal 50 mina), and 24–27 kg (nominal 1-talent), S. C. Bakhuizen, "La grande batterie de Goritsa et l'artillerie défensive," in Leriche and Tréziny (1986) 315–321 at 321 n. 2. Balls of 9 kg, 12 kg, and 24 kg here stand on the margins and could go into "calibers" either side.

7. See e.g. James 2004: 213–214. Although ancient and modern authors are concerned about missile length, the mass is what most determines performance, for sharps as well as for stones. James notes some of the difficulties obtaining weight data.

8. Calahorra: Frías, Martínez, and Sábada "Nuevo testimonio de las guerras sertorianas en Calahorra: un depósito de proyectiles de catapulta," *Kalakorikos* 8 (2003) 9–30, esp. table on p. 13, figures 8–12, and discussion on p. 11–14. Wallsend: M. P. Wright, M. W. C. Hassall, R. S. O. Tomlin, "Roman Britain in 1975," *Britannia* 7 (1976) 290–392 at 389.

9. Reinschmidt Unpublished: 9–10.

10. James 1994: 94 and figures 1 and 2.

11. Wilkins 2000: 93.

12. Appian *Illyrian Wars* 2.11. On fire-arrows and incendiary catapult bolts see S. James, "Archaeological evidence for Roman incendiary projectiles," *Saalburg Jahrbüch* 39 (1983)

142–143.

13. I have in mind something similar to Cotterell and Kamminga's "the standard spring length was nine times its diameter," 1990: 191. This is true of palintones, but not euthytones.

14. (i) the height of the side stanchions, plus (ii) the thickness of the hole-carrier and (iii) its equivalent below the stanchions, which together form the frame holding the spring, plus (iv) the height of the washers top and (v) bottom.

15. The Caminreal washer, like those from Auerberg and Sounion, has no rebate for the lever. By contrast, the Cremona 2 washer has a rebate more than one-third its total height. The extent to which rebates in the washer, or absence of same, may make a difference to the total length of the spring is perhaps best illustrated graphically. See Vicente et al. 1997: 182 for a diagram showing the currently known variety of washer types and sizes.

16. Reinschmidt Unpublished: 6, 8.

17. Reinschmidt Unpublished: 23–24 and figure there.

18. Reinschmidt Unpublished: 18. The arms will also crash into the stanchions if the levers have been twisted to tighten the springs *after* the arms have been inserted.

19. The terms are *katatonus* (short-sprung) and *anatonus* (long-sprung). Wilkins (2000: 79) translates these as under-sprung and over-sprung, respectively. I have avoided this translation because the former English term seems to me to imply inadequate power and the latter excess, whereas the short-spring *katatonus* is so hard to withdraw that it requires longer arms, and the long springs of the *anatonus* are said to be relatively lax, which imply the opposite. I use the term over-sprung to mean a catapult that has a spring-diameter greater than its other dimensions suggest it should have—a machine whose spring-holes have been enlarged after setting the ratios for the other parts of the catapult.

20. Vicente, Pilar Punter, and Ezquerra "La catapult tardo-republicana y otro equipamiento military de 'La Caridad' (Caminreal, Teruel)," *Journal of Roman Military Equipment Studies* 8 (1997) 167–199, at 180 fig. 22.

21. 1971: 161 n. 31 and 57–58 n. 36.

22. Remains of this device have apparently been found in Vani, O. Lordkipanidzé, "The fortification of ancient Colchis," in Leriche and Treziny (1986) 179–184, at 183 n. 13. It is at this stage that most modern reconstructions part company with ancient instructions, relying instead on turning the levers to tension the ropes, which results in twisted instead of straight springs when the machine is braced. As a result, the arms crash into the stanchions on such reconstructions, whereas they would not do on a pre-tensioned catapult.

23. Marsden (1971: 57–58 n. 36) offers a much more complicated solution based on areas rather than diameters, and deducts the area occupied by the lever, with a different result, which seems to overestimate the cordage. I prefer to offer the sort of calculation I think Ktesibios and Philon had in mind for this. I do not deduct the lever because it is variable and sits totally or partially above the spring-hole (depending on the model), and I believe that all references to the spring-hole mean the hole as measured across the bottom not the top (which is sometimes splayed and usually slightly wider across the top than the bottom), without this sporadic complication. I assume that hair *can* be stretched

to this degree without breaking, though I do not know what figure is now put on its resilience. Ktesibios claims it was similar to sinew: "the spring-cord on the arms is also made from women's hair. For being slender and long and maintained with a lot of oil, when plaited it acquires much vigour, such that its strength is not out of tune with sinew," *Bel.* 112.4–6 Heron.

24. The magister ballistarii of the Ermine Street Guard reports another, eminently practical, method to ensure that both arms are equally tensioned: measure the angle of the bowstring to the stock with a set square: they should be perpendicular to one another.

25. Landels 1978: 110. Cotterell and Kamminga give the mechanical properties of sinew (and other materials, including modern spring steel) 1990: 68 table 3.1. Note sinew's high maximum elastic strain (8.3 percent) and resilience per unit mass (3,200 J/kg).

26. D. Stephenson discovered this in 1996 in research for his undergraduate dissertation. Coulston has a relatively good discussion of sinew, "Roman archery equipment," in M. C. Bishop (ed.) *The Production and Distribution of Roman Military Equipment* BAR IS 275 (1985) 220–336 at 253. It has since been made by others, Wilkins 2003: 36, but is a very time-consuming and tedious procedure.

27. Reinschmidt Unpublished: 18; Baatz "Waffenwirkung antiker Katapulte," in Baatz 1997: 138.

28. Wilkins 2000: 93. The former shot a bolt weighing 200g, the latter 102 g.

29. Reinschmidt Unpublished: 15–17.

30. Reinschmidt Unpublished: 15–17.

31. Denny 2003: 373–374.

32. *The Book of the Crossbow* (1903) 199.

33. Reinschmidt Unpublished: 19.

34. "Scorpio and Cheiroballistra," 2000: 79–80.

35. The MSS have 1 1/3 but Marsden corrected and has since been shown right.

36. It is not clear why this is so; perhaps because there is a large collection of uncatalogued shot at Hatra (photograph in Campbell 2003: 20). This could have been ammunition for other machines, or even, since Hatra was repeatedly besieged by the Romans, incoming shot rather than the defenders'.

37. Wilkins 2003: 65 fig. 51.

38. Scott 1985: 160.

39. W. Scheidel, "Finances, figures and fiction," *Classical Quarterly* 46 (1996) 222–238.

40. The length of the bowstring can serve as a simple (slightly generous) guide to the width of a machine since it is 2.1 times the length of the arm, and the arm is 6D on the palintone and 7D on the euthytone.

41. Payne-Gallwey gives 1/2 oz. as a standard size for a sporting stonebow (e.g. p. 199). He adds that if it is shot at a metal target at 20 yards, more than half the bullet will be flattened. Cf. the flattened *glandes* that are sometimes found. Held at 45°, such a bullet from such a bow could reach a maximum range of 300 yards. He also notes that this weapon is absolutely quiet (which is why it was favored for hunting, and see now Karger

et al. 2004: 334), p. 199. The lightest *glandes* currently known (they have not been well studied) is 18 g.

42. SEG 31.1617.

43. Philon says this is sufficient to destroy enemy artillery, *Pol.* C6 & D17.

44. Philon says this is sufficient to destroy siege engines, *Pol.* C67–68.

45. So Marsden 1969: 145.

46. Marsden 1969: 145 has 12 feet.

47. But this component should be only 12 cubits = 5.54m = 18′2″ in Philon *Pol.* A71, 85.1–2.

48. Marsden 1: 123 calculated 3 feet here since he was using 19D as the stock length for euthytones as well as palintones.

49. Marsden 1: 123 has 4′6″ (using 19D).

50. Marsden 1: 123 has 6 feet (19D).

51. We occasionally hear of larger sharps, e.g. those 12 Roman feet long (7 1/2 cubits, 11 1/2 British feet or 3.5 meters) shot at Caesar's forces during the siege of Massilia in 49 B.C., but these would have been shot, almost certainly, from palintones (*Civil War* 2.2). *Sudis* (stake, spike, pike, pile) seems to be used regularly by Latin historians for large sharps.

52. = 288 daktyls.

53. = 96 daktyls.

54. Compare Baatz and Feugere (1981) esp. 209.

55. The cylinders sit on top of the rings, as washers on a Philonian or Vitruvian catapult. Heron gives their height as slightly less than the spring diameter (2 daktyls versus 2 1/3 daktyls).

## Appendix 2

1. Schramm 1918: 40–46.

2. Baatz in G. Ulbert *Der Auerberg I,* 1994, 173–187.

3. Initial publication: J. Cabré *Archivo Espagnol de Arte y Arqueologia* 1 (1925) at 14. Remains of a second machine were discovered in 1942. Much more detail: F. G. Díez, "Las catapultas de Azaila (Teruel)" *Gladius* 5 (2002) 293–302. The remains represent at least three catapults, and shot from the site indicates more calibers. It seems that six washers were found, but most were lost during the Spanish Civil War. There is currently one washer, one counterplate, and lots of plating, but the rest of the finds are being sought.

4. Baatz in B. Cunliffe *The Temple of Sulis Minerva at Bath* 2, 1988, 7–9, and pl. 5–7.

5. Vicente et al. 1997.

6. Baatz *Mitt DAI Rom* 87 (1980) 283–299. Reprinted in Baatz 1994. Vicente et al. 1997.

7. I. Kayumov and A. Minchev (2013). "The kambestrion and other Roman military equipmen from Thracia" in M. Sanander, A. Rendic-Miočevic, D. Tončinic, I. Radman-Livaja (eds.) *XVII ROMEC* Zagreb 2010, Zagreb.

8. Baatz *Mitt DAI Athen* 97 (1982) 211–233. Reprinted in Baatz 1994. Twenty-one washers (of an original twenty-eight) from seven different catapults found here.

9. Gudea and Baatz (1974) 50–72; Baatz (1978) "Recent finds."

10. Baatz 1978; Hatra 2 was represented by finds on display in the Mosul museum when Baatz wrote this paper.

11. Steidl, B. "Römische Waffen und Ausrüstungsteile" in G. Seitz ed. *Im Dienste Roms* 2006: 307–317.

12. Feugère and Baatz 1981. Reprinted in Baatz 1994.

13. Baatz *Arch Anz.* (1985) 679–691. Reprinted in Baatz 1994.

14. Gudea and Baatz (1974) 50–72; Reprinted in Baatz 1994. Baatz (1978) "Recent finds."

15. W. Gaitzsch, "Hellenistische Geschützteile aus Pergamon," in C. Dobiat *Festschrift O.-H. Frey*, 1994: 235–42.

16. Baatz, *Saalburg-Jahrb.* 44 (1988) 59–64. Reprinted in Baatz 1994.

17. C. Boube-Piccot, *Les bronzes antiques du Maroc IV* (1994) no. 465, pp. 188–195 and pl. 49, 96–98. See also pp. 25–27.

18. H. Williams, *Échos du monde antique* 11 (1992) 181–187. For potential historical contexts for this catapult, see H. Goette, "Cape Sounion and the Macedonian occupation" (2003), which includes photos of the bastion and inscribed ballista balls found at the site, C. Habicht, "Athens after the Chremonidean War: some second thoughts," same volume, and D. Gill, "Arsinoe in the Peloponnese: the Ptolemaic base on the Methana peninsula" (forthcoming), which discusses the Ptolemaic camps set up during the Khremonidean War at Koroni, a little north of Sounion, and Patroklos's Island, just offshore.

19. Baatz, "Katapultfunde 1914–1988," in Baatz 1994: 7.

20. C. Boube-Piccot, *Les bronzes antiques du Maroc IV* (1994) nos. 466 and 467, pp. 195–197 and pl. 50, 99–100. See also pp. 25–27. Boube-Piccot believes that both washers came from the same machine, but they need not have. Provenance data is missing for the second and they do not have consecutive museum inventory numbers, suggesting that they were not catalogued together. There is a wide variety of calibers attested in the range of shot and bolt sizes from Volubilis, e.g., shot from 36 g to 12.75 kg (inv. nos. 463 and 485, respectively, and 484 is larger but no weight is recorded), so there were obviously many more catapults present, in attack or defense of the site.

21. *PM* 12 February 2005, Geschichte section (newspaper report of find). Interim publication: H.-J. Schalles, "Eine frühkaiserzeitliche Torsionswaffe aus der Kiesgrube Xanten-Wardt," in H. G. Horn et.al. *Archäologie in Nordrhein-Westfalen*, 2005: 378–381. I owe this reference to Jon Coulston. See now H.–J. Schalles, *Die fruhkaiserzeitliche Manuballista aus Xanten-Wardt,* Xantener Berichte 18, Zabern, Mainz, 2010: 35–43.

# *Bibliography*

## Ancient Works


Aineias Taktikos *How to Survive Under Siege*. Translated with a historical commentary and introduction by D. Whitehead, London, 2002. Also in Loeb.

Ammianus Marcellinus *Res Gestae* J. C. Rolfe, Loeb, 1935–39.

Anonymous *Chronicon Paschale* trans. with notes and introduction by M. and M. Whitby, Liverpool, 1989.

Anonymous *De obsidione toleranda* (On withstanding siege) ed. H. Van der Berg, Leiden, 1947.

Anonymous *De rebus bellicis* (Fourth century A.D.) ed., trans., and commentary, by R. Ireland, *Anonymi Auctoris Libellus De Rebus Bellicis*, published as Part 2 of M. W. C. Hassall, ed. *De Rebus Bellicis* BAR IS 63 (1979). Also E. A. Thompson, *A Roman Reformer and Inventor* text, trans., and introduction to the *De Rebus Bellicis*, Oxford, 1952, reprint Chicago 1996.

Anonymous *Miracles of Saint Demetrius* ed. P. Lemerle, *Les plus anciens recueils des miracles de Saint Démétrius* vol. 1 (text), Paris, 1979.

Anonymous *Notitia Dignitatum* (c. 395–408 A.D.), selections in translation in R. Rees, *Diocletian and the Tetrarchy*, Edinburgh, 2004.

Anonymous Byzantine Περὶ Στρατηγίας *(On Strategy)* (C6 A.D.) G. T. Dennis, *Three Byzantine Military Treatises*, Washington, D.C., 1985.

Anonymous *Taktikon* (*Tactics*) (Tenth century A.D.) G. T. Dennis, *Three Byzantine Military Treatises*, Washington, D.C., 1985.

Anonymous Valesianus, *History of King Theodoric*, Loeb (with Ammianus) J. C. Rolfe, Loeb, 1939.

Appian *Civil War, Illyrian Wars, Mithridatic Wars* H. White, Loeb, 1912–13.

Arrian *Anabasis (Campaigns of Alexander)* P. A. Brunt, Loeb, 1976–83.

Arrian *Taktika* and *Expedition Against the Alans* J. G. DeVoto, Chicago, 1993.

Athenaios Mekhanikos *On Machines* D. Whitehead and P. Blyth, Historia-Einzelshrift 182, Stuttgart, 2004.

Biton *Construction of War Engines and Catapults*, in E. W. Marsden (1971) *Greek and Roman Artillery: Technical Treatises* Oxford.

Caesar *Civil Wars* A. G. Peskett, Loeb, 1914.

Caesar *Gallic War* H. J. Edwards, Loeb, 1917.

[Caesar] *Alexandrian War, African War, Spanish War* A. G. Way, Loeb, 1955.

Cassiodorus *Variae* Sam Barnish, partial English translation, Liverpool, 1992. A full trans. by the same author is forthcoming, Ann Arbor, Michigan.

Cassius Dio *Roman History* E. Cary, Loeb, 1914–27.

Celsus *De Medicina* W. G. Spencer, Loeb, 1935–38.

Curtius *History of Alexander* J. C. Rolfe, Loeb, 1946.

Diodoros Siculus *Library of History* C. H. Oldfather, R. M. Gees, C. L. Sherman, F. R. Walton, and C. B. Wells, Loeb, 1933–67.

Dionysios Halikarnassos, *Roman Antiquities* E. Cary, Loeb, 1937–50.

Eutocius *Commentary on the Sphere and the Cylinder* J. L. Heiberg and E. Stamatis, *Archimedis opera omnia cum commentariis Eutocii* Leipzig, 1915.

Florus *Epitome of Roman History* E. S. Forster, Loeb, 1929.

Herodian *History of the Empire After Marcus* C. R. Whittaker, Loeb 1969–70.

Herodotos *Histories* A. D. Godley, Loeb, 1920–25.

Heron *Ktesibios' Belopoiika* in E. W. Marsden (1971) *Greek and Roman Artillery: Technical Treatises* Oxford.

[Hyginus Gromaticus] *De munitionibus castrorum* M. Lenoir, Les Belles Lettres, Paris, 1979.

Isokrates *Epistles I (Dionysios), 2 (Philip 1), 9 (Arkhidamos)*, L. van Hook, Loeb, 1945.

Josephus *Jewish War* H. St. J. Thackeray, L. Feldman, R, Marcus, and A. Wikgren, Loeb, 1926–65.

Justin *Epitome of Trogus' History of Philip* J. C. Yardley, Atlanta, 1994.

Ktesibios *Belopoiika* in E. W. Marsden (1971) *Greek and Roman Artillery: Technical Treatises* Oxford, under the name of Heron's *Belopoeica*.

Livy *History of Rome* B. O. Foster, F. G. Moore, E. T. Sage, and A. C. Schlesinger, Loeb, 1919–59.

Lucan *Civil War* (also known as the *Pharsalia*) J. D. Duff, Loeb, 1928.

Lysias *Oration* 33, W. Lamb, Loeb, 1930.

Malalas, John (sixth century) *The Chronicle: A Translation*, by E. Jeffreys, M. Jeffreys and R. Scott, Australian Association for Byzantine Studies 4, Melbourne, 1986.

Maurikios (Maurice) *Das Strategikon des Maurikios* (early seventh century A.D.), Greek text ed. G. Dennis (with German translation by E. Gamillscheg) Vienna, 1981; English translation, Dennis *Maurice's Strategikon*, Philadelphia, 1984.

Nepos *Life of Hannibal* E. S. Forster, Loeb, 1929.

Nikephoros Ouranos *Taktika* E. McGeer "The Taktika of Nicephoros Ouranos," *Dumbarton Oaks Papers* 45 (1991) 129–40; ibid. *Sowing the Dragon's Teeth: Byzantine Warfare in the Tenth Century* Dumbarton Oaks Studies 33, 1996.

Onasander *Strategikon* The Illinois Greek Club, Loeb, 1923.

Panopolis Papyri, in R. Rees, *Diocletian and the Tetrarchy*, Edinburgh, 2004.

Philon *Belopoiika* in E. W. Marsden (1971) *Greek and Roman Artillery: Technical Treatises* Oxford.

Philon *Poliorketika* (*Siege-craft*) and *Paraskeuastika* (*Preparations for a Siege*) are cited with reference to Garlan's edition, *Recherches de poliorcétique grecque,* 1974, which has Greek text, facing French translation, and extensive commentary with plans and diagrams, pp. 279–404. In this edition the text is divided into four sections, labeled A, B, C, and D. Each sentence within a section is numbered; A 5 means section A sentence 5, for example. For the convenience of classical scholars, I give parallel reference to the Thevenot edition. A. W. Lawrence has translated parts of the *Poliorketika* into English, to form part II of his *Greek Aims in Fortification*. These Philonian texts are largely concerned with building fortifications and defense outworks in an age that knew stone-throwing catapults sufficiently powerful to threaten walls.

Pliny the Elder *Natural History* H. Rackham, D. E. Eichholz, and W. H. S. Jones, Loeb, 1938–62.

Plutarch *Life of Aemilius Paullus,* B. Perrin, Loeb, 1918.

Plutarch *Life of Alexander,* B. Perrin, Loeb, 1919.

Plutarch *Life of Antony,* B. Perrin, Loeb, 1920.

Plutarch *Life of Marcellus,* B. Perrin, Loeb, 1917.

Plutarch *Life of Pyrrhus,* B. Perrin, Loeb, 1920.

Polyainos *Stratagems* P. Krentz and E. Wheeler, Chicago, 1994.

Polybios *The Histories* W. R. Paton, Loeb, 1922–27.

Proklos *On Providence, Fate and that which is in our power, to Theodoros, the Mechanic* D. Isaac, *Proclus trois études sur la providence* Paris, 1979. An English translation by C. Steel is forthcoming.

Prokopios *Wars* H. B. Dewing, Loeb, 1914–28.

Seneca *Natural Questions* T. H. Corcoran, Loeb, 1971–72.

Strabo *Geography* H. L. Jones, Loeb, 1917–36.

Suda *Lexicon* A. Adler, Stuttgart, 1928–38. Currently being translated and made available online, http://www.stoa.org/sol.

Tacitus *Annals* J. Jackson, Loeb, 1931–37.

Tacitus *Histories* C. H. Moore, Loeb, 1925–31.

Theopompos *Theopompus the historian,* G. S. Shrimpton, Montreal, 1991.

Tyrtaios *Tyrtaeus* C. Prato, Ateneo, Rome, 1968.

Vegetius *Epitome of Military Science* 2nd revised ed., N. P. Milner, Liverpool, 1996.

[Vergil] *Aetna* J. Wight Duff and A. M. Duff, Loeb, 1934.

Vitruvius *On Architecture*, M. H. Morgan, Harvard, 1914. Vitruvius was himself a catapult builder: *De arch.* 1. praef. 2.

Xenophon *Kyropaedia* W. Miller, Loeb, 1914.

## Inscriptions

CIL I. $2^2$ I 847–888 Inscribed glandes.

CIL XIII 6677 *Negotiator gladiarius.*

Epidaurus: *The Epidaurian Miracle Inscriptions*, text, translation, and commentary, L. R. LiDonnici, Atlanta, 1995.

GM = Maier, F. G. (1959) *Griechische Mauerbauinschriften*, Heidelberg.

IG II$^2$ 120 (Khalkotheka stores) Molonos archon, 362/1. Lines 36–37 = "Boxes of arrows [–]; boxes of catapults two."

IG II$^2$ 468 (from the acropolis) dated to 306/5. Very fragmentary. Mentions stone-throwers (*petroboloi*), the generals, *oxubeleis*, spears/lances, and lead.

IG II$^2$ 554 (Honours for Euxenides son of Eupolis of Phaselis), dated 306/5. A crown awarded to this metic because (lines 13–16) "...he embarked on twelve ships and now has given sinew for the catapults..."

IG II$^2$ 665 (Ephebic honours) Nikias archon, 282. The important part is partially restored: lines 27–28 = "[And the katap]al[ta]phetes Mnesitheos son of Mnesitheos of Kopros."

IG II$^2$ 900 (Ephebic honours) καταπαλταφέτης (catapult instructor).

IG II$^2$ 944, 1006, 1008, 1009, 1011, 1028 (Ephebic honours for their instructor in catapult use) dating between 158–101 B.C.; 1006 (Ephebes repair an old catapult and work on an engine).

IG II$^2$ 1422 (from the acropolis). A very, very fragmentary inscription. Two boxes each of arrows and catapults have been restored, as has the archon's name, giving a date of 371/370.

IG II$^2$ 1440 B (treasurers of Athena), about 350. The "two boxes of arrows and two boxes of catapults" here are conjectural; nothing survives on this part of the stone.

IG II$^2$ 1467 face B column 2 lines 48–56. Date: about 330. (Items in the Khalkotheka store, letter forms of the Lykourgan period, 338–326.) This surviving piece of a larger inscription is the earliest evidence for torsion springs, in this case made of hair. Three of the eight catapults listed here before the stone breaks off are said to be incomplete and unusable. The text and translation are given in Marsden 1969: 56.

IG II$^2$ 1469 face B (Treasurers of Athena and the other gods) Arkhippos archon; 321/0. This stone is quite badly damaged, so much so that Marsden and Garlan can offer slightly different texts, restorations, and numbers of catapult bolts (1,800 v. 1,900). Marsden follows the *IG*. Garlan (216 n. 1) follows Tréheux. We are in any case dealing with a large number of catapult bits and bolts.

IG II$^2$ 1475 B lines 30–35 (Treasurers of Athena and the other gods), about 318/7. Most of this side of the stone has been entirely obliterated. The word "catapult" is restored four times in six lines from a few letters each time; "hair spring" is a certain restoration; "sinew spring" is conjectured twice. The text and translation are given in Marsden 1969: 69.

IG II$^2$ 1487 B lines 84–90 (Treasurers of Athena and the other gods), Koroibos archon, 306/5. This is probably the most important Athenian inscription that mentions catapults: full and detailed and in relatively good condition. The catapult-maker Bromios is named. Text and translation in Marsden 1969: 70.

IG II$^2$ 1488 (Treasurers of Athena and the other gods), end of the century. This is a very fragmentary inscription, missing left and right sides, with a few odd words surviving down the middle. "Katapulti-" on the first line, the number "601" on the second, "spans" on the third, "240" and "of those" on the fourth, "14" and "darts" on the fifth, "500" and "of lead" on the sixth, "boxes" on the seventh, "guards" on the eighth, and "darts" again on the ninth and last surviving line. All the numbers are minima, since they are missing their beginnings or ends. It would appear to be an inventory of missiles for catapults and hand weapons.

IG II$^2$ 1627 B lines 328–341 (Naval inventory, Peiraios) 330/29. A large and largely complete inscription, running to nearly two hundred lines of text in each of three columns. In "the large building near the gates" were quite a lot of catapult bits – for example, stocks, sliders, bases, bows in leather cases (τόξα ἐσκευτομένα), windlass-drums, and bolts.

IG II$^2$ 1628 D 510–21 (another long and well-preserved naval inventory) 326/5. The same items, in the same quantities, as in 1627.

IG II$^2$ 1629 E 985–998 (an even longer and well-preserved naval inventory) 325/4. The same items, in the same quantities, as in 1627.

IG II$^2$ 1631 B 220–229 (another very long but not so well-preserved naval inventory), 323/2. Different only insofar as the numbers seem to have grown in two cases. Probably accounting errors, inscribing, or typographic errors.

IG II$^2$ 3234 (gravestone for Heraleidas the Mysian, catapult-discharger)

IG XII.5.647 (catapult-shooting competitions)

IG XII.8.156 (catapults and missiles) from Samothrace, 228–225 B.C.

IG XII.9.259 (from Eretria, Euboia) an engine craftsman, *organa kheirotekhnes.*

IG XIV 1717 (gravestone for Euemeros, *organopoios*).

*Inscriptiones Tyrae Olbiae Chersonesi Tauricae aliorum locorum a Danubio usque ad regnum Bosporanum*, vol. 1, 195, ed. B. Latyschev, Hildesheim, 1965 (Anaxagoras' extraordinary shot).

Panakton invoice 1992–300, M. Munn "The first excavations at Panakton on the Attic-Boiotian frontier," in *Boiotia Antiqua* 6, ed. J. M. Fossey, 1996: 47–58 at 52. Firmly dated to 343–342, specifies numbers of catapult missiles at Panakton fort.

RIB 1912 Ballistarium near Hadrian's wall.

R&O=Rhodes, P. J., and Osborne, R. (2003) *Greek Historical Inscriptions 404–323 B.C.*, Oxford (selected texts plus translations and commentaries in English).

SEG 19.96, 32.125, 32.129, and *Hesperia* 15.197 appear to refer to two or three generations of a family specialising in catapult-instruction.

SEG 27.966 Glans of Andron, possibly made by him.

SEG 31.1607 Glans made by Andron.

SEG 31.1617 Glans weighing 69.76 gr. inscribed Epam(inondas), said to have come from Thebes.

SEG 35.581 Funerary·epigram for Dikaiogenes, killed by a "cruel" kestros, c. 200–150 B.C.

SEG 47.1635 Glans from Miletos inscribed Dromas Ainian.

*Sylloge Inscriptionum Graecarum*, Dittenberger, 3rd ed., vol. 3, 958.23–32. (competitions on Keos in catapult-shooting, archery and javelin throwing).

## Modern Works

Adam, J.-P. (1982) *L'architecture militaire grecque*, Paris.

———. (1992) "Approche et défense des portes dans le monde hellénisé," in S. Van de Maele and J. M Fossey, eds. *Fortificationes Antiquae*, Amsterdam, 5–43.

Adan-Bayewitz, D. (1997) "Jotapata, Josephus, and the siege of 67: preliminary report on the 1992–94 seasons," *Journal of Roman Archaeology* 10: 131–165.

Adcock, F. E. (1957) *The Greek and Macedonian Art of War*, Berkeley.

Alexander, J. (1979) "The illustrations of the Anonymous *De Rebus Bellicis*," M. W. C. Hassall, ed. *De Rebus Bellicis* BAR IS 63 (1979) 11–15.

Allason-Jones, L. (2007) "Small objects," in W. S. Hanson, *Elginhaugh: A Flavian Fort and Its Annexe* Britannia Monograph Series, London.

Allen, F. D. (1896) "On 'Os Colvmnatvm,' (Plaut. M. G. 211) and ancient instruments of confinement," *Harvard Studies in Classical Philology* 7: 37–64.

Allen, P. (1979) "The Justinianic plague," *Byzantion* 49: 5–20.

Alsatian, G., "The physics of the sling," online at www.slinging.org/29.html.

Angour, A. D. (1998) *The Dynamics of Innovation: Newness and Novelty in the Athens of Aristophanes*, Ph.D. diss., University College London.

Anstee, J. (1998) "'Tours de force': an experimental catapult / ballista," *Studia Danubiana*, Bucharest, 131–139.

Baatz, D. (1966) "Zur Geschützbewaffnung römischer Auxiliartruppen in der frühen und mittleren Kaiserzeit," *Bonner Jahrbücher* 166: 194–207.

———. (1977) "The Hatra Ballista," *Sumer* 33: 141–151.

———. (1978) "Recent finds of ancient artillery," *Britannia* 9: 1–17.

———. (1980) "Ein Katapult der Legio IV Macedonica aus Cremona," *Röm. Mitt.* 87: 283–299 and plates 90–94, reprinted in Baatz 1994.

———. (1982) "Hellenistische Katapulte aus Ephyra (Epirus)," *Mitt. DAI Athen* 97: 211–233, reprinted in Baatz 1994.

———. (1985) "Katapultteile aus dem Schiffswrack von Mahdia (Tunesien)," *Archäologischer Anzeiger* 679–691.

———. (1991) "Die römische Jagdarmbrust," *Archäologisches Korrespondenzblatt*, 21: 283–290.

———. (1994) *Bauten und Katapulte des römischen Heeres*, Mavors 11, Stuttgart (reprints many relevant articles, and some new).

———. (1999) "Katapulte und mechanische Handwaffen des spätrömischen Heeres," *Journal of Roman Military Equipment Studies* 10: 5–19.

Bakhuizen, S. C. (1986) "La grande batterie de Goritsa et l'artillerie défensive," in Leriche and Tréziny 315–321.

Baldwin B. (1981) "Eusebius and the siege of Thessalonica," *RhM* 124: 291–296.

Balfour, H. (1890) "On the structure and affinities of the composite bow," *Journal of Anthropological Institute*, 19: 220–250.

Balty, J. C. (1988) "Apamea in Syria in the second and third centuries A.D.," *Journal of Roman Studies* 78: 91–104.

Barag, D., and Hershkovitz, M., et.al. (1994) *Masada: the Yigael Yadin excavations 1963–5: final reports vol. 4*, Jerusalem.

Barthel, W. (1914) "Die Catapulta von Emporion," *Frankfurter Zeitung und Handelsblatt* 118.2: 1 (29 April).

Beazley, J. D. (1966) "Un realista greco," in *Adunanze straordinarie per il conferimento dei premi della Fondazione A. Feltrinelli* 1.3: 53–60.

Belkin, M. (1978) "Wound ballistics," *Progress in Surgery* 16: 7–24.

Beltrán, F. (1990) "La pietas de Sertorio," *Gerion* 8: 211–226.

Bennett, J. (1991) "Plumbatae from Pitsunda (Pityus) Georgia," *Journal of Roman Military Equipment Studies* 2: 59–63.

Berryman, S. (2003) "Ancient automata and mechanical explanation," *Phronesis* 48: 344–369.

Birley, A. R. (1991) "Vindolanda: notes on some new writing tablets," *Zeitschrift für papyrologie und epigraphy* 88: 87–102.

———. (1996) *Vindolanda Research Reports. New Series 4 Small finds. Fasc. 1 The Weapons*, Hexham.

Birley, E. (1984/1985) "Ballista and Trebellius Pollio," *Bonner Hist.-Aug.-Coll.* 55–60.

Bishop, M. C., ed. (1983) *Roman Military Equipment*, Sheffield.

Bishop, M. C. (1985a) "The military *fabrica* and the production of arms in the early Principate," in Bishop (ed.), *The Production and Distribution of Roman Military Equipment*, BAR S275, Oxford, 1–42.

Bishop, M. C., ed. (1985b) *The Production and Distribution of Roman Military Equipment* BAR S275, Oxford.

Bishop, M. C., and Coulston, J. C. N. (1993 and 2nd revised edition 2006) *Roman Military Equipment*, London.

Blaine, B. B. (1994) "Medieval military technology" (review of K DeVries' book of same name, 1992), *Speculum* 69: 452–454.

Bloedow, E. F. (1998) "The siege of Tyre in 332 B.C.: Alexander at the crossroads of his career," *Parola del Passato* 53: 255–293.

Bockius, R. (1996) "A Roman depiction of a warship equipped with two catapults?" in H. Tzalas ed., *Tropis VI: 6th international symposium on ship construction in antiquity*, Lamia, 89–97.

Boehringer, E. (1937) *Die hellenistischen Arsenale. Altertümer von Pergamon* 10, Berlin.

Bohec, Y. le (2000) *The Imperial Roman Army*, trans. R. Bate, London (originally published as *L'Armée Romaine, sous le Haut-Empire*, 1989).

Borza, E. N. (1990) *In the Shadow of Olympus: The Emergence of Macedon*, Princeton.

Bosman, A. V. A. J. (1995) "Pouring lead in the pouring rain: making lead slingshot under battle conditions," *Journal of Roman Military Equipment Studies* 6: 99–103.

Boube-Piccot, C. (1994) *Les bronzes antiques du Maroc IV: L'équipement militaire et l'armement*, Paris.

Bowman, A. K., and Thomas, J. D. (1996) "New writing tablets from Vindolanda," *Britannia* 27: 299–328.

Bradbury, J. (1992) *The Medieval Siege*, Woodbridge.

Brennan, P. (1980) "Combined legionary detachments as artillery units in late Roman Danubian bridgehead dispositions," *Chiron* 10: 553–567.

Bret, P. (1996) "The organisation of gunpowder production in France 1775–1830," in Buchanan ed. 261–274.

Brok, M. F. A. (1977) "Bombast oder Kunstfertigkeit. Ammians Beschreibung der ballista (23, 4, 1–3)," *Rheinisches Museum* 120: 331–345.

Bruce I. A. F. (1988) "Diodorus on the siege of Chalcedon," *Ancient History Bulletin* 2: 54–56.

Brun, P. (1994) "Les fortifications d'Hyllarima, Philon de Byzance et Pleistarchos," *Revue des études anciennes* 96: 193–204.

Buchanan, B. J., ed. (1996) *Gunpowder: The History of an International Technology*, Bath.

Buchanan, B. J. (1996) "Meeting standards: Bristol powder makers in the eighteenth century," in Buchanan ed., 237–252.

Cabré, J. (1925) "Los bronces de Azaila," *Archivo Español de Arte y Arqueologia* 1: 297–315.

Cameron, A. (1979) "The date of the Anonymous *De Rebus Bellicis*," in M. W. C. Hassall, ed. *De Rebus Bellicis* BAR IS 63: 1–10.

Campbell D. B. (1986) "Auxiliary artillery revisited," *Bonner Jahrbücher* 186: 117–132.

———. (1988) "Dating the siege of Masada," *Zeitschrift für Papyrologie und Epigraphik* 73: 156–158.

———. (2003) *Greek and Roman Artillery 399B.C.–A.D. 363*, Oxford.

Cartledge, P. (1987) *Agesilaos*, Baltimore.

Cavalier, K. R. (2000) "An old saw," in C. C. Mattusch, A. Brauer, and S. E. Knudsen eds., *From the parts to the whole* vol. 1, JRA Supp. 39, Rhode Island, 81–85.

Caven, B. (1990) *Dionysius I*, New Haven.

Chaniotis, A. (2003) "The divinity of Hellenistic Rulers," in A. Erskine ed., *A Companion to the Hellenistic World*, Oxford, 431–445.

Chevedden, P. (1995) "Artillery in late antiquity," in I. Corfis and M. Wolfe (eds) *The Medieval City under Siege* Woodbridge, 131–173.

Ciasca, A. (1986) "Fortificazioni di Mozia (Sicilia): dati tecnici e proposta preliminare di periodizzazione," in Leriche and Tréziny, eds., 221–227.

Cole, P. J. (1981) "The catapult bolts of *IG* II 1422," *Phoenix* 25: 216–219.

Connolly, P. (1981) *Greece and Rome at War*, London.

Cotterell, B., and Kamminga, J. (1990) *The Mechanics of Pre-Industrial Technology*, Cambridge.

Coulston, J. C. (1985) "Roman Archery Equipment," in M. Bishop (ed.), *The production and Distribution of Roman Military Equipment*, BAR S275, Oxford, 220–348.

———. (1989) "The value of Trajan's Column as a source for military history," in van Driel-Murray 1989: 31–44.

Crosby, A. W. (2002) *Throwing Fire*, Cambridge.

Crow, J. G. (1991) "A review of current research on the turrets and curtain of Hadrian's wall," *Britannia* 22: 51–63.

Cuomo, S. (2002) "The machine and the city: Hero of Alexandria's Belopoeica," in T. E. Rihll and C. Tuplin, eds., *Science and Mathematics in Ancient Greek Culture*, Oxford, 165–177.

———. (2004) "The sinews of war: ancient catapults," *Science* 303: 771–772.

———. (2007) *Technology and culture in Greek and Roman antiquity*, Cambridge.

Davies, R. W. (1970) Review of Marsden, Historical Development, *Journal of Roman Studies* 60: 225–226.

De Morgan, J., Jequier, G., et Lampre, G. (1900) *Fouilles à Suse en 1897–8 et 1898–9, Mémoires Recherches Archaéologiques*, Tome 1, Paris.

Debord, P. (1999) *L'Asie Mineure au IV^e Siècle (412–323 a.C.)*, Bordeaux.

Degering, H. (1905) *Die Orgel, ihre Erfindung und ihre Geschichte bis zur Karolingerzeit*, Münster.

Dennis, G. T. (1981) "Flies, Mice, and the Byzantine crossbow," *Byzantine and Modern Greek Studies* 7: 1–6.

———. (1998) "Byzantine heavy artillery: the helepolis," *Greek, Roman and Byzantine Studies* 39: 99–115.

Denny, M. (2003) "Bow and catapult internal dynamics," *European Journal of Physics* 24: 367–378.

DeVries, K. (1992) *Medieval military technology,* Ontario.

———. (1997) "Catapults are not atomic bombs: towards a redefinition of 'effectiveness' in premodern military technology," *War in History* 4: 454–470.

Diels, H. (1924) *Antike Technik,* 3rd ed., Leipzig and Berlin.

Díez, F. G. (2002) "Las catapultas de Azaila (Teruel)," *Gladius* 5: 293–302.

Dillon, Viscount (1894) "Military Archery in the Middle Ages," chapter 7 in A. Lever ed. *Archery*, London.

Dohrenwend, R. E. (1994) "The sling: forgotten firepower of antiquity," *Canadian Journal of Arms Collecting* 32.3: 85–91.

Donaldson, G. H. (1989) "*Tormenta, Auxilia* and *Ballistaria* in the environs of Hadrian's wall," *Archaeologia Aeliana* 17: 217–219.

———. (1990) "A reinterpretation of RIB 1912 from Birdoswald," *Britannia* 21: 207–214.

Douglas, D. (1981) "The siege of Uxellodunum in 51 B.C.: a historical mystery," *Classicum* (Sydney University Latin Department) 7: 3–10.

Drachmann, A. G. (1963) *The Mechanical Technology of Greek and Roman Antiquity*, Copenhagen.

———. (1977) "Biton, and the development of the catapult," *Prismata, Naturwissenschaftsgesichtliche Studien. Festschrift für Willy Hartner,* 119–131.

Driel-Murray, C. van, ed. (1989) *Roman Military Equipment*, BAR IS 476.

Ducrey, P. (1986) "Les fortifications grecques: rôle, function, efficacite," in Leriche and Tréziny, eds. *La fortification dans l'histoire du monde grec*, 133–142.

Eggermont P. H. L. (1975) *Alexander's campaigns in Sind and Baluchistan and the siege of the Brahmin town of Harmatelia* Leuven.

Empereur, J-Y. (1981) "Collection Paul Canellopoulos (XVII)," *Bulletin de correspondance hellénique* 105: 537–568.

Erdmann, E. (1977) *Ausgrabungen in Alt-Paphos auf Cypern 1 Nordost-Tor und persische Belagerungsrampe I Waffen und Kleinfunde*, Deutsches Archäoligisches Institute.

Eriksson, A., Georén, B., and Öström, M. (2000) "Work-place Homicide by Bow and Arrow," *Journal of Forensic Science* 45: 911–916.

Esper, T. (1965) "The replacement of the longbow by firearms in the English army," *Technology and Culture* 6: 382–393.

Evans, E. (1994) "Military architects and building design in Roman Britain," *Britannia* 25: 143–164.

Everson, P., and Cocroft, W. (1996) "The royal gunpowder factory at Waltham Abbey: the field archaeology of gunpowder manufacture," in Buchanan ed., 377–394.

Feugère, M. (2002) *The Weapons of the Romans*, trans. D. G. Smith, Stroud (originally published as *Les armes des romains* Paris 1993).

Fisher, N. (1993) *Slavery in Classical Greece*, London.

Fleming, S. (1988) "Classical artillery," *Archaeology* 41: 56–57.

Fleury, P. (1981) "Vitruve et la nomenclature des machines de jet Romaines," *Revue des Etudes Latines* 59: 216–234.

———. (1993) *La Méchanique de Vitruve*, Caen.

Foley, V., Palmer, G., and Soedel, W. (1985) "The crossbow," *Scientific American* 252.1: 80–86.

Follette, L. La (1986) "The Chalkotheke on the Athenian Acropolis," *Hesperia* 55: 75–87.

Forage, P. C. (1996) Review of Needham, Science and Civilisation vol. 5. 6, *Harvard Journal of Asiatic Studies* 56: 508–521.

Fossey, J. M. (1992) "The development of some defensive networks in eastern central Greece during the classical period," in van de Maele and Fossey, 109–132.

Fougères (1877) "Glans," in C. Daremberg and E. Saglio, eds., *Dictionnaire des Antiquités*, Graz, 1608–1611.

France, J. (1994) *Victory in the East: a military history of the First Crusade*, Cambridge.

Franklin, J. C. (2002) "Harmony in Greek and Indo-Iranian Cosmology," *Journal of Indo-European Studies* 30: 1–25.

Frederiksen, M. (1984) *Campania*, edited with additions by N. Purcell, British School at Rome.

Frías, J. V., Martínez, J. L. C., and Sábada, J. L. R. (2003) "Nuevo testimonio de las guerras sertorianas en calahorra: un depósito de proyectiles de catapulta," *Kalakorikos* 8: 9–30.

Furneaux, R. (1973) *The Roman Siege of Jerusalem*, London.

Gabrielsen, V. (1997) *The Naval Aristocracy of Hellenistic Rhodes*, Aarhus.

Gaitzsch, W. (1994) "Hellenistische Geschützteile aus Pergamon," in C. Dobiat (ed) *Festschrift für Otto-Herman Frey*, 235–242.

Garlan, Y. (1973) "Cités, armées et stratégie a l'époque hellénistique d'après l'oeuvre de Philon de Byzance," *Historia* 22: 16–33.

———. (1974) *Recherches de poliorcétique grecque*, Paris.

———. (1975) *War in the Ancient World: A Social History*, trans J. Lloyd, London.

———. (1988) *Slavery in Ancient Greece* revised ed. trans. J. Lloyd, Ithaca.

Gerkan, A. von (1935) *Milet 2.3: Die Stadtmauern*, Berlin.

Gille, B., ed. (1978) *Histoire des techniques*, Tours.

Gille, B. (1980) *Les mécaniciens grecs*, Paris.

Gilliver, C. M. (1999) *The Roman Art of War*, Stroud.

Goette, H. R. (2003) "Cape Sounion and the Macedonian occupation," in O. Palagia and S. V. Tracy, eds., *The Macedonians in Athens 322–229 B.C.*, Oxford, 152–161.

Goldsworthy, A. (2003) *The Complete Roman Army*, London.

Gorman, V. B. (2001) *Miletos, the Ornament of Ionia*, Michigan.

Gourley, B. and Williams, H. (2005) "The fortifications of Stymphalos," *Mouseion* Series III 5 forthcoming.

Greaves, A. M. (2002) *Miletos: A History*, London.

Greep, S. J. (1987) "Lead sling-shot from Windbridge farm, St. Albans and the use of the sling by the Roman army in Britain," *Britannia* 18: 183–200.

Gregory, T. (2005) *A History of Byzantium*, Oxford.

Grmek, M. D. (1991) *Diseases in the Ancient Greek World* trans. M. Muellner and L. Muellner, Baltimore.

Griffiths, W. B. (1989) "The sling and its place in the Roman Imperial Army," in van Driel-Murray (ed.) 255–279.

———. (1994) "Reconstructing Roman slings," *Arbeia Journal* 3: 1–11.

Habicht, C. (2003) "Athens after the Chremonidean War: some second thoughts," in O. Palagia and S. V. Tracy, eds., *The Macedonians in Athens 322–229 B.C.*, Oxford, 52–55.

Hacker, B. C. (1968) "Greek catapults and catapult technology," *Technology and Culture* 9: 34–50.

Haldon, J. (1999) *Warfare, State and Society in the Byzantine World 565–1204*, Routledge.

Hall, B. S. (1973) "Crossbows and Crosswords" (review of Marsden), *Isis* 64: 527–533.

———. (1996) "The corning of gunpowder and the development of firearms in the renaissance," in Buchanan ed. 87–120.

———. (1997) "Weapons of war and late Medieval cities: Technological innovation and tactical changes," in Bradford Smith and Wolfe (eds.), *Technology and Resource Use in Medieval Europe: Cathedrals, Mills and Mines* Aldershot, 185–208.

Hamilton-Smith, K., and Corlett, E. (1989) "Some possible technical features of a Salamis *trieres*," in S. Fisher (ed.), *Innovation in Shipping and Trade*, Exeter, 9–36.

Hammond, N. G. L. (1986) *History of Greece* 3rd ed., Oxford.

———. (1992) Review of W. K. Pritchett, *Greek State at War V, Classical Review* 42: 375–377.

Harris, H. A. (1963) "Greek javelin throwing," *Greece and Rome* 10: 26–36.

Hassall, M. W. C., ed. (1979) *De Rebus Bellicis* BAR IS 63, Oxford.

Hassall, M. (1999) "Perspectives on Greek and Roman catapults," *Archaeology International*, London, 23–26.

Hauben, H. (1974) "IG II 492 and the siege of Athens in 304 B.C.," *ZPE* 14: 10.

Hawkins, W. (1847) "Observations on the use of the sling as a warlike weapon among the ancients," *Archaeologia* 32: 96–107.

Heath, T. L. (1981) *A History of Greek Mathematics* New York (corrected reprint of the 1921 Oxford ed.)

Herrmann, L. (1964) "Sur le sic vos non vobis et Ballista," *Studies Ullman* 1: 223–228.

Hoch, P. K. (1986) "Emigrés in science and technology transfer," *Phys. Technology* 17: 225–229.

Hodges, H. (1979) "The Anonymous in the later Middle Ages," in M. W. C. Hassall, ed. *De Rebus Bellicis* BAR IS 63: 119–126.

Holley, A. E. (1994) "The ballista balls from Masada," in D. Barag et al., eds., *Masada IV*, Jerusalem, 349–366.

Hornblower, S. (1982) *Mausolus*, Oxford.

Hoxha, G. (2001) "Philon von Byzanz und die spätantiken Befestigungen in Albanien," *Archäologisches Korrespondenzblatt* 31: 601–616.

Irby-Massie, G. L., and Keyser, P. T. (2002) *Greek Science of the Hellenistic Era: A Sourcebook*, London.

Iriarte, A. (2000) "Pseudo-Heron's *cheiroballistra*: a(nother) reconstruction: 1. Theoretics," *Journal of Roman Military Equipment Studies* 11: 47–75.

Jackson, A. (1975) Review of *La guerre dans l'antiquité* by Y. Garlan, *Journal of Hellenic Studies* 95: 238.

James, S. (1983) "Archaeological evidence for Roman incendiary projectiles," *Saalburg Jahrbuch* 39: 142–143.

James, S. T., and Taylor, J. H. (1994) "Parts of Roman artillery projectiles from Qasr Ibrim, Egypt," *Saalburg Jahrbuch* 47: 93–98.

James, S. (2004) *The Excavations at Dura-Europos Conducted by Yale University and the French Academy of Inscriptions and Letters 1928–1937* Final Report VII: The Arms and Armour and Other Military Equipment, British Museum, London.

Jones, R., and Robinson, D. (2005) "The structural development of the House of the Vestals (VI.i.6–8, 24–26)," in P. G. Guzzo and M. P. Guidobaldi (eds.), *Nuove ricerche archeologiche a Pompei e Ercolano*, Soprintendenza Archeologica di Pompei 10, Naples, pp. 257–269.

Kaegi, W. E. (1990) "Procopius the military historian," *Byzantinische Forschungen* 15: 53–85.

Karger, B., Sudhues, H., Kneubuehl, B. P., and Brinkmann, B. (1998) "Experimental arrow wounds: ballistics and traumatology," *Journal of Trauma* 45.3: 495–501.

Karger, B., Sudhues, H., and Brinkmann, B. (2001) "Arrow wounds: major stimulus in the history of surgery," *World Journal of Surgery* 25: 1550–1555.

Karger, B., Bratzke, H., Graß, H., Lasczkowski, G., Lessig, R., Monticelli, F., Wiese, J., Zweihoff, R. F. (2004) "Crossbow homicides," *International Journal of Legal Medicine* 118: 332–336.

Katz, B. R. (1976) "The siege of Rome in 87 B.C.," *Classical Philology* 71: 328–336.

Keay, S. (2003) "Recent archaeological work in Roman Iberia 1990–2002," *Journal of Roman Studies* 93 (2003) 146–211.

Kelso, I. (2003) "Artillery as a classicizing digression," *Historia* 52: 122–125.

Kenny, E. J. A. (1932) "The date of Ctesibius," *Classical Quarterly* 26: 190–192.

Kern, P. B. (1999) *Ancient Siege Warfare*, London.

Keyser, P. T. (1994) "The use of artillery by Philip II and Alexander the Great," *Ancient World* 25: 27–59.

Kiernan, B. (2004) "The first genocide: Carthage, 146 B.C.," *Diogenes* 203: 27–39.

Kingsley, P. (1995a) *Ancient Philosophy, Mystery, and Magic*, Oxford.

———. (1995b) "Artillery and prophesy: Sicily in the reign of Dionysius I," *Prometheus* 21: 15–23.

Klee, K. (1937) Paulys *Real-Encyclopaedie der classischen Altertumswissenschaft* 6A.2: 1329.

Kooi, B. W. (1991) "The 'cut and try' method in the design of the bow," *Engineering Optimization in Design Processes*, H. Eschenauer, C. Mattheck, and N. Olhoff, eds., Berlin, 283–292.

Kooi, B. W., and Bergman, C. A. (1997) "An approach to the study of ancient archery using mathematical modelling," *Antiquity* 71: 124–134.

Kooi, B. W., and Sparenberg, J. A. (1997) "On the mechanics of the arrow: Archer's Paradox," *Journal of Engineering Mathematics* 31: 285–303.

Korfmann, M. (1973) "The sling as a weapon," *Scientific American* 229: 34–42.

Kosmetatou, E. (2003) "The Attalids of Pergamon," in Erskine (ed) 159–174.

Kramer, G. W. (1996) "Das Feuerwerkbuch: its importance in the early history of black powder," in Buchanan ed. 45–56.

Kromayer, J., and Veith, G. (1928) *Heerwesen und Kriegführung der Griechen und Römer*, Munich.

Kylander, M. E. et al. (2005) "Refining the pre-industrial atrmospheric Pb isotope evolution curve in Europe using an 8000 year old peat core from NW Spain," *Earth and Planetary Science Letters* 240: 467–485.

Landels, J. G. (1978) *Engineering in the Ancient World*, London.

Lapatin, K. D. S. (1997) "Pheidias ελεφαντουργος," *American Journal of Archaeology* 101: 663–682.

Lascaratos, J. (1997) "The wounding of Alexander the Great in Cyropolis (329 B.C.): the first reported case of the syndrome of transient cortical blindness?" *History of Ophthalmology* 42.3: 283–287.

Laurenzi, L. (1964) "Perchè Annibale non assediò Roma: considerazioni archeologiche," *Studi Annibalici*, Cortona, 141–154.

Lauter, H. (1992) "Some remarks on fortified settlements in the Attic countryside," in van de Maele and Fossey, 77–91.

Lawrence, A. W. (1979) *Greek Aims in Fortification*, Oxford.

Lawson, H. (2005) *The Hatra Ballista* M.A. dissertation, Swansea.

Lendle, O. (1983) *Texts und Untersuchungen zum technischen Bereich der Antiken Poliorketik*, Weisbaden.

Leriche, P. (1986) "Fortification grecques: bilan de la recherché au proche et moyen orient," in Leriche and Tréziny, 39–49.

Leriche, P., and Tréziny, H., eds. (1986) *La fortification dans l'histoire du monde grec*, CNRS.

Lewis, D. M. (1977) *Sparta and Persia*, Leiden.

Lewis, M. J. T. (1999) "When was Biton?" *Mnemosyne* 52: 159–168.

———. (2000) "Theoretical hydraulics, automata and water clocks," in O. Wikander, ed., *Handbook of Ancient Water Technology* Brill, Leiden, 343–370.

———. (2001) *Surveying Instruments of Greece and Rome*, Cambridge.

Lewis, R. G. (1971) "The siege of Praeneste, 82 B.C.," *Papers of the British School of Rome* 39: 32–39.

Lienhard, J. (2000) *The Engines of Our Ingenuity*, Oxford.

Lightfoot, C. S. (1988) "Facts and fiction. The third siege of Nisibis (A.D. 350)," *Historia* 37: 105–125.

Lindsay, J. (1974) *Blast Power and Ballistics: Concepts of Force and Energy in the Ancient World*, London.

Lissarrague, F. (1990) *L'Autre Guerrier: Archers, peltastes, cavaliers dans l'imagerie Attique*, Paris and Rome.

Lloris, M. Beltrán. (1976) *Arqueologia e historia de las ciudades antiques del Cabezo de Azaila*, Museo Arqueológico Nacional no. 3034, Zaragoza.

Lomas, K. (1995) "The Greeks in the West and the Hellenization of Italy," in *The Greek World* ed. A. Powell, London, 347–367.

López, F., et. al. (2000) "Pentrating craniocerebral injury from an underwater fishing harpoon," *Child's Nervous System* 16: 117–119.

Lordkipanidze, O. (1986) "The fortification of ancient Colchis," in Leriche and Treziny, 179–184.

MacGregor, A. (1975–6) "Two antler crossbow nuts and some notes on the early development of the crossbow," *Proceedings of the Society of Antiquaries of Scotland* 107: 317–321.

Macnamara, E. (1964) "The Causeway and Lagoon 1962," in *Motya*, by B. S. J. Isserlin et al., *Annual of Leeds University Oriental Society* IV [1962–63], Leiden, 86–104.

Maddin, R., Hauptmann, A., and Baatz, D. (1991) "A metallographic examination of some iron tools from the Saalburg Museum," *Saalburg Jahrbuch* 46: 5–23.

Maele, S. van de (1992) "Le réseau mégarien de défense territoriale contre l'Attique à l'époque classique (Ve et IV s. av. J-C)," in Maele and Fossey, eds., 93–107.

Maele, S. van de, and Fossey, J. M., eds. (1992) *Fortificationes Antiquae*, Amsterdam.

Maier, F. G., and Karageorgis, V. (1984) *Paphos*, Nicosia.

Manganaro, G. (1999) *Sikelika: Studi di antichità e di epigrafia della Sicilia greca*, Pisa and Rome.

Manganaro, G., "Onomastica greca su anelli, pesi da telaio e glandes in Sicilia," *ZPE* 133 (2000) 123–134.

Marsden, E. W. (1969) *Greek and Roman Artillery: Historical Development*, Oxford.

———. (1971) *Greek and Roman Artillery: Technical Treatises*, Oxford.

Masquelez, M. (1877) "Cestrosphendoné," in C. Daremberg and E. Saglio, eds., *Dictionnaire des Antiquités*, 1089–1090.

Massey, D. (1994) "Roman archery tested," *Military Illustrated* 74: 36–39.

Matthews, J. (1989) *The Roman Empire of Ammianus Marcellinus*, London.

Mayor, A. (2003) *Greek Fire, Poison Arrows and Scorpion Bombs: Biological and Chemical Warfare in the Ancient World*, Woodstock.

McDermott, W. C. (1942) "Glandes plumbeae," *Classical Journal* 38: 35–37.

McGeer, E. (1995a) "Byzantine siege warfare in theory and practice," in Corfis and Wolfe, eds., *The Medieval City Under Siege*, Woodbridge, 123–129.

———. (1995b) *Sowing the Dragon's Teeth*, Washington, D.C.

McInerney, J. (1999) *The Folds of Parnassos*, Austin.

McNicoll, A. (1978) "Some developments in Hellenistic siege warfare with special reference to Asia Minor," *Proc. 10th International Congress Archaeologique*, Ankara, 405–420.

———. (1986) "Developments in techniques of siegecraft and fortification in the Greek world ca. 400–100 B.C.," in Leriche and Tréziny, eds., 305–313.

———. (1997) *Hellenistic Fortifications from the Aegean to the Euphrates*, with revisions and a new chapter by N. P. Milner, Oxford.

McNicoll, A. and Winikoff, T. (1983) "A Hellenistic fortress in Lycia–the Isian tower?" *American Journal of Archaeology* 87: 311–323.

Miks, C. (2001) "Die Χειροβαλλιστρα des Heron. Überlegungen zu einer Geschützentwicklung der Kaiserzeit" *Salburg Jahrbuch* 51: 153–234.

Morin, M. (2002) "The earliest European firearms," in D. Nicolle, ed., *A Companion to Medieval Arms and Armour*, Woodbridge, 55–62.

Morton, A. (2006) "Finding the corkscrew," *Studies in the History and Philosophy of Science* 37: 114–117.

Morrison, J., and Williams, R. T. (1968) *Greek Oared Ships 900–322 B.C.*, Cambridge.

Murray, W. M. (1997) Report on the Actium Project, online at http://luna.cas.usf.edu/~murray/actium/brochure.html, accessed 29 July 2005.

Mußmann, O. (1996) "Gunpowder production in the electorate and the kingdom of Hanover," in Buchanan ed. 329–350.

Needham, J., Ho, P-Y, Lu, G-D, and Wang, L. (1986) *Science and Civilisation in China*, vol. 5 part 7, Cambridge.

Needham, J. and Yates, R. (1994) *Science and Civilisation in China*, vol. 5 part 6, Cambridge.

Nickel, H. (2002) "The mutual influence of Europe and Asia," in Nicolle, ed., *A Companion to Medieval Arms and Armour*, 107–125.

Nicolle, D. (1987) "Armes et Armures dans les Épopées des Croisades," in K-H. Bender (ed) *Les Épopées de La Croisade*, Stuttgart, 17–34.

Nicolle, D. (ed) (2002) *A companion to medieval arms and armour*, Woodbridge.

Oates, D. (1968, reprint 2005) *Studies in the Ancient History of Northern Iraq*, London.

Ober, J. (1987) "Early artillery towers," *American Journal of Archaeology* 91: 569–604.

———. (1992) "Towards a typology of Greek artillery towers: the first and second generations (c. 375–275 B.C.)," in Van de Maele and Fossey, 147–169.

Oldstein, J. (1985) "Manufacture and supply of the Roman army with bronze fittings," in Bishop, ed., *The Production and Distribution of Roman Military Equipment*, Oxford, 82–94.

Oliver, R. P. (1955) "A note on the De Rebvs Bellicis," *Classical Philology* 50: 113–118.

Özyigit, Ö. (1994) "The city walls of Phokaia," *Revue des Études Anciennes* 96: 77–109.

Parke, H. W. (1933) *Greek Mercenary Soldiers*, Oxford.

Partington, J. R. (1999) *A History of Greek fire and Gunpowder*, 2nd ed. with introduction by B. S. Hall (1st ed. 1960), Baltimore.

Patterson, O. (1982), *Slavery and Social Death*, Cambridge, Mass.

Payne-Gallwey, R. (1903) *The Book of the Crossbow*, London (reprinted Mineola, N.Y.,1995, with Payne-Gallwey's *Appendix* of 1907).

Peddie, J. (1994) *The Roman War Machine*, Stroud.

Petersen, E. (1896) *Die Marcus-Säule*, Munich.

Pimouguet-Pedarros, I. (2000) "L'apparition des premiers engines balistiques dans le monde grec et hellénisé: un état de la question," *Revue des études anciennes* 102: 5–26.

Ponting, C. (2005) *Gunpowder*, London.

Pritchett, W. K. (1991) *The Greek State at War*, Part 5, Berkeley.

Rausing, G. (1997) *The Bow: Some Notes on Its Origin and Development*, Manchester (1st edition Sweden 1967).

Reinschmidt, K. F. (1999) "On torsion catapults," unpublished MS. I am grateful to the author for providing me with a copy of this document.

Richardson, T. (1998a) "Ballistic testing of historical weapons," *Royal Armouries Yearbook* 3: 50–52.

———. (1998b) "The ballistics of the sling," *Royal Armouries Yearbook* 3: 44.

Richmond, I. (1962) "The Roman siege-works of Masada, Israel," *Journal of Roman Studies* 52: 142–155.

Rincón, G., and María, D. (1994) "El dístico de Ballista y sus variants en la Vita Vergiliana de Focas," *VIII congreso español de estudios clásicos II*, Madrid, 849–853.

Rihll, T. E. (2006) "On catapult sizes and artillery towers," *Annual of the British School at Athens* 101 (to appear).

Robinson, D. M. (1941) *Excavations at Olynthus XI Metal and miscellaneous finds*, London.

Rodgers, W. L. (1937, reprinted 1964) *Greek and Roman Naval Warfare*, Annapolis.

Rogers, C. J. (1998) "The efficacy of the English longbow: a reply to Kelly DeVries," *War in History* 5: 233–242.

Rogers, R. (1992) *Latin Siege Warfare in the Twelfth Century*, Oxford.

Roland, A. (1995) "Science, technology and war," *Technology and Culture* 36 Supplement: 83–100.

Romane, J. P. (1987) "Alexander's siege of Tyre," *Ancient World* 16: 79–90.

———. (1988) "Alexander's siege of Gaza, 332 B.C.," *Ancient World* 18: 21–30.

Rona-Tas, A. (1999) *Hungarians and Europe in the Early Middle Ages*, trans. N. Bodoczky, Budapest.

Rorres, C., and Harris, H. G. (2001) "A formidable war machine," online at http://math.nyu.edu/~crorres/Archimedes/Claw/harris/rorres_harris.pdf http://math.nyu.edu/~crorres/Archimedes/Claw/Models.html.

Rosman, K. et al. (1997) "Lead from Carthaginian and Roman Spanish mines isotopically identified in Greenland ice dated from 600 B.C. to 300 A.D.," *Environmental Science and Technology* 31: 3413–3416.

Rostovtzeff, M. I., Bellinger, A. R., Hopkins, C., and Welles, C. B. (1936) *Excavations at Dura-Europos, Preliminary Report of 6th Season of Work Oct 32 to March 33*, New Haven.

Roth, J. (1995) "The length of the siege of Masada," *SCI* 14: 87–110.

Russo, F. (2002) *Tormenta: venti secoli di artiglieria meccanica*, Rome.

———. (2004) *L'artiglieria delle legioni romance*, Rome. Despite the best efforts of Inter-Library loans and myself, I was only able to consult a portion of this work and then just before this book went to press. I want to thank Serafina Cuomo for drawing my attention to its existence.

Sáez, R. (2005) "La maquinaria bélica en Hispania," *Aquila Legiones*, Murcia, in press.

Sage, M.M. (1996) *Warfare in Ancient Greece: A Sourcebook*, London.

Salazar, C. F. (1998) "Getting the Point: Paul of Aegina on Arrow Wounds," *Sudhoffs Archiv* 82: 170–187.

———. (2000) *The Treatment of War Wounds in Greco-Roman Antiquity*, Leiden.

Schramm, E. (1980) *Die antiken Geschütze der Saalberg*, 2nd ed. with introduction by D. Baatz, Bad Homburg (1st ed. 1918).

Schramm, E. (1928a) "Poliorketik," in J. Kromayer and G. Veith, eds., *Heerwesen und Kriegsführung der Griechen und Römer = Handbuch der Altertumwissenschaft* 4.3.2, 209–247.

Scott, I. R. (1985) "First century military daggers and the manufacture and supply of weapons for the Roman army," in Bishop, M. C., (ed.) *The Production and Distribution of Roman Military Equipment* BAR S275, Oxford, 160–219.

Sellier, K. G. and Kneubuehl, B. P. (1994) *Wound Ballistics,* Amsterdam.

Shipley, G. (1987) *A History of Samos 800–188 B.C.*, Oxford.

Shipp, G. P. (1960) "Ballista," *Glotta* 39: 149–152.

Shrimpton, G. S. (1984) "Strategy and tactics in the preliminaries to the siege of Potidaea (Thuc. 1.61–5)," *Symbolae Osloensis* 59: 7–12.

Simms, D. L. (1987) "Archimedes and the invention of artillery and gunpowder," *Technology and Culture* 28: 67–79.

Sleewsyk, A. W. (1996) "The lineage of the triacontor," in H. Tzalas, ed., *Tropis VI: 6th international symposium on ship construction in antiquity*, Lamia, 517–527.

Snodgrass, A. (1964a) *Early Greek Armour and Weapons from the End of the Bronze Age to 600 B.C.*, Edinburgh.

———. (1964b) "The Metalwork," in *Motya*, by B. S. J. Isserlin et al., *Annual of Leeds University Oriental Society* IV [1962–63], Leiden, 127–131.

———. (1971) review of Marsden 1969, *Classical Review* 21: 106–108.

Soar, H. (2004), *The Crooked Stick*, Yardley, Pa.

Soedel W., and Foley, V. (1979) "Ancient catapults," *Scientific American* 240.3: 120–128.

Sommer, M. (2003) *Hatra: Geschichte und Kultur einer Karawanenstadt im Römisch-Parthischen Mesopotamien*, Mainz.

Southern, P. (1989) "The numeri of the Roman imperial army," *Britannia* 20: 81–140.

Southern, P., and Dixon, K. R. (1996) *The Late Roman Army*, London.

Spieser, J. M. (1982) "Philon de Byzance et les fortifications paléochrétiennes," *La fortification dans l'histoire du monde grec* P. Leriche, ed., Paris, 363–368.

Thévenot, M. (no date, before 1693) *Mathematici Veteres* (poliorketic texts of Athenaios, Apollodoros, Biton, Heron, and Philon), re-edited and published by C. Wescher in 1867.

Tilley, A. (2004) *Seafaring on the Ancient Mediterranean*, BAR IS 1268, Oxford

"Trajan's artillery: the archaeology of a Roman technological revolution," *Current World Archaeology* 3 (2003) 41–48.

Vernidub, I. I. (1996) "One hundred years of the Russian smokeless (nitrocellulose) powder industry," in Buchanan ed. 395–400.

Vicente Redón, J. D., Pilar Punter, M., and Ezquerra, B. (1997) "La catapult tardo-republicana y otro equipamiento military de 'La Caridad' (Caminreal, Teruel)," *Journal of Roman Military Equipment Studies* 8: 167–199.

Vogel, S. (1998) *Cats' Paws and Catapults: Mechanical Worlds of Nature and People*, New York.

Vryonis, S. Jn. (1981) "The evolution of Slavic society and the Slavic invasions of Greece," *Hesperia* 50: 378–390.

Wees, H., van (2004) *Greek Warfare: Myths and Realities*, London.

Weiß, P. (1997) "Milet 1994–5: Schleuderbleie und Marktgewichte," *Archäologischer Anzeiger* 143–153.

Wells, H. B. (1973) "Knowledge of spring steel in Hellenistic times," *Journal of the Arms and Armour Society* 7: 249–256.

Wescher, C. (1867) *Poliorcétique des Grecs*, Paris.

Wheeler, R. E. M. (1943) *Maiden Castle, Dorset*, Oxford.

White, K. D. (1976) "Technology in classical antiquity: some problems," *Museum Africum* 5: 23–35.

White, L., Jr. (1962) "The act of invention: causes, contexts, continuities and consequences," *Technology and Culture* 3: 486–500.

Whitehorn, J. N. (1946) "The catapult and the ballista," *Greece and Rome* 15: 49–60.

Wild, H. W. (1996) "Black powder in mining – its introduction, early use, and diffusion over Europe," in Buchanan ed. 203–217.

Wild, J. P., "Fourth century underwear with special reference to the Thoracomachus," in M. W. C. Hassall, ed. *De Rebus Bellicis*, BAR IS 63 (1979) 105–110.

Wilkes, J. J. (2005) "The Roman Danube: an archaeological survey," *Journal of Roman Studies* 95: 124–225.

Wilkins, A. (2003) *Roman Artillery*, Haverfordwest.

Wilkins, A., and Morgan, L. (2000) "Scorpio and Cheiroballistra," *Journal of Roman Military Equipment Studies* 11: 77–101.

Williams, H. (1992) "A Hellenistic catapult washer from Sounion?" *Échos du Monde Classique/Classical Views* 36: 181–188.

Winter, F. E. (1971) *Greek Fortifications*, Toronto.

———. (1992) "Philon of Byzantion and the Hellenistic fortifications of Rhodos," in van de Maele and Fossey eds., 185–209.

———. (1997) "The use of artillery in fourth-century and Hellenistic towers," *Échos du Monde Classique/Classical Views* 16: 247–292.

Wright, M. P., Hassall, M. W. C., and Tomlin, R. S. O. (1976) "Roman Britain in 1975," *Britannia* 7: 290–392.

Wyse, W. (1900) "On the meaning of σφενδονη in Aesch. *Ag.* 997," *Classical Review* 14: 5.

Zertal, A. (1995) "The Roman siege-system at Khirbet al-Hamam (Narbata)," *The Roman and Byzantine Near East*, 71–94.

Ziman, J. M. (1974) "Ideas move around inside people," *Fifth J. D. Bernal Lecture* Birkbeck College, London.

Zinsser, H. (1935) *Rats, Lice and History*, Boston.

# Index

## Index Locorum

## General Index

# Acknowledgments

MY THANKS BEGIN WITH MY OWN INSTITUTION, the University of Wales Swansea, for allowing me a one-term sabbatical in 2005, during which time five chapters took shape. Nina and her colleagues in the Interlibrary loans office of Swansea University Library obtained for me most of the rare and unusual items in the bibliography. The British Library held all but one of the remaining such works, and its staff was of much help. Many archaeologists and scholars have helped me with material that they worked on directly; I thank especially Jo Berry for information on the ballista balls recently found at Pompeii; Robin Birley for help on some of the Vindolanda material; Jon Coulston for help with the Xanten find; Serafina Cuomo for sending me draft copy of her chapter on catapults, to appear in her forthcoming book on ancient technology; Carol van Driel-Murray on sling pouches; Bill Griffiths and Ben Gourley for extensive correspondence about sling things; Bill Hanson for information on the Elginhaugh finds, and for comments on an early draft of Chapter 5; Darejan Karacheva and Michael Vickers for help with the Vani material; Ruben Sáez for help on the Azaila material; Alex Tilley for help tracking down the Rhine river boat; Hector Williams and Chris Hagerman for help on Stymphalos; and John France, who helped me understand the late Roman/early medieval world, and read and commented on the entire manuscript in draft. Thanks to Jenny Hall for discussion and correspondence about the bolts, hipposandals, and other items in the Museum of London; and to Mark Lewis of the Legionary Museum at Caerleon for help with its material. Thanks to staff at the British Museum, especially Peter Higgs for provenance information on their slingshot. Thanks to Ken Reinschmidt, for sending me a copy of his unpublished paper on palintone mechanics, and answering carefully and patiently my questions on it. Thanks to members of the Ermine Street Guard for discussion of their machines, and to George at Legion XXIV for correspondence on theirs. Thanks to the Royal Engineers' Museum in Chatham, which displays latter-day versions of technologies whose origins are ancient, such as caltrops and two-pound guns, and thereby demonstrates that some of these inventions had a very long life indeed. Thanks also to my publisher, Bruce Franklin, for outstanding efficiency, good sense, and good taste. Thanks to Noreen O'Connor for similar efficiency and effectiveness in editing. Thanks to my husband, John Tucker, for help and support too great to express during the twenty months this book has been in the making. I dedicate this book to my children, for all the games we didn't play while I was working on it.